# DYNAMICS OF PARTICLES AND RIGID BODIES

**Wiley-ASME Press Series List**

| | | |
|---|---|---|
| Dynamics of Particles and Rigid Bodies – A Self-Learning Approach | Daqaq | July 2018 |
| Stress in ASME Pressure Vessels, Boilers, and Nuclear Components | Jawad | October 2017 |
| Robust Adaptive Control for Fractional-Order Systems with Disturbance and Saturation | Shi | November 2017 |
| Robot Manipulator Redundancy Resolution | Zhang | October 2017 |
| Combined Cooling, Heating, and Power Systems: Modeling, Optimization, and Operation | Shi | August 2017 |
| Applications of Mathematical Heat Transfer and Fluid Flow Models in Engineering and Medicine | Dorfman | February 2017 |
| Bioprocessing Piping and Equipment Design: A Companion Guide for the ASME BPE Standard | Huitt | December 2016 |
| Nonlinear Regression Modeling for Engineering Applications | Rhinehart | September 2016 |
| Fundamentals of Mechanical Vibrations | Cai | May 2016 |
| Introduction to Dynamics and Control of Mechanical Engineering Systems | To | March 2016 |

# DYNAMICS OF PARTICLES AND RIGID BODIES

## A SELF-LEARNING APPROACH

**Mohammed F. Daqaq**

*Global Network Associate Professor*
*Division of Engineering*
*New York University, Abu Dhabi*

*Affiliate Associate Professor*
*Tandon School of Engineering, New York University, NY.*

This Work is a co-publication between ASME Press and John Wiley & Sons Ltd.

This edition first published 2019


*Registered Offices*
John Wiley & Sons, Inc., 111 River Street, Hoboken, NJ 07030, USA
John Wiley & Sons Ltd, The Atrium, Southern Gate, Chichester, West Sussex, PO19 8SQ, UK

*Editorial Office*
The Atrium, Southern Gate, Chichester, West Sussex, PO19 8SQ, UK

For details of our global editorial offices, customer services, and more information about Wiley products visit us at www.wiley.com.

Wiley also publishes its books in a variety of electronic formats and by print-on-demand. Some content that appears in standard print versions of this book may not be available in other formats.



*Library of Congress Cataloging-in-Publication Data:*

Names: Daqaq, Mohammed F., 1979- author.
Title: Dynamics of particles and rigid bodies : a self-learning approach / by Mohammed F. Daqaq.
Description: First edition. | Hoboken, NJ : John Wiley & Sons, 2018. | Series: Wiley-ASME Press Series | Includes bibliographical references and index. |
Identifiers: LCCN 2018012505 (print) | LCCN 2018023172 (ebook) | ISBN 9781119463184 (pdf) | ISBN 9781119463191 (epub) | ISBN 9781119463146 (pbk.)
Subjects: LCSH: Dynamics, Rigid. | Dynamics of a particle.
Classification: LCC QA861 (ebook) | LCC QA861 .D37 2018 (print) | DDC 531/.11–dc23
LC record available at https://lccn.loc.gov/2018012505

Cover Design: Wiley
Cover Image: Courtesy of Hyperloop One

Set in 10/12pt TimesLTStd by SPi Global, Chennai, India

10 9 8 7 6 5 4 3 2 1

*To Farid, Hana, Shaima,*

*Rayyan and Laith . . .*

# Contents in Brief

# Contents

# List of Figures

# Preface

The first graduate courses on dynamical systems, often referred to as "Intermediate Dynamics", introduce students to the fundamentals necessary to model the motion of particles and rigid bodies. On top of what they learn in introductory undergraduate dynamics courses, intermediate dynamics courses typically introduce the concept of rotating frames to simplify the kinematic description of particles and rigid bodies. Using this concept, they extend the application of Newton's equations and the principles of impulse and momentum to describe three-dimensional motion of rigid bodies, eventually leading to the derivation of Euler's rotational equations. They also present many new concepts including, but not limited to, constraints, generalized coordinates, generalized forces, virtual displacement, the principle of virtual work, Hamilton's extended principle, Lagrange's equations, and Hamilton's canonical equations.

During my tenure at Clemson University, I taught the Intermediate Dynamics course numerous times, and through this I had the opportunity to interact with students hailing from different cultural and academic backgrounds. Some of those students had a very strong background in introductory dynamics, while unfortunately many others had not had the chance to properly learn the fundamentals of dynamics upon which the *Intermediate Dynamics* course was based. Many students in the latter group struggled silently. Due to cultural beliefs and, sometimes, the language barrier, they would nod their heads in agreement every time I asked whether they understood the material I had just explained in class.

After the first midterm examinations, my own delusion of being a good teacher would often be crushed and I would be left to face a hard reality: many of the students, who were constantly nodding their heads, scored very low on the midterm. I made it a habit to bring these students to my office to understand their difficulties and to learn how I could help them improve their understanding. They would often cite their weak background knowledge, and say that the concepts seemed very clear when explained in class. However, when homework time came around, they would struggle to use what they had learned in class to solve the assigned problems. Often, they would try to learn the material from the references provided in the syllabus, but they would have a very difficult time understanding it. They said that many of the textbooks were directed at experts and used difficult language, or that the authors took shortcuts that made it hard for students to follow. They also mentioned that many of the references did not have enough of the solved examples that would help them understand the material on their own.

Ever since I was a graduate student, I knew that dynamics can only be learned properly by solving as many examples as possible. Struggling with problems for hours at a time is what really allowed me to comprehend dynamics and be able to make a career out of it. Based on my experience as a student, a teacher, and my interaction with the students over the years, I came to realize that a modern textbook on intermediate dynamics had become necessary. It is also critical at a time when online courses are becoming common and students often rely on self-study. In my opinion, such a book needs to contain:

- detailed derivations combined with simple, yet comprehensive textual descriptions of these derivations;
- highlights of the important definitions and concepts, serving serve as a constant reminder of the important concepts to be learned in a given chapter;
- a large number of comprehensively solved examples of gradually increasing difficulty;
- methods that help the teacher improve the interactions in the classroom in order to help him/her assess the students' true level of understanding.

While writing this book, I have tried to the best of my ability to adhere to the aforementioned points. I have made it a habit to explain derivations in a very simple, yet comprehensive manner. I have presented highlights emphasizing the important concepts whenever they appear. I have also highlighted the important equations regularly throughout the text. I have solved a large number of examples – 120 to be exact – and presented the solutions in detail along with illustrations. These examples should aid the student in understanding the theoretical concepts at a deep level.

In addition, throughout the book, I have presented a set of exercises, which I call "Flipped classroom exercises". Flipping the classroom is a modern teaching technique which has been shown to be very effective in teaching topics requiring mathematical derivations. In a typical flipped classroom, the instructor will briefly discuss the material for ten minutes, do some of the important derivations, and solve one example. Subsequently, the instructor hands the students one or more problems to solve on their own, but provides some guidelines that help the students by dividing the problem into several sub-problems. The student will then spend the rest of the class trying to solve as many of these problems as possible. The problems are carefully designed to have an increasing difficulty level. At the end of the class, the full solution of the problems is given to the student. This approach allows the instructor to assess the level of each student in the class and allows the student to assess his/her own level of understanding and to pinpoint the exact concepts they need to inquire about.

The book presents a whole chapter on electromechanical modeling of systems involving particles and rigid-body motion. Such topics, which are essential in modern-day systems, have rarely been addressed systematically in a book on intermediate dynamics. Furthermore, in order to establish the critical connection between modeling of dynamical systems, which is the core of this book, and the analysis of their responses, which is usually covered in courses like perturbation theory and non-linear dynamics, I allocated the final chapter of this book to the introduction of some introductory analysis tools. This includes finding equilibrium points and analyzing their stability, establishing the phase-plane of the dynamics, and constructing bifurcation diagrams and basins of attraction of equilibrium points. The examples used in this chapter are mostly a continuation to the models developed in the previous chapters of the book.

In summary, I believe that I was fortunate to learn dynamics at the hands of a passionate teacher who made the topic very enjoyable and the material accessible. Nonetheless, I know that many students will not have such an opportunity. As such, I wrote this book hoping that many students would find it a useful companion in their journey of learning dynamics.

M.F. Daqaq
*Abu Dhabi, UAE*
*November 2017*

# Acknowledgement

Many people I have met in my life contributed, in some way or another, to my personal growth, and consequently to this humble manuscript. To each and everyone of you, I extend my sincerest gratitude. Nevertheless, there are certain individuals who left a long lasting impression on my life and career. I truly believe that, without those individuals, this work would have not been completed, at least not in this stage of my career. In what follows, I would like to acknowledge each one of you.

First and foremost, I thank my parents Farid and Hana. I know that you love me beyond words and I know that you have sacrificed everything to make sure that I had what I needed to be the person I am today. I pray everyday that you will be around forever as I cannot imagine my life without you. I love you.

I also thank my older sisters Alia, Ruba, and Manal. I felt your endless love and warmth ever since I became conscious of this world and continue to feel it everyday of my life. I am extremely fortunate to have sisters like you. To my younger brother Nihad, I am very proud of you and your achievements. I love you beyond words and ask you to forgive the "Big Brother" attitude I sometimes express. Thank you for being always there to support me.

No words can do justice to describe my endless love Shaima. I would like to thank you an infinite number of times for being there everyday to support me, to love me, to push me forward, to be on my side, to make me feel loved, and to make me and the kids your everyday priority. I love you.

To Rayyan and Laith: the joy you have brought to my world is unparalleled. You have changed every bit of me to the better. You have shown me the true meaning of love. For this I thank you. I am eager to tell you that I was able to finish this work despite all your efforts to distract me.

Three friends have contributed to my personal growth more than anyone else. Abdelghani Shehabelddin "Abu Yasser" for being a true friend and a brother for the past twenty one years. I have learned many things from you. Prof. Khaled Alhazza "Abu Mohammad" for being my best friend during my M.Sc. and Ph.D. Your generosity is unparalleled; I will never forget our daily conversation at Barnes and Nobles and our weekly trips to Washington, DC. Dr. Samir Arabasi for your willingness to go with me everywhere, and for the times, good and bad, we spent together in Blacksburg. You have always been a true brother and, for this, I thank you.

Four people have contributed to my academic growth more than anyone else. The influence of these people on my career is immeasurable and can never be paid back. Prof. Mohammad Al-Nimr for believing in me and introducing me to the elements of research. You have shown me the path to success. Dr. Scott L. Hendricks for making dynamics so simple and enjoyable.

Your courses were the best. Your ability to teach dynamics in such a smooth manner has always amazed me. This book is an extension to your teaching style and many examples were taken from your notes. I am so fortunate to have taken your classes. Prof. Ziyad Masoud, without whom I must admit that my Ph.D. research would have stalled. You have always been so dedicated, kind, generous, fair, and supportive. I have learned from you that the human element comes before any career gains. You have sacrificed a lot to make me successful and, for this, I thank you. Last, but not least, my late mentor, Prof. Ali H. Nayfeh from whom I learned most of the things I know today. In addition to his unprecedented technical knowledge, he was the most humble, hard working, professional, and meticulous researcher I have worked with. He taught me that, every minor detail is important and to never give up when it comes to achieving one's dreams. You showed, not only me, but all Palestinians who struggle every day to make ends meet, that dreams can be realized even with little opportunities.

Finally, I would like to extend my deepest gratitude to my previous students and dear friends, Dr. Ali Alhadidi, Dr. Ravindra Masana, and Dr. Rafael Sanchez Crespo, for proofreading the manuscript and solving the examples to make sure there were no mistakes or typos. A word of appreciation also goes to Dr. Mohammed Al-Qasaimeh for reading a sample chapter of the book and providing his feedback.

M.F.D.

# Introduction

In the realm of physics and engineering, "dynamics" refers to the study of the motion of material bodies, rigid and elastic, under the influence of their surroundings. A rigid body is a material body that undergoes translational and rotational motions. However, it is assumed that the relative distance between any two points on a rigid body remains constant regardless of the type and magnitude of external forces acting on it. In other words, unlike elastic bodies, a rigid body does not undergo measurable strain levels when subjected to external stresses.

When a rigid body undergoes only translational motion, or when its resistance to rotational motion is negligible, a rigid body is treated as a particle. In such a case, the mass of the rigid body, regardless of its shape, can be reduced to a single point and the body's motion can be fully described by the motion of this point.

This book is concerned *only* with the dynamics of particles and rigid bodies, a topic critical in our understanding of how systems, natural or man made, move under the influence of forces. Such understanding has already enabled us to decipher some of the key forces that sustain the motion of the universe and to develop key enabling technologies that improve our lives, pushing the boundaries of what is considered achievable. Terrestrial, aerial, and waterborne vehicles, energy systems, robots, machinery, bionics, and nano/micro-electromechanical systems are only a few examples of technologies whose evolution would not have been possible without our knowledge of dynamics. Today, dynamics of particles and rigid bodies play an important role in almost every aspect of our daily life, and hence it is imperative for every physicist and engineer to understand its fundamentals.

## I.1 Brief History

The word "dynamics" originated from the Greek word *Dynamis*, which means power. Dynamics, as a science, can also be traced back to the Greeks, in particular, to the Greek philosopher Aristotle (384–322 BC; Figure I.1) whose studies on the motion of bodies are the earliest documented evidence of the human curiosity about motion [1]. Similar to other Greek philosophers, Aristotle's understanding of dynamics was influenced by the four classical elements: fire, air, water, and earth. Each of these was thought to have a specific hierarchical position in nature: fire stands at the top, followed by air, then water, then earth. Aristotle argued that when one of these elements is taken out of its place, it naturally wants to return back to it. This explains why a stone thrown in water sinks to the bottom, while a bubble of air inside a fluid rises up to the surface.

**Figure I.1** Aristotle.
Source: https://en.wikipedia.org/wiki/Aristotle.

Based on this understanding, Aristotle divided motion into two parts: *natural* to refer to any motion that follows the hierarchy of the classical elements, and *violent* to refer to motion that opposes it. For example, a stone that slides down a hill is performing a natural motion whereas one thrown upwards is undergoing a violent motion.

Aristotle's limited understanding, which was based on the four elements, led him to many fallacious conclusions regarding the motion of physical bodies. He claimed that heavier objects fall faster than lighter ones and that the speed at which an object falls is proportional to its weight. Nevertheless, he correctly surmised that, in a void, all objects, regardless of their weight, fall at the same speed. This, unfortunately however, led him to incorrectly conclude that a void could not exist.

Most of Aristotle's ideas were later demolished by the Alexandrian philosopher Philoponus (490–570 AD) [2]. Philoponus's work led him to conclude that the rate at which objects fall is not proportional to their weight, as claimed by Aristotle. However, he still argued that heavier objects fall faster than lighter ones in a void. Philoponus also deciphered some of Aristotle's work on violent versus natural motion, alluding tangentially to the concepts of the potential and kinetic energy. He stated that, when an object is violently moved, it is subjected to a finite supply of forcing impetus, which makes the body move as long as it lasts. When this forcing impetus is exhausted, the body stops moving. He also indicated that the forcing impetus must be an internal property of the body.

During the Middle Ages, two Muslim Andalusian philosophers namely Avempace (anglicized from the Arabic, Ibn-Bajja) (1085–1138) and Averroes (anglicized from the Arabic, Ibn-Rushd) (1126–1198) played an important role in preserving and expanding upon Philoponus's studies on the motion of bodies [3]. This is clearly evident in their conclusion that Aristotle's work concerning the inverse proportionality between the velocity of a body and the density of the medium it moves through cannot be true. They argued that, if it were true, then a body could move instantaneously in a void.

Contrary to common belief, the understanding of the motion of bodies also evolved during the Middle Ages [4]. In fact, it was in those times that kinematics was transformed from a qualitative philosophy into a mathematical science. The concepts of instantaneous velocity, mean velocity, and acceleration – although without referring to the forces behind them – were introduced and studied by many researchers including the group known as the "Oxford Calculators", notably the scholar William Heytesbury, and the French priest Jean Buridan.

It was not until Galileo (1564–1642) that the forces behind motion were studied in the modern sense [5]. Galileo developed his understanding of forces by studying Archimedes' hydro-static principle. He noted that the same buoyancy forces that cause a body to rest in liquid are responsible for how fast it moves in the liquid. Another important contribution to dynamics was introduced by Huygens (1629–1695), who was the first to allude to the principle of conservation of momentum, through his investigation of the motion of impacting particles.

Sir Isaac Newton (1643–1728; Figure I.2) is, unarguably, the father of vectorial dynamics as we know it today. In his book, the *PhilosophæNaturalis Principia Mathematica*, usually referred to as the *Principia*, he introduced the concepts of force, acceleration, inertia, mass, and linear momentum [6]. He also stated (in the English of the time) his three famous laws, upon which all of our utilization and understanding of vectorial mechanics is based:

**Figure I.2** Isaac Newton. Source: https://en.wikipedia.org/wiki/Isaac_Newton.

- *First Law*: Every body perseveres in its state of rest, or of uniform motion in a right line, unless it be compelled to change that state by forces impressed upon it thereon.
- *Second Law*: The alteration of motion is proportional to the motive force impressed; and is made in the direction of the right line in which that force is impressed.
- *Third Law*: To every action there is always opposed an equal reaction: or the mutual actions of two bodies upon each other are always equal, and directed to contrary parts.

In *The Principia*, Newton also introduced the general law of gravitational attraction used today to study orbital mechanics.

What Newton did not discuss in *The Principia*, was developed by many of his compatriots and successors, such as Daniel Bernoulli (1700–1782), Euler (1707–1783; Figure I.3), D'Alembert (1717–1783), Lagrange (1736–1813), Jacobi (1804–1851), and Hamilton (1805–1865). This work includes, but is not limited to, the treatment of rigid bodies in rotation, and hence the concept of angular momentum, the motion of systems including more than two degrees of freedom, elastic bodies, and the concepts of work and energy. It is also worth noting that many historians argue that credit for the application of Newton's second law along the three Cartesian coordinates is actually due to Euler [1].

**Figure I.3** Leonhard Euler. Source: https://en.wikipedia.org/wiki/Leonhard_Euler.

Many also argue that Jacob Bernoulli was the first to introduce the concept of angular momentum, even before *The Prinicipia* was published [8]. He was also the first to develop the equation of motion of an elastic body, after investigating the motion of strings using the balance of forces and moments across a string element.

Euler, on the other hand, was the first to tackle the three-body problem and to introduce the concept of three-body rotations about orthogonal axes, which he reasoned must pass through

**Figure I.4** Joseph-Louis Lagrange.
Source: http://ro.math.wikia.com/wiki/Teorema_lui_Lagrange.

the center of mass of the rotating body. In 1776, he published his famous equations on the conservation of angular momentum known today as Euler's rotational equations [9].

The concepts of work and energy and the field of analytical dynamics were first alluded to by Leibniz (1647–1716). Subsequently, upon the formalization of the calculus of variation by the likes of d'Alembert, Euler, and John Bernoulli (1667–1748), the principle of virtual work was established by Jacob Bernoulli as a way to understand static equilibrium, and then extended to bodies in motion by d'Alembert.

The principle of least action, which is used to derive the equations of motion of different systems, is due to Maupertuis (1698–1759), who stated that "nature is thrifty in all actions" and applied this concept to mechanics by suggesting the minimization of the kinetic energy of the system to obtain the equations of motion. Along similar lines, Euler suggested applying the principle of least action to the linear momentum in order to obtain the same equations.

In 1788, Lagrange (1736–1813; Figure I.4) introduced his equations of motion for mechanical bodies, leading to what is known today as Lagrangian mechanics [10]. Hamilton in 1833 applied the variational principle to the classical Lagrangian function and obtained Hamilton's variational principle, which later led to the formulation of Hamiltonian mechanics.

## I.2 Introductory Concepts

This section presents many of the fundamental concepts used throughout this book. It is assumed that the reader will have been exposed to these concepts in introductory courses on statics, dynamics, vector and matrix algebra, multi-variable calculus, and differential equations. Nonetheless, this section reintroduces some important concepts briefly in order to rejuvenate them in the reader's mind.

### *I.2.1 Matrices and Linear Algebra*

An $n$ by $m$ matrix is an array of objects arranged in $n$ rows and $m$ columns as follows:

$$\mathbf{A} = \begin{pmatrix} a_{11} & a_{12} & \dots & a_{1m} \\ a_{21} & a_{22} & \dots & a_{2m} \\ \vdots & \vdots & \ddots & \vdots \\ a_{n1} & a_{n2} & \dots & a_{nm} \end{pmatrix}, \tag{I.1}$$

where the $a_{ij}$ are denoted as the matrix entries. A matrix is said to be square when $n = m$. Matrix algebra satisfies the following important properties/identities:

1. Two matrices **A** and **B** are said to be equal if and only if they have the same number of rows and columns and have equal entries; that is, $[a_{ij}] = [b_{ij}]$ for each $i$ and $j$.
2. The sum of two $n$ by $m$ matrices $\mathbf{A} = [a_{ij}]$ and $\mathbf{B} = [b_{ij}]$ is equal to another $n$ by $m$ matrix **C** such that $c_{ij} = a_{ij} + b_{ij}$ for each $i$ and $j$.
3. If a matrix $\mathbf{A} = [a_{ij}]$ is multiplied by a scalar $\alpha$ then $\alpha\mathbf{A} = [\alpha a_{ij}]$.
4. The product of an $n$ by $r$ matrix **A** and an $r$ by $m$ matrix **B** is an $n$ by $m$ matrix **C** whose $c_{ij}$ elements are given by

$$c_{ij} = a_{i1}b_{1j} + a_{i2}b_{2j} + \ldots + a_{ir}b_{rj}. \tag{I.2}$$

5. For two matrices **A** and **B** having the same dimensions, $\mathbf{A} + \mathbf{B} = \mathbf{B} + \mathbf{A}$.
6. For three matrices **A**, **B**, and **C**; $\mathbf{A}(\mathbf{B} + \mathbf{C}) = \mathbf{AB} + \mathbf{AC}$ provided **A** is an $n$ by $k$ matrix and **B** and **C** are $k$ by $m$ matrices.
7. For three matrices **A**, **B**, and **C**; $\mathbf{A}(\mathbf{BC}) = (\mathbf{AB})\mathbf{C}$ provided **A** is an $n$ by $k$ matrix, **B** is a $k$ by $r$ matrix and **C** is an $r$ by $m$ matrix.

#### I.2.1.1 Transpose of a matrix

The transpose of an $n$ by $m$ matrix $\mathbf{A} = [a_{ij}]$ is an $m$ by $n$ matrix $\mathbf{A}^T = [a_{ji}]$. The transpose of a matrix satisfies the following properties:

1. $(\mathbf{A}^T)^T = \mathbf{A}$.
2. $(\mathbf{AB})^T = \mathbf{B}^T\mathbf{A}^T$.
3. A matrix **A** is said to be symmetric if $\mathbf{A}^T = \mathbf{A}$.
4. A matrix **A** is said to be skew-symmetric if $\mathbf{A}^T = -\mathbf{A}$.
5. A matrix **A** is said to be orthogonal if $\mathbf{AA}^T = \mathcal{I}$, where $\mathcal{I}$ is denoted as the identity matrix

$$\mathcal{I} = \begin{pmatrix} 1 & 0 & 0 & 0 \\ 0 & 1 & \ldots & 0 \\ \vdots & \vdots & \ddots & \vdots \\ 0 & 0 & \ldots & 1 \end{pmatrix}. \tag{I.3}$$

#### I.2.1.2 Inverse of a matrix

A $n$ by $n$ matrix **A** is said to have an inverse $\mathbf{A}^{-1}$ when $\mathbf{AA}^{-1} = \mathcal{I}$. A matrix that has an inverse is called a non-singular matrix. The following are some of the properties of the matrix inverse:

1. If **A** and **B** are non-singular square matrices then **AB** is also a non-singular matrix and satisfies $(\mathbf{AB})^{-1} = \mathbf{B}^{-1}\mathbf{A}^{-1}$.

2. If **A** is non-singular then so is $\mathbf{A}^{-1}$.
3. If either **A** or **B** is singular then **AB** and **BA** are both singular.
4. For an orthogonal matrix $\mathbf{A}^T = \mathbf{A}^{-1}$.

#### I.2.1.3 Determinant

The determinant of a 2 by 2 matrix **A** is denoted as $|\mathbf{A}|$ and is given by $|\mathbf{A}| = a_{11}a_{22} - a_{12}a_{21}$. The determinant of a 3 by 3 matrix $|\mathbf{A}|$ is given by

$$|\mathbf{A}| = a_{11}\begin{vmatrix} a_{22} & a_{23} \\ a_{32} & a_{33} \end{vmatrix} - a_{12}\begin{vmatrix} a_{21} & a_{23} \\ a_{31} & a_{33} \end{vmatrix} + a_{13}\begin{vmatrix} a_{21} & a_{22} \\ a_{31} & a_{32} \end{vmatrix} \tag{I.4}$$

In general, the determinant of any $n$ by $n$ matrix is given by

$$|\mathbf{A}| = (-1)^{i+j} a_{ij} \mathcal{Q}_{ij} \tag{I.5}$$

where $\mathcal{Q}_{ij}$ are called the minors of $|\mathbf{A}|$ and are obtained by finding the determinant of the matrix formed by deleting the $i$th row and $j$th column of $|\mathbf{A}|$. Note that a matrix is singular if its determinant is zero. The determinant satisfies the following properties:

1. $|\alpha\mathbf{A}| = \alpha|\mathbf{A}|$.
2. If **A** has a zero row or column, then $|\mathbf{A}| = 0$.
3. If **A** has a repeated row (or column), or has one row (or column) that is a constant multiplication of another row (or column), then $|\mathbf{A}| = 0$.
4. If **A** is an $n$ by $n$ matrix and **u** is an $n$ by 1 vector, then $\mathbf{Au} = 0$ if and only if $\mathbf{u} = 0$ (trivial solution) or $|\mathbf{A}| = 0$ (nontrivial solution).

#### I.2.1.4 Eigenvalues and eigenvectors

A scalar $\lambda$, real or complex, is said to be a right eigenvalue of the matrix **A** if and only if it satisfies the following equation:

$$\mathbf{Au} = \lambda\mathbf{u}, \tag{I.6}$$

where **u** is a non-zero $n$ by 1 vector known as the eigenvector of **A**. An easy way to find the eigenvalues of a matrix **A** is by manipulating the previous equation such that

$$(\mathbf{A} - \lambda\mathcal{I})\mathbf{u} = 0. \tag{I.7}$$

For the previous equation to have non-trivial solutions, $\mathbf{u} \neq 0$, then by property 4 of the determinants of a matrix, we can write

$$|\mathbf{A} - \lambda\mathcal{I}| = 0. \tag{I.8}$$

For each distinct eigenvalue $\lambda_i$ of the matrix **A**, there exists an eigenvector $\mathbf{u}_i$. The eigenvectors $\mathbf{u}_i$ of the matrix **A** are said to be linearly independent if they satisfy

$$\alpha_1\mathbf{u}_1 + \alpha_2\mathbf{u}_2 + \ldots + \alpha_n\mathbf{u}_n = 0, \tag{I.9}$$

*only* when all of the $\alpha_i = 0$, where $\alpha_i$ are scalars.

#### I.2.1.5 Diagonalization

A matrix $\mathbf{A} = [a_{ij}]$ is said to be diagonal if $a_{ij} = 0$ when $i \neq j$. A non-diagonal $n$ by $n$ matrix can be diagonalized if there exists an $n$ by $n$ matrix $\mathbf{P}$ such that $\mathbf{B} = \mathbf{P}^{-1}\mathbf{AP}$ is a diagonal matrix. In such a case, we say that $\mathbf{P}$ diagonalizes $\mathbf{A}$.

The condition that guarantees that existence of a matrix $\mathbf{P}$ that diagonalizes $\mathbf{A}$ is that $\mathbf{A}$ has $n$ linearly independent eigenvectors, $\mathbf{u}_i$. In such a case, $\mathbf{P}$ can be formed by using the eigenvectors of $\mathbf{A}$ as the columns of $\mathbf{P}$. The resulting diagonal matrix, $\mathbf{B} = \mathbf{P}^{-1}\mathbf{AP}$ will have the eigenvalues of $\mathbf{A}$ as its diagonal elements.

The following are some properties associated with diagonalization:

1. Any $n$ by $n$ matrix with $n$ distinct eigenvalues is always diagonalizable because the presence of $n$ distinct eigenvalues guarantees the existence of $n$ linearly-independent eigenvectors.
2. Any $n$ by $n$ real symmetric matrix is diagonalizable even when it does not possess $n$ distinct eigenvalues. By construction, real symmetric matrices have $n$ linearly-independent eigenvectors.
3. Any $n$ by $n$ real symmetric matrix $\mathbf{A}$ is diagonalizable by an orthogonal matrix, $\mathbf{P}$. This implies that $\mathbf{P}^{-1}\mathbf{AP} = \mathbf{P}^T\mathbf{AP}$ for any real symmetric matrix $\mathbf{A}$.

### *I.2.2 Vectors*

Vectors are used in physics to describe quantities that have both magnitude and direction. In a Cartesian coordinate, a vector is described by three components, each of which indicates its magnitude along the unit vectors $(\hat{\mathbf{i}}, \hat{\mathbf{j}}, \hat{\mathbf{k}})$ forming the Cartesian coordinate system. For instance, as shown in Figure I.5, the vector $\mathbf{u}$ can be expressed as

$$\mathbf{u} = u_x\hat{\mathbf{i}} + u_y\hat{\mathbf{j}} + u_z\hat{\mathbf{k}}. \tag{I.10}$$

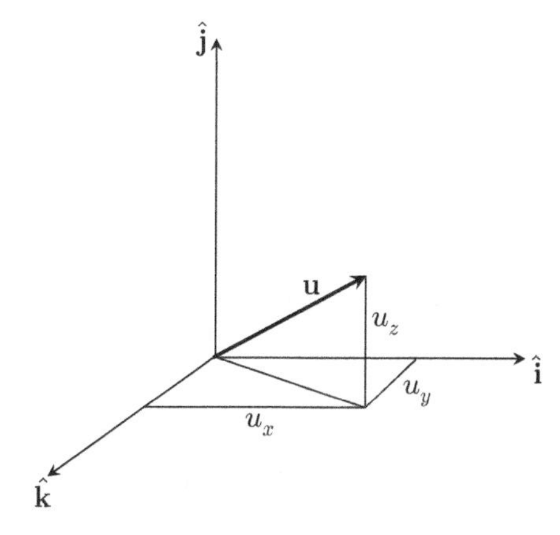


**Figure I.5** Projection of a vector into its different components along the unit vectors of a Cartesian coordinate system.

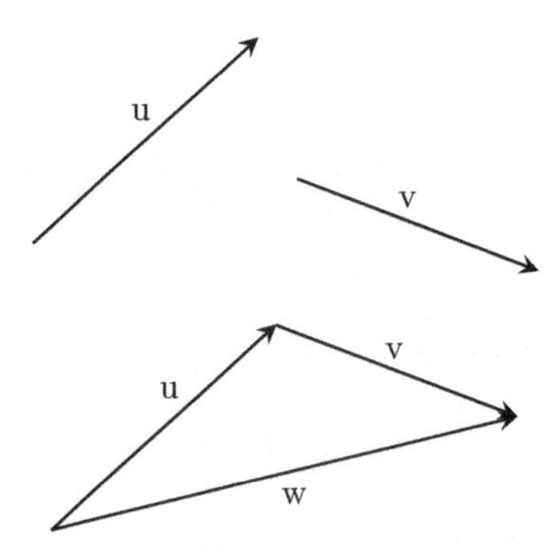


**Figure I.6** Graphical representation of the sum of two vectors.

The magnitude of the vector is defined as

$$\text{Magnitude of } \mathbf{u} = \|\mathbf{u}\| = \sqrt{u_x^2 + u_y^2 + u_z^2}. \tag{I.11}$$

As shown in Figure I.6, the sum of two vectors, **u** and **v**, is another vector **w** whose magnitude and direction depends on the magnitude and direction of **u** and **v**; that is:

$$\begin{aligned}\mathbf{w} = \mathbf{u} + \mathbf{v} &= (u_x\hat{\mathbf{i}} + u_y\hat{\mathbf{j}} + u_z\hat{\mathbf{k}}) + (v_x\hat{\mathbf{i}} + v_y\hat{\mathbf{j}} + v_z\hat{\mathbf{k}}) \\ &= (u_x + v_x)\hat{\mathbf{i}} + (u_y + v_y)\hat{\mathbf{j}} + (u_z + v_z)\hat{\mathbf{k}}. \end{aligned} \tag{I.12}$$

The summation of vectors is commutative; that is, $\mathbf{u}+\mathbf{v}=\mathbf{v}+\mathbf{u}$ and associative; that is, $(\mathbf{u}+\mathbf{v})+\mathbf{w}=\mathbf{u}+(\mathbf{v}+\mathbf{w})$.

Multiplication involving vectors can be classified as follows:

1. *Multiplication by a scalar quantity,* $\alpha$: In this case the product is defined by

$$\mathbf{v} = \alpha\mathbf{u} = \alpha u_x\hat{\mathbf{i}} + \alpha u_y\hat{\mathbf{j}} + \alpha u_z\hat{\mathbf{k}}, \tag{I.13}$$

   in which case $\|\mathbf{v}\| = \alpha\|\mathbf{u}\|$.
2. *The dot product*: The dot product of two vectors **u** and **v** is a scalar defined as

$$\mathbf{u} \cdot \mathbf{v} = \|\mathbf{u}\|\|\mathbf{v}\| \cos\theta, \tag{I.14}$$

   where $\theta$ is the angle between **u** and **v**. We say that two vectors are orthogonal when $\mathbf{u} \cdot \mathbf{v} = 0$. In general, the dot product can be expressed as

$$\mathbf{u} \cdot \mathbf{v} = u_x v_x + u_y v_y + u_z v_z. \tag{I.15}$$

   It follows that $\mathbf{u} \cdot \mathbf{u} = \|u\|^2$. The dot product satisfies the following properties:

   (a) $\mathbf{u} \cdot \mathbf{v} = \mathbf{v} \cdot \mathbf{u}$

(b) $(\mathbf{u}+\mathbf{v})\cdot\mathbf{w} = (\mathbf{u}\cdot\mathbf{w}) + (\mathbf{v}\cdot\mathbf{w})$
(c) $\mathbf{u}\cdot\mathbf{u} = 0$ only when $\mathbf{u} = 0$.

3. *The cross product*: The cross product of two vectors **u** and **v** is another vector **w** which is perpendicular to both **u** and **v** and is defined as

$$\mathbf{w} = \mathbf{u}\times\mathbf{v} = \|\mathbf{u}\|\,\|\mathbf{v}\|\sin\theta\hat{n}, \tag{I.16}$$

where $\theta$ is the angle between **u** and **v** and $\hat{n}$ is the vector perpendicular to both **u** and **v**. We say that two vectors are parallel when $\mathbf{u}\times\mathbf{v} = 0$. In terms of the Cartesian unit vectors, the cross product can be written as

$$\mathbf{u}\times\mathbf{v} = (u_y v_z - u_z v_y)\hat{\mathbf{i}} + (u_z v_x - u_x v_z)\hat{\mathbf{j}} + (u_x v_y - u_y v_x)\hat{\mathbf{k}}, \tag{I.17}$$

or

$$\mathbf{u}\times\mathbf{v} = \begin{vmatrix} \hat{\mathbf{i}} & \hat{\mathbf{j}} & \hat{\mathbf{k}} \\ u_x & u_y & u_z \\ v_x & v_y & v_z \end{vmatrix} \tag{I.18}$$

The cross product satisfies the following properties:

(a) $\mathbf{u}\times\mathbf{v} = -\mathbf{v}\times\mathbf{u}$
(b) $\mathbf{u}\times(\mathbf{v}+\mathbf{w}) = \mathbf{u}\times\mathbf{v}+\mathbf{u}\times\mathbf{w}$
(c) $\alpha(\mathbf{u}\times\mathbf{v}) = \alpha\mathbf{u}\times\mathbf{v} = \mathbf{u}\times\alpha\mathbf{v}$
(d) $\mathbf{u}\cdot(\mathbf{v}\times\mathbf{w}) = (\mathbf{u}\times\mathbf{v})\cdot\mathbf{w} = (\mathbf{w}\times\mathbf{u})\cdot\mathbf{v}$
(e) $\mathbf{u}\times(\mathbf{v}\times\mathbf{w}) = \mathbf{v}(\mathbf{u}\cdot\mathbf{w}) - \mathbf{w}(\mathbf{u}\cdot\mathbf{v})$.
(f) The cross product of two vectors can also be expressed as the product of a skew-symmetric matrix and a vector as following:

$$\mathbf{u}\times\mathbf{v} = \begin{pmatrix} 0 & -u_z & u_y \\ u_z & 0 & -u_x \\ -u_y & u_x & 0 \end{pmatrix} \begin{bmatrix} v_x \\ v_y \\ v_z \end{bmatrix}.$$

(g) $(\mathbf{u}\times\mathbf{v})\cdot(\mathbf{w}\times\mathbf{z}) = (\mathbf{u}\cdot\mathbf{w})(\mathbf{v}\cdot\mathbf{z}) - (\mathbf{v}\cdot\mathbf{w})(\mathbf{u}\cdot\mathbf{z})$
(h) $(\mathbf{u}\times\mathbf{v})\times(\mathbf{w}\times\mathbf{z}) = (\mathbf{u}\cdot(\mathbf{v}\times\mathbf{z}))\mathbf{w} - (\mathbf{u}\cdot(\mathbf{v}\times\mathbf{w}))\mathbf{z}$

## *I.2.3 Vector Calculus*

### I.2.3.1 Gradient

Consider a scalar function $\Phi(x, y, z)$. The gradient of $\Phi$ is a vector defined as

$$\nabla\Phi = \frac{\partial\Phi}{\partial x}\hat{\mathbf{i}} + \frac{\partial\Phi}{\partial y}\hat{\mathbf{j}} + \frac{\partial\Phi}{\partial z}\hat{\mathbf{k}}, \tag{I.19}$$

where $\nabla = \frac{\partial}{\partial x}\hat{\mathbf{i}} + \frac{\partial}{\partial y}\hat{\mathbf{j}} + \frac{\partial}{\partial z}\hat{\mathbf{k}}$ is known as the *del operator*. For a surface $\Phi(x, y, z) = 0$, the gradient of the surface evaluated at a point $P_0 : (x_0, y_0, z_0)$ represents the vector normal to the surface, $\Phi = 0$ at $P_0$.

#### I.2.3.2 Divergence

The divergence of a vector field $\mathbf{u}(x, y, z) = f(x, y, z)\hat{\mathbf{i}} + g(x, y, z)\hat{\mathbf{j}} + h(x, y, z)\hat{\mathbf{k}}$ is a scalar quantity defined as

$$\begin{aligned} \nabla \cdot \mathbf{u} &= \left(\frac{\partial}{\partial x}\hat{\mathbf{i}} + \frac{\partial}{\partial y}\hat{\mathbf{j}} + \frac{\partial}{\partial z}\hat{\mathbf{k}}\right) \cdot (f(x, y, z)\hat{\mathbf{i}} + g(x, y, z)\hat{\mathbf{j}} + h(x, y, z)\hat{\mathbf{k}}), \\ \nabla \cdot \mathbf{u} &= \frac{\partial f}{\partial x} + \frac{\partial g}{\partial y} + \frac{\partial h}{\partial z}. \end{aligned} \tag{I.20}$$

In general, the divergence of the vector field evaluated at a given point is a measure of the net flow into or out of that point in the field.

#### I.2.3.3 Curl

The curl of a vector field $\mathbf{u}(x, y, z) = f(x, y, z)\hat{\mathbf{i}} + g(x, y, z)\hat{\mathbf{j}} + h(x, y, z)\hat{\mathbf{k}}$ is a vectorial quantity defined as

$$\begin{aligned} \nabla \times \mathbf{u} &= \left(\frac{\partial}{\partial x}\hat{\mathbf{i}} + \frac{\partial}{\partial y}\hat{\mathbf{j}} + \frac{\partial}{\partial z}\hat{\mathbf{k}}\right) \times (f(x, y, z)\hat{\mathbf{i}} + g(x, y, z)\hat{\mathbf{j}} + h(x, y, z)\hat{\mathbf{k}}), \\ \nabla \times \mathbf{u} &= \left(\frac{\partial h}{\partial y} - \frac{\partial g}{\partial z}\right)\hat{\mathbf{i}} + \left(\frac{\partial f}{\partial z} - \frac{\partial h}{\partial x}\right)\hat{\mathbf{j}} + \left(\frac{\partial g}{\partial x} - \frac{\partial f}{\partial y}\right)\hat{\mathbf{k}} \end{aligned} \tag{I.21}$$

The curl at a given point in the vector field is a measure of the degree of rotation of the flow field around that point.

The gradient, divergence, and curl satisfy the following identities:

1. $\nabla(\Phi + \Psi) = \nabla\Phi + \nabla\Psi$.
2. $\nabla(\Phi\Psi) = \Psi\nabla\Phi + \Phi\nabla\Psi$.
3. $\nabla(\mathbf{u} \cdot \mathbf{v}) = (\mathbf{u} \cdot \nabla)\mathbf{v} + (\mathbf{v} \cdot \nabla)\mathbf{u} + \mathbf{u} \times (\nabla \times \mathbf{v}) + \mathbf{v} \times (\nabla \times \mathbf{u})$.
4. $\nabla \cdot (\mathbf{u} \times \mathbf{v}) = \mathbf{v} \cdot (\nabla \times \mathbf{u}) - \mathbf{u} \cdot (\nabla \times \mathbf{v})$.
5. $\nabla \times (\nabla\Phi) = 0$. The curl of the gradient of a scalar field is zero. This is because $\nabla$ is parallel to $\nabla\Phi$.
6. $\nabla \cdot (\nabla \times \mathbf{u}) = 0$. The divergence of the curl of a vector field is zero. This is because $\nabla$ is perpendicular to $\nabla \times \mathbf{u}$.
7. $\nabla(\nabla \cdot \mathbf{u}) - \nabla \times (\nabla \times \mathbf{u}) = \nabla^2\mathbf{u}$.
8. $\nabla \cdot (\Phi\nabla\Psi) = \Phi\nabla^2\Psi + \nabla\Phi \cdot \nabla\Psi$.

#### I.2.3.4 Gauss divergence theorem

The Gauss divergence theorem states that the flow in or out (flux) of a closed surface $\sigma$ is equal to the divergence of the vector field $\mathbf{u}$ away from the points in the volume $V$ enclosed by $\sigma$; that is

$$\iint_\sigma \mathbf{u} \cdot \mathrm{d}\mathbf{A} = \iiint_V \nabla \cdot \mathbf{u}\, \mathrm{d}V, \tag{I.22}$$

where, as shown in Figure I.7, $\mathrm{d}\mathbf{A}$ is an element normal to the surface $\sigma$.

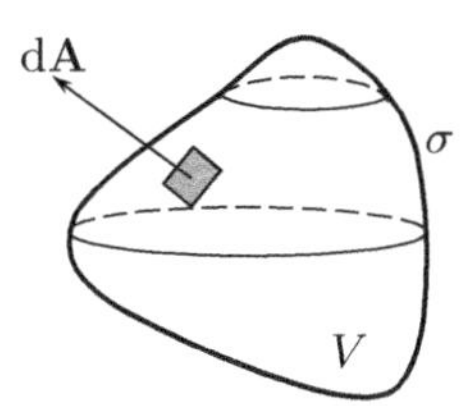


**Figure I.7** Graphical representation of the Gauss divergence theorem.

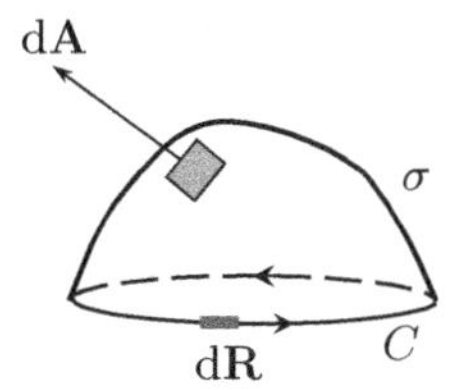


**Figure I.8** Graphical representation of Stokes' theorem.

#### I.2.3.5 Stokes' theorem

Stokes theorem states that the circulation of the vector field $\mathbf{u}$ around a closed loop $C$ is equal to the flux of the curl of the vector field through a surface bounded by the curve; that is

$$\oint_C \mathbf{u} \cdot \mathrm{d}\mathbf{R} = \iint_\sigma (\nabla \times \mathbf{u}) \cdot \mathrm{d}\mathbf{A}, \tag{I.23}$$

where, as shown in Figure I.8, $\mathrm{d}\mathbf{A}$ is an area element normal to the surface $\sigma$ and $\mathrm{d}\mathbf{R}$ is a differential length along the closed loop $C$.

### *I.2.4 Basic Quantities used in Dynamics*

The following are some basic definitions of the scalar and vectorial quantities that are commonly used to describe the motion of dynamical systems.

- Mass (symbol: $m$, $M$, SI unit: kg): a scalar quantity used to measure the resistance of a body to movement. A larger mass implies larger resistance to motion.
- Time (symbol: $t$, SI unit: s: a scalar quantity that is used to measure the irreversible succession of events in a dynamical system.
- Position (symbol: $\mathbf{r}$, SI unit: m): a vectorial quantity used to locate the position of a point in space with respect to another point.
- Distance (symbol: $s$, SI unit: m): a scalar quantity used to measure the magnitude of a vector $\mathbf{r}$ connecting two points in space; $s = \|\mathbf{r}\|$.
- Velocity (symbol: $\mathbf{v}$, SI unit: m/s): a vectorial quantity used to measure the time rate of change of a position vector; $\mathbf{v} = \dot{\mathbf{r}}$.

- Speed (symbol: $\dot{s}$, SI unit: m/s): a scalar quantity used to measure the magnitude of the velocity vector; $\dot{s} = \|\mathbf{v}\|$.
- Acceleration (symbol: $\mathbf{a}$, SI unit: m/s$^2$): a vectorial quantity used to measure the time rate of change of the velocity vector; $\mathbf{a} = \dot{\mathbf{v}}$.
- Gravitational acceleration (symbol: $\mathbf{g}$, SI unit: m/s$^2$): a vectorial quantity that measures the acceleration of a free-falling body in the earth's atmosphere. The value of $\mathbf{g}$ differs ever so slightly at different latitudes. Nonetheless, a value of 9.81 m/s$^2$ is usually used by scientists.
- Angle (symbol: none specific ($\theta$ commonly used), SI unit: rad): a vectorial quantity that measures the amount of turn between two straight lines that have a common end point; that is, it is the orientation of one vector with respect to the other.
- Angular velocity (symbol: $\omega$, SI unit: rad/s): a vectorial quantity that measures the time rate of change of the angle between two vectors; $\omega = \dot{\theta}$.
- Angular acceleration (symbol: $\alpha$, SI unit: rad/s$^2$): a vectorial quantity that measures the time rate of change of the angular velocity; $\alpha = \dot{\omega}$.
- Force (symbol: $\mathbf{F}$, SI unit: Newton N): a vectorial quantity that measures the action of one body on another. According to Newton's second law of dynamics applied to a body of constant mass, the force acting on a body can be related to its ensuing acceleration via $\mathbf{F} = m\mathbf{a}$.
- Weight (symbol: $\mathbf{W}$, SI unit: N): the force exerted by the gravitational field on a given body; $\mathbf{W}=m\mathbf{g}$.
- Moment (symbol: $\mathbf{M}$, SI unit: N.m): a vectorial quantity that measures the turning effect of a force about a point. The moment is defined as $\mathbf{M} = \mathbf{r} \times \mathbf{F}$ where $\mathbf{r}$ is the position vector from the point about which the moment is measured to the point at which the force is exerted.
- Mass moment of inertia (symbol: $I$, SI unit: kg.m$^2$): a scalar quantity that measures a body's resistance to rotation about a given axis.

## I.3 Book Organization

The study of bodies in motion under the influence of their surroundings can be divided into three different parts:

- Part 1 concerns modeling, a process through which the equations used to describe the motion of the system are obtained. For particles and rigid bodies, the motion is sufficiently described by one or more, linear or non-linear, ordinary differential equations.
- Part 2 concerns the analysis, a process through which the equations of motion are analyzed to understand the influence of the different design parameters on the motion of the system.
- Part 3 concerns controlling the motion, a process through which special controllers are designed to force the dynamical system to behave in a desired manner.

This book focuses on Part 1 but also presents some of the introductory analysis tools commonly used in Part 2. Specifically, this book is organized as follows:

1. Chapter 1 presents the concepts of inertial and rotating frames and their utilization to study the kinematics of particles.
2. Chapter 2 presents Newton's laws of dynamics and their implementation to obtain the equations of motion of particles. This is commonly referred to as the "vectorial" approach to dynamics.

3. Chapter 3 implements the vectorial approach to obtain the equations of motion of rigid bodies. Both planar and non-planar motion are considered. In the process, the concepts of center of mass, moment of inertia, and principal axis of rotation are presented and Euler's rotational equations are derived.
4. Chapter 4 classifies the types of constraints to which a dynamical system can be subjected and discusses the difference between an actual and a virtual displacement.
5. Chapter 5 presents the analytical approach to dynamics. In this chapter, Lagrange's equations for conservative and non-conservative systems as well as Hamilton's principle are derived and applied to model the motion of particles.
6. Chapter 6 extends the analytical approach to model the motion of rigid bodies. Both planar and non-planar motion are discussed.
7. Chapter 7 presents the concepts of the impulse and the linear and angular momenta, their conservation and utilization to derive the equations of motion of particles and rigid bodies.
8. Chapter 8 studies the motion of charged bodies in electrostatic and electromagnetic fields. To this end, the Coulomb force, Lorentz force, Maxwell's equations, Ampere's law, Gauss's law, and Faraday's law of induction are presented and used in conjunction with Newton's second law and Lagrange's equation to obtain the equations of motion for charged bodies.
9. Chapter 9 presents some of the most-widely utilized tools to analyze the behavior of dynamical systems. This includes finding equilibrium solutions of ordinary differential equations and assessing their stability, establishing the phase-space representation of the dynamics, and constructing bifurcation diagrams and basins of attractions for equilibrium solutions. Examples of analyzing the motion of particles and rigid bodies using these tools are presented to establish the critical connection between modeling and analysis of dynamical systems.

## References

1. Aristotle (330 BC) *Physics.* Kessinger Publishing Co., 2004 edn.
2. Philoponus, J. "Aristotle's physics", in *Ancient Commentators on Aristotle*, Gerald Duckworth and Co. Ltd., 2006 edn.
3. Moody A. (1951) "Galileo and Avempace: The dynamics of the leaning tower experiment (I)", *Journal of the History of Ideas,* 12(2), 163–193.
4. Truesdell, C. (1968) *Essays in The History of Mechanics*, Springer-Verlag.
5. Galileo (1638) *Dialogue Concerning Two New Sciences.* Elsevier.
6. Newton, I. (1687) *Philosophae Naturalis Principia Mathematica*, Prometheus Books, 3rd edn, 1995, translated by Andrew Motte.
7. Euler, L. (1750) "Discovery of a new principle of mechanics". *Memoires de L'Academie Royale des Sciences*, **Par. XXIII**, 196.
8. Bernoulli J. (1703) "Démonstration générale du centre de balancement ou d'oscillation, tirée de la nature du evier. *Mémoires de l'Academie Royale des Sciences*, pp. 78–84.
9. Euler, L. (1776) "A new method for generating the motion of a rigid body". *Novi Commentarii Academiae Scientiarum Petropolitanae*, 20, pp. 208–238, E479.
10. Lagrange, J. L. (1788) *Mécanique Analytique.* Boston Studies in the Philosophy of Science, Springer-Verlag, 2001 edn.

# About the Companion Website

Don't forget to visit the companion website for this book:

**www.wiley.com/go/daqaq/dynamics**

There you will find valuable material designed to enhance your learning, including:

- Solution manuals

Scan this QR code to visit the companion website

# 1

# Kinematics of Particles

Kinematics is the description of the motion of material bodies without referring to their inertia or the forces that caused their motion. In particular, in kinematics we are interested in defining the velocity, acceleration, angular velocity, and angular acceleration of a given body. To achieve this objective, this chapter introduces the important concept of inertial and non-inertial frames of reference and uses them to illustrate how to fully describe the kinematics of particles.

## 1.1 Inertial Frames

We embark on this journey of learning dynamics by learning about inertial frames. In describing the position, velocity, and acceleration of a moving point in space, it is essential to define a frame of reference. A common example used in undergraduate textbooks of dynamics to explain the concept of the reference frame is that of two cars moving in the same direction at the same speed. An observer standing at the side of the road sees both cars moving at a certain speed relative to him, while an observer riding in one of the cars will see the other car as stationary. The difference in the observed speed stems from the observer's point of view, referred to in dynamics as the observer's "reference frame".

In his early work, Newton realized the importance of the frame of reference in deriving his first and second laws of dynamics. Therefore, he used the fixed stars as his reference frame. However, it was later shown using the theory of relativity that this choice of reference can yield discrepancies, especially for systems moving at a very high speed close to the speed of light.

In an attempt to overcome this problem, an inertial frame of reference, also known as a Galilean frame, was introduced to define the acceleration of points in space. An inertial frame was defined as a frame that can only undergo pure translation at a constant velocity without any rotation with respect to an *absolute* space. However, as our understanding of the universe evolved, it became apparent that the notion of absolute space does not really exist, because everything we know is moving and rotating with respect to something else. As such, referring to an absolute space to define an inertial frame is fundamentally incorrect. Today, most scientists

*Dynamics of Particles and Rigid Bodies: A Self-Learning Approach*, First Edition. Mohammed F. Daqaq.

Companion website: www.wiley.com/go/daqaq/dynamics

working in the field of classical mechanics define an inertial frame of reference as one in which the motion of a particle not subject to forces is in a straight line at constant speed. In other words, an inertial frame is a frame in which the motion of particles follows Newton's first law of dynamics.

 **Inertial Frame**

An inertial frame of reference is one in which the motion of a particle not subject to forces is in a straight line at constant speed. In other words, an inertial frame is a frame in which the motion of particles follows Newton's first law of dynamics.

The existence of an inertial frame is extremely important because Newton's first and second laws hold true *only* if such a frame exists. In particular, Newton's first law of dynamics, which states that any free motion of a particle has a constant magnitude and direction, is true *only* when the observer of this particle is not rotating or accelerating. Similarly, his second and widely celebrated law of dynamics concerning particles, given by

$$\mathbf{F} = m\mathbf{a}, \tag{1.1}$$

which means that the net force, $\mathbf{F}$, acting on a particle is equal to its mass, $m$, times its acceleration, $\mathbf{a}$, is true in this simple widely-used form *only* when the acceleration is measured with respect to an observer standing in an inertial frame. Otherwise, one has to account for other fictitious forces resulting from other types of acceleration, such as the Coriolis, centrifugal, and tangential forces.

## 1.2 Rotating Frames

In addition to the inertial frame of reference, rotating frames, which do not obey Newton's laws in their simple form, are also used quite frequently in kinematics to describe positions, velocities, and accelerations. This is usually done to reduce complexities that could arise when describing the kinematics of bodies involved in complex rotational motions. To show the importance of rotating frames, we consider the kinematics of a ball attached to a rigid cable of fixed length, $l$, forming a simple pendulum, as shown in Figure 1.1. The position, velocity, and acceleration of the ball, $P$, with respect to an observer standing in an inertial Cartesian frame, $N$-frame, with unit vectors $(\hat{n}_1, \hat{n}_2, \hat{n}_3)$ located at $O$ can be easily obtained by defining the position vector **OP** then differentiating it as follows:

$$\begin{aligned}
\text{Position: } \mathbf{OP} &= l\sin\theta\hat{n}_1 - l\cos\theta\hat{n}_2,\\
\text{Velocity: } \mathbf{v} = \frac{\mathrm{d}\mathbf{OP}}{\mathrm{d}t} &= l\cos\theta\dot{\theta}\hat{n}_1 + l\sin\theta\dot{\theta}\hat{n}_2,\\
\text{Acceleration: } \mathbf{a} = \frac{\mathrm{d}\mathbf{v}}{\mathrm{d}t} &= (l\ddot{\theta}\cos\theta - l\dot{\theta}^2\sin\theta)\hat{n}_1 + (l\ddot{\theta}\sin\theta + l\dot{\theta}^2\cos\theta)\hat{n}_2.
\end{aligned}$$

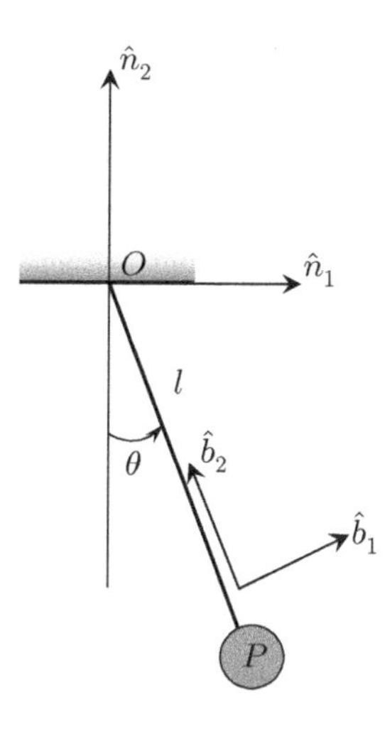


**Figure 1.1** Schematic of a simple pendulum.

It is evident that the complexity of the kinematic description of motion increases substantially as more and more derivatives are taken, despite the very simple nature of the pendulum motion. However, it turns out that it is possible to simplify these expressions considerably, if we describe the motion of the pendulum in a rotating frame. For instance, let us form a new Cartesian frame and call it the $B$-frame, with unit vectors $(\hat{b}_1, \hat{b}_2, \hat{b}_3)$. This rotates with the pendulum such that the $\hat{b}_1$ unit vector is always perpendicular to the cable and the $\hat{b}_2$ unit vector is always parallel to it, as shown in Figure 1.1. As we will learn later in this chapter, we can find the velocity and acceleration of point $P$ with respect to point $O$ and express it in the rotating frame as

$$\text{Position: } \mathbf{OP} = -l\hat{b}_2,$$

$$\text{Velocity: } \mathbf{v} = l\dot{\theta}\hat{b}_1,$$

$$\text{Acceleration: } \mathbf{a} = l\ddot{\theta}\hat{b}_l + l\dot{\theta}^2\hat{b}_2.$$

The reader should not be concerned yet about how we obtained the velocity and acceleration expressions; we use them here only to see how the acceleration expression in the rotating $B$-frame is much simpler than the one obtained using the inertial, $N$-frame. As such, for more complex motions involving multiple rotations, describing kinematics in an inertial frame is often the wrong approach to take.

Another reason for describing motion in a rotating frame stems from the simplicity of measuring angular velocities, angular accelerations, and the mass moment of inertia of rigid bodies in such frames. The importance of rotating frames in dynamics will become clearer as the reader delves into the subsequent chapters of this book.

 **Rotating Frame**

Rotating frames are frames that can rotate with respect to the inertial frame of reference. Such frames are often used to simplify the kinematic description of particles and rigid bodies.

## 1.3 Rotation Matrices

Since both of the inertial and rotating frames are critical in kinematics, it is quiet important to learn how to switch back and forth between two different frames in a simple way. To this end, we consider Figure 1.2a, which depicts two different frames: the $N$-frame and the $B$-frame. The $B$-frame is formed by rotating the $N$-frame around an unknown axis. Our goal in this section is to find a set of equations that allows us to easily go back and forth between these two frames. As will be shown next, this set of equations can be used to construct a matrix known as the *rotation* or *transformation* matrix.

To relate the $B$-frame to the $N$-frame, we consider the general rotation shown in Figure 1.2a and decompose it into three successive rotations around the unit vectors of the Cartesian coordinate system. The first rotation, shown in Figure 1.2b, is a rotation of angle $\theta$ about the $\hat{n}_1$ unit vector. This rotation creates a new intermediate Cartesian coordinate system, which is denoted as the $E$-frame, such that $\hat{e}_1 \equiv \hat{n}_1$.

Next, we take a second rotation, as shown in Figure 1.2c, this time about the $\hat{e}_2$ unit vector through an angle $\phi$ forming a second intermediate frame, denoted as the $C$-frame, such that $\hat{c}_2 \equiv \hat{e}_2$. Finally, as shown in Figure 1.2d, the $C$-frame is rotated by an angle $\psi$ about the $\hat{c}_3$ unit vector to form the $B$-frame such that $\hat{b}_3 \equiv \hat{c}_3$. As such, it is clear now that going from the original $N$-frame to the $B$-frame can be done by performing three successive rotations:[1]

- a rotation $\theta$ around $\hat{n}_1$ to form the $E$ frame;
- a rotation $\phi$ around $\hat{e}_2$ to form the $C$ frame;
- a rotation $\psi$ around $\hat{c}_3$ to form the $B$ frame.

It is now possible to relate the orientation of the $B$-frame to the original $N$-frame by using these three successive rotations. First, we refer back to Figure 1.2b to relate the $E$-frame to the $N$-frame. Using simple vector projections, we can write

$$\begin{aligned} \hat{e}_1 &= \hat{n}_1, \\ \hat{e}_2 &= \cos\theta \hat{n}_2 + \sin\theta \hat{n}_3, \\ \hat{e}_3 &= -\sin\theta \hat{n}_2 + \cos\theta \hat{n}_3. \end{aligned}$$

Note that $\hat{e}_1 = \hat{n}_1$ because the first rotation occurs about the $\hat{n}_1$ unit vector, and hence the $\hat{e}_1$ unit vector remains in the same direction as the $\hat{n}_1$ unit vector. The relationship between $\hat{e}_1$, $\hat{e}_2$ and $\hat{n}_1$, $\hat{n}_2$ can be clarified by referring to Figure 1.3, which shows a planar projection

[1] Note that this is not a unique transformation. One could go from the $N$-frame to the $B$-frame by performing many other different rotations using different angles.

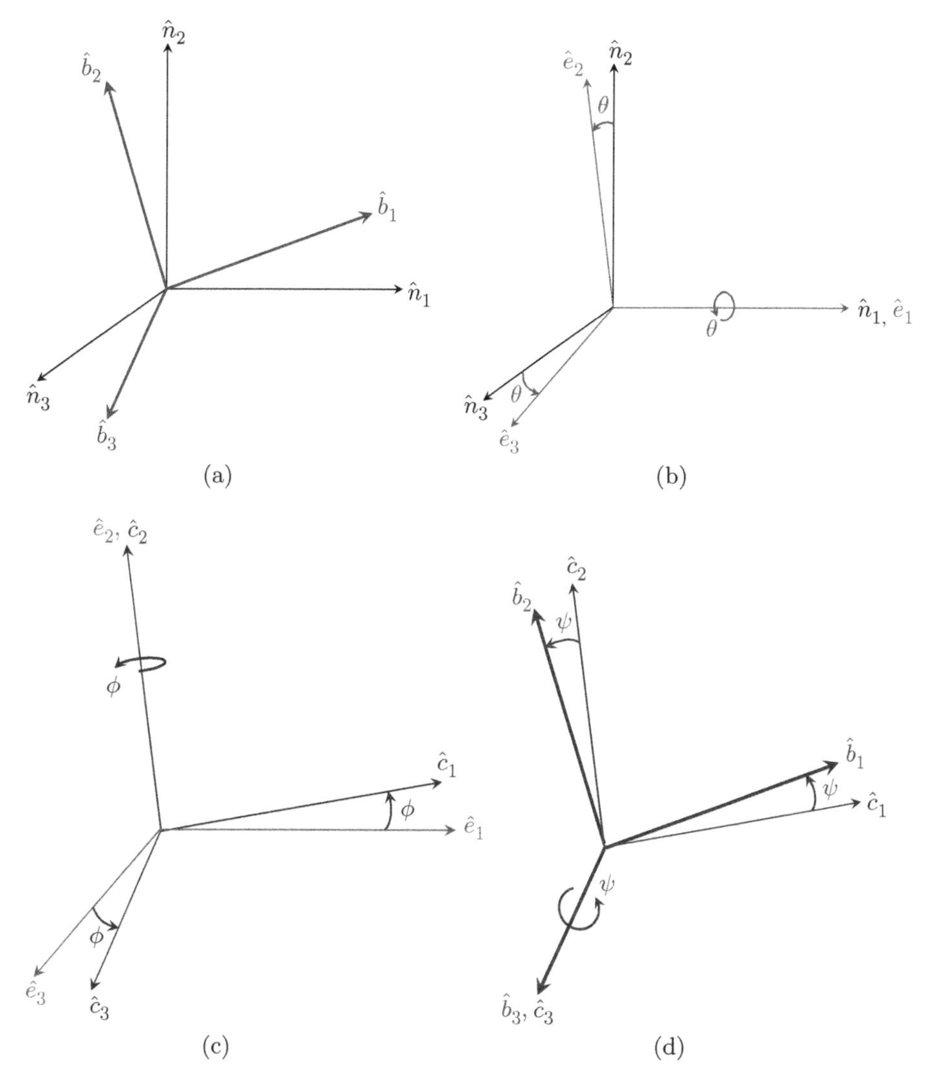


**Figure 1.2** A 1-2-3 rotation performed to orient the $N$-frame in the unit coordinates of the $B$-frame. (*See color plate section for the color representation of this figure.*)

of the $(\hat{n}_2 - \hat{n}_3)$ plane shown in Figure 1.2b. Using Figure 1.3 it can be shown that $\hat{e}_2 = \cos\theta\hat{n}_2 + \sin\theta\hat{n}_3$. Similarly, using Figure 1.3 one can project the unit vector $\hat{e}_3$ in the $N$-frame by letting $\hat{e}_3 = -\sin\theta\hat{n}_2 + \cos\theta\hat{n}_3$.

The previous projections can be described in matrix form as

$$\begin{bmatrix} \hat{e}_1 \\ \hat{e}_2 \\ \hat{e}_3 \end{bmatrix} = \begin{pmatrix} 1 & 0 & 0 \\ 0 & \cos\theta & \sin\theta \\ 0 & -\sin\theta & \cos\theta \end{pmatrix} \begin{bmatrix} \hat{n}_1 \\ \hat{n}_2 \\ \hat{n}_3 \end{bmatrix}.$$

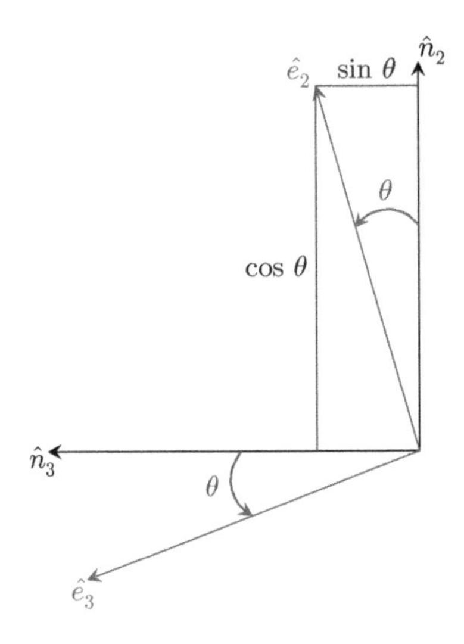


**Figure 1.3** Relationship between the unit vectors of the $E$-frame and the unit vectors of the $N$-frame. (*See color plate section for the color representation of this figure.*)

where the matrix relating the $E$−frame to the $N$-frame is known as a rotation matrix. Furthermore, since this rotation matrix is formed by a single rotation about the $\hat{n}_1$ axis, it is commonly referred to as a 1-rotation.

Following similar reasoning, we can relate the $C$-frame to the $E$-frame via a 2-rotation of angle $\phi$; such that

$$\begin{bmatrix} \hat{c}_1 \\ \hat{c}_2 \\ \hat{c}_3 \end{bmatrix} = \begin{pmatrix} \cos\phi & 0 & -\sin\phi \\ 0 & 1 & 0 \\ \sin\phi & 0 & \cos\phi \end{pmatrix} \begin{bmatrix} \hat{e}_1 \\ \hat{e}_2 \\ \hat{e}_3 \end{bmatrix},$$

and the $B$-frame to the $C$-frame via a 3-rotation of angle $\psi$

$$\begin{bmatrix} \hat{b}_1 \\ \hat{b}_2 \\ \hat{b}_3 \end{bmatrix} = \begin{pmatrix} \cos\psi & \sin\psi & 0 \\ -\sin\psi & \cos\psi & 0 \\ 0 & 0 & 1 \end{pmatrix} \begin{bmatrix} \hat{c}_1 \\ \hat{c}_2 \\ \hat{c}_3 \end{bmatrix}.$$

To describe the rotation from the original $N$-frame to the final $B$-frame we combine the previous rotations, letting

$$\begin{bmatrix} \hat{b}_1 \\ \hat{b}_2 \\ \hat{b}_3 \end{bmatrix} = \begin{pmatrix} \cos\psi & \sin\psi & 0 \\ -\sin\psi & \cos\psi & 0 \\ 0 & 0 & 1 \end{pmatrix} \begin{pmatrix} \cos\phi & 0 & -\sin\phi \\ 0 & 1 & 0 \\ \sin\phi & 0 & \cos\phi \end{pmatrix} \begin{pmatrix} 1 & 0 & 0 \\ 0 & \cos\theta & \sin\theta \\ 0 & -\sin\theta & \cos\theta \end{pmatrix} \begin{bmatrix} \hat{n}_1 \\ \hat{n}_2 \\ \hat{n}_3 \end{bmatrix}.$$

Multiplying the previous matrices yields the following general transformation matrix that can be used to go from the $N$-frame to the $B$-frame via a 1-2-3 rotation.

$$L = \begin{pmatrix} \cos\phi\cos\psi & \cos\theta\sin\psi + \sin\theta\sin\phi\cos\psi & \sin\theta\sin\psi - \cos\theta\cos\phi\sin\psi \\ -\cos\phi\sin\psi & \cos\theta\cos\psi - \sin\theta\sin\phi\sin\psi & \sin\theta\cos\psi + \cos\theta\sin\phi\sin\psi \\ \sin\phi & -\sin\theta\cos\phi & \cos\theta\cos\phi \end{pmatrix}.$$

**Example 1.1 Properties of a Rotation Matrix**

Prove that any rotation matrix has the following two properties:

- $L^{-1} = L^T$
- $|L| = \pm 1$.

1. A rotation matrix is a matrix that preserves the length of the vectors involved in the rotation and the angle between them. In other words, a Cartesian coordinate system formed by three normal unit vectors undergoing a rotation remains a Cartesian coordinate system with unit vectors normal to each other. A matrix that preserves length and direction between vectors is known as an orthogonal matrix, and satisfies the property $LL^T = \mathcal{I}$, where $\mathcal{I}$ is the identity matrix. It follows that $L^{-1} = L^T$ for a rotation matrix (see Property 5 in Section I.2.1.1).
2. Using $LL^T = \mathcal{I}$, and taking the determinant of both sides, we obtain:

$$|LL^T| = |\mathcal{I}|, \qquad |LL^T| = 1.$$

Using $|LL^T| = |L|\ |L^T| = |L|^2$ because $|L^T| = |L|$, we obtain:

$$|L|^2 = 1, \qquad |L| = \pm 1.$$

**Properties of a Rotation Matrix**

A rotation matrix $L$ is an orthogonal matrix that satisfies the following conditions:

1. $L^{-1} = L^T$
2. $|L| = \pm 1$.

When using the right-hand rule to define the successive rotations, it can be shown that $|L| = +1$.

**Flipped Classroom Exercise 1.1**

Find the rotation matrix necessary to take you from a certain frame, $N$, to another frame $B$ by performing a successive 2-1-3 rotation using angles ($\theta$, $\phi$, $\psi$).

To answer this exercise, follow the following steps:

1. Which rotation takes place first? What is the rotation matrix associated with it?
2. Which rotation takes place second? What is the rotation matrix associated with it?
3. Which rotation takes place third? What is the rotation matrix associated with it?
4. Multiply the rotation matrices obtained in steps 1, 2, and 3. Since the 2-rotation occurs first, the matrix obtained in step 1 must be on the far left. Show that the transformation matrix from $N$ to $B$ can be written as

$$L = \begin{pmatrix} \cos\phi\cos\psi + \sin\theta\sin\phi\sin\psi & \cos\theta\sin\psi & -\cos\psi\sin\phi + \cos\phi\sin\theta\sin\psi \\ \cos\psi\sin\theta\sin\phi - \cos\phi\sin\psi & \cos\theta\cos\psi & \cos\phi\cos\psi\sin\theta + \sin\phi\sin\psi \\ \cos\theta\sin\phi & -\sin\theta & \cos\theta\cos\phi \end{pmatrix}$$

## 1.4 Velocity of a Particle in a Three-dimensional Space

The kinematics of a point moving in space is fully described using three vectorial quantities:

- its position
- its velocity (being the time rate of change of the position)
- its acceleration (being the time rate of change of the velocity).

When the position vector is defined in an inertial frame, the velocity and acceleration can be easily obtained by differentiating the position vector with respect to time. On the other hand, the process is not as simple when the position vector is defined in a rotating frame. *This is because the unit vectors that are used to describe directions in the rotating frame are continuously changing their orientation with respect to the inertial frame.*

In what follows, we explain in detail how to find the velocity and acceleration of a point whose position is described in a rotating frame when the observer is standing in an inertial frame of reference. To this end, we consider a hypothetical situation in which a student named Joe is trying to understand kinematics. Joe is standing at point $O$ and observing point $P$, as shown in Figure 1.4. The position vector **OP** is given by $\mathbf{r} = r_1\hat{n}_1 + r_2\hat{n}_2 + r_3\hat{n}_3$. Here, $r_i$ are the lengths of the vector **OP** projected in the directions of $\hat{n}_i$, where $\hat{n}_i$ are the unit vectors of the inertial Cartesian frame denoted as the $N$-frame.

When point $P$ is moving such that the vector **OP** does not change orientation with respect to the $N$-frame, then, from Joe's perspective, the vector **OP** only changes length. As such, the velocity of point $P$ with respect to point $O$, as measured in the inertial $N$-frame, is the derivative of the vector **r** with respect to time, $t$; that is,

$$^{N}\mathbf{v}^{P/O} = \dot{r}_1\hat{n}_1 + \dot{r}_2\hat{n}_2 + \dot{r}_3\hat{n}_3, \qquad (1.2)$$

where the dot is a derivative with respect to time. At this point, it is worth decoding the notation on the left-hand side of Equation (1.2), as similar notation will be used throughout this book. The right-hand superscript, $P/O$, on the vector **v**, means $P$ with respect to $O$. The left-hand superscript, $N$ on the vector **v** refers to the frame of the observer. As such, $^{N}\mathbf{v}^{P/O}$ reads as "the velocity of point $P$ with respect to point $O$ as observed in the $N$ frame".

Throughout this book, the following notation will be used to describe kinematic quantities:

- Bold face **v** and **a** represent, respectively, the velocity and acceleration vectors. Such vectors will always appear with superscripts on their right- and left-hand sides. For

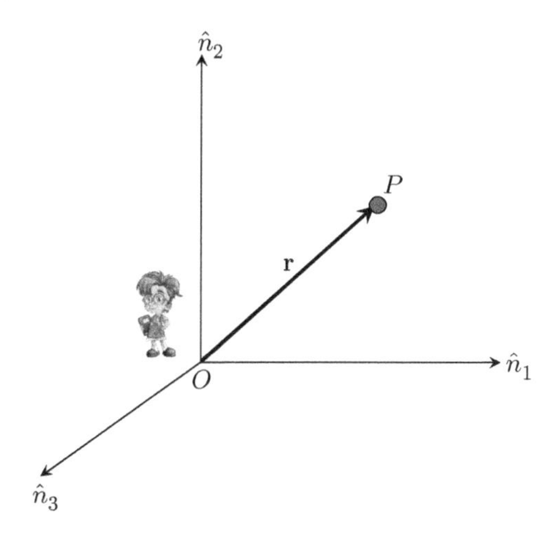


**Figure 1.4** Kinematics of a particle in an inertial frame.

instance, you will often see velocities described in the form ${}^N\mathbf{v}^{P/O}$. The right-hand superscript, $P/O$, on the vector **v**, means $P$ with respect to $O$. The left-hand superscript, $N$ on the vector **v** refers to the frame of the observer. As such, ${}^N\mathbf{v}^{P/O}$ reads as "the velocity of point $P$ with respect to point $O$ as observed in the $N$ frame". Along similar lines, ${}^B\mathbf{a}^{Q/P}$ reads as "the acceleration of point $Q$ with respect to point $P$ as observed in the $B$-frame".

- Bold face $\boldsymbol{\omega}$ and $\boldsymbol{\alpha}$ represent, respectively, the angular velocity and angular acceleration vectors. Such vectors will always appear with superscripts on their right- and left-hand sides. For instance, you will often see angular velocities described in the form ${}^N\boldsymbol{\omega}^B$, which reads as "the angular velocity of the $B$-frame with respect to the $N$-frame". Similarly, ${}^C\boldsymbol{\alpha}^A$, reads as "the angular velocity of the $A$-frame with respect to the $C$-frame".

Now, Joe allows the vector **r** to change direction, but he also decides to sit in a three-degrees-of-freedom chair, which changes its orientation (pitch, roll, yaw), such that at any instant his orientation is always in the direction of **r**. In other words, again Joe can *only* observe changes in length. To reflect the fact that Joe's frame of reference is now rotating in space, we define a rotating frame called the $B$-frame at point $O$, as shown in Figure 1.5. In this rotating frame, the position and velocity of point $P$ with respect to $O$ can be written as

$$\begin{aligned}\mathbf{OP} &= r_1\hat{b}_1 + r_2\hat{b}_2 + r_3\hat{b}_3,\\ {}^B\mathbf{v}^{P/O} &= \dot{r}_1\hat{b}_1 + \dot{r}_2\hat{b}_2 + \dot{r}_3\hat{b}_3.\end{aligned} \tag{1.3}$$

Next, Joe decides to step out of the three degrees-of-freedom chair and observe the velocity of particle, $P$, from the inertial frame. In this case, Joe observes that the vector $r$ is simultaneously changing its length and direction. To describe the velocity of point $P$ from Joe's perspective,

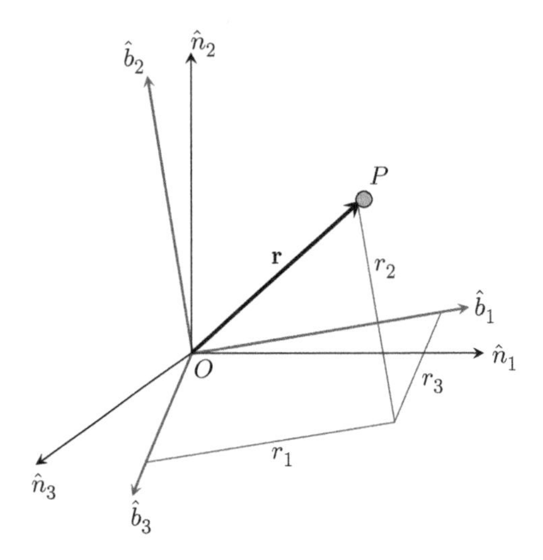


**Figure 1.5** Kinematics of a particle in a rotating frame.

we differentiate the position vector **OP** once with respect to time, taking into account that the unit vectors $\hat{b}_i$ are also changing orientation with respect to the inertial frame. This yields

$$^N\mathbf{v}^{P/O} = \underbrace{\dot{r}_1\hat{b}_1 + \dot{r}_2\hat{b}_2 + \dot{r}_3\hat{b}_3}_{\text{Translation}} + \underbrace{r_1\dot{\hat{b}}_1 + r_2\dot{\hat{b}}_2 + r_3\dot{\hat{b}}_3}_{\text{Rotation}}, \tag{1.4}$$

where the first three terms on the right-hand side represent the translational component of the velocity, while the last three terms represent the rotational component.

In Equation (1.4), the derivatives of the unit vectors $\hat{b}_i$ emerge to reflect the fact that the $B$-frame is rotating with respect to the inertial $N$-frame. The derivative of the unit vectors depends on the rate at which the rotation occurs; in other words, the angular velocity between the two frames. To this end, in order to capture the change of orientation of the $B$-frame with respect to the $N$-frame, we define the angular velocity vector $^N\boldsymbol{\omega}^B$, which describes the angular rate at which the $B$-frame rotates with respect to the $N$-frame.

Next, we turn our attention to describing the derivative of the unit vectors in terms of the angular velocity $^N\boldsymbol{\omega}^B$. For simplicity, consider Figure 1.6, where we have assumed that the $B$-frame is rotating around one axis, chosen here to be the third axis $\hat{b}_3$, at an angular velocity $^N\boldsymbol{\omega}^B = \omega_3\hat{b}_3$. The rotation of the $B$-frame at this angular velocity causes the tip of the unit vectors $\hat{b}_1$ and $\hat{b}_2$ to rotate with a velocity equal to their length – unity in this case – times the angular velocity $\omega_3$. The direction of their rotation will be perpendicular to the unit vectors themselves; that is, in the direction of $\hat{b}_2$ for unit vector $\hat{b}_1$, and in the direction of $-\hat{b}_1$ for unit vector $\hat{b}_2$. Mathematically, we can write

$$\begin{aligned} \frac{\mathrm{d}\hat{b}_1}{\mathrm{d}t} &= \omega_3\hat{b}_2, \\ \frac{\mathrm{d}\hat{b}_2}{\mathrm{d}t} &= -\omega_3\hat{b}_1, \\ \frac{\mathrm{d}\hat{b}_3}{\mathrm{d}t} &= 0. \end{aligned} \tag{1.5}$$

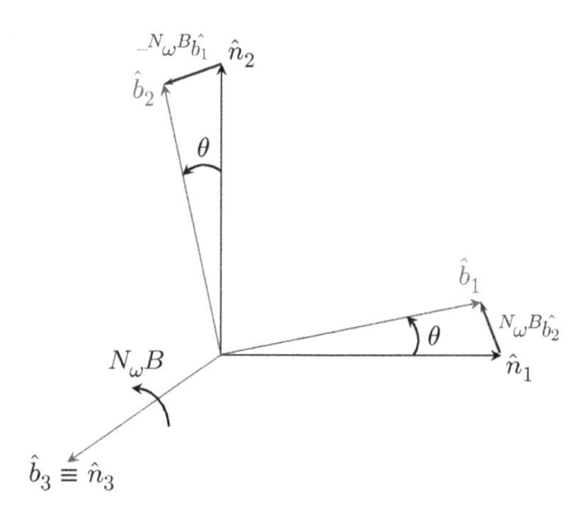


**Figure 1.6** Time rate of change of the unit vectors as a result of their rotation.

Substituting Equation (1.15) into Equation (1.4), we obtain

$$^N\mathbf{v}^{P/O} = (\dot{r}_1\hat{b}_1 + \dot{r}_2\hat{b}_2 + \dot{r}_3\hat{b}_3) + \omega_3(r_1\hat{b}_2 - r_2\hat{b}_1). \tag{1.6}$$

Note that the term $\omega_3(r_1\hat{b}_2 - r_2\hat{b}_1)$ is nothing but $\omega_3\hat{b}_3 \times \mathbf{r}$, where $\times$ refers to the cross product. Thus, we can write

$$^N\mathbf{v}^{P/O} = {}^B\mathbf{v}^{P/O} + \omega_3\hat{b}_3 \times \mathbf{r}, \tag{1.7}$$

where the first term on the right-hand side represents the change in the velocity of point $P$ with respect to $O$ as observed in the rotating frame.

By using the concept of successive rotations described earlier, the same conclusion can be achieved, even when the rotation is not restricted to one axis; that is, when ${}^N\boldsymbol{\omega}^B = \omega_1\hat{b}_1 + \omega_2\hat{b}_1 + \omega_3\hat{b}_3$. In other words, one can generalize Equation (1.7), such that

$$\boxed{{}^N\mathbf{v}^{P/O} = {}^B\mathbf{v}^{P/O} + {}^N\boldsymbol{\omega}^B \times \mathbf{r}} \tag{1.8}$$

In words, Equation (1.8) states that "the velocity of point $P$ with respect to point $O$ as observed in the $N$-frame is equal to the velocity of point $P$ with respect to point $O$ as observed in the rotating $B$-frame (translation) plus a rotational component resulting from the angular velocity of the $B$-frame with respect to the $N$-frame".

Equation (1.8) can also be generalized to any kinematic quantity defined in two different frames (velocity, acceleration, angular velocity, or angular acceleration). In general, the temporal derivative of the vector described in one frame can be expressed as its derivative in another frame plus the cross product of the angular velocity vector between the two frames and the vector itself.

 **Differentiating in a Rotating Frame**

When taking the time derivative of a vector, $\mathbf{r}$, described in a rotating frame – say the $B$-frame – with respect to an observer standing in another frame – say the $C$-frame – do not forget to add the term ${}^C\boldsymbol{\omega}^B \times \mathbf{r}$:

$$^C\frac{\mathrm{d}}{\mathrm{d}t}(\mathbf{r}) = {}^B\frac{\mathrm{d}}{\mathrm{d}t}(\mathbf{r}) + {}^C\boldsymbol{\omega}^B \times \mathbf{r}$$

**Example 1.2 Velocity of a Particle 1**

Consider a particle $P$ whose position with respect to a fixed point $O$ is given by $\mathbf{r} = 2t\hat{n}_1 + 3t^3\hat{n}_2$ m, where $\hat{n}_i$ are the unit vectors of a Cartesian inertial frame. Find the velocity of the particle after 2 s, as measured by an observer standing in the $N$-frame.

Since the components of the vector $\mathbf{r}$ are all described in the $N$-frame and the observer is also in the $N-$frame, the velocity is just a direct derivative of $\mathbf{r}$:

$$^N\mathbf{v}^{P/O} = 2\hat{n}_1 + 9t^2\hat{n}_2 = 2\hat{n}_1 + 36\hat{n}_2 \text{ m/s}.$$

**Example 1.3 Velocity of a Particle 2**

Consider a particle $P$ whose position with respect to a fixed point $O$ is given by $\mathbf{r} = 2t\hat{b}_1 + 3t^3\hat{b}_2$ m, where the $\hat{b}_i$ are the unit vectors of a Cartesian rotating frame, $B$, which rotates with respect to a stationary frame $N$ at a constant angular velocity ${}^N\boldsymbol{\omega}^B = 2\hat{b}_1 + 3\hat{b}_2$ rad/s. Find the velocity of the particle after 2 s as measured by:

(a)an observer standing in the $N$-frame;
(b)an observer standing in the $B$-frame.

***(a) Observer standing in the N-frame*** Note that the position vector is defined in the rotating frame. Hence, when taking a derivative with respect to the inertial frame, we need to take into account the rotation of the unit vectors. As such, for each component of the vector $\mathbf{r}$, we have to take a direct derivative as well as account for the rotation resulting from the angular velocity; that is,

$$\begin{aligned} {}^N\mathbf{v}^{P/O} &= 2\hat{b}_1 + 9t^2\hat{b}_2 + {}^N\boldsymbol{\omega}^{\mathbf{B}} \times (2t\hat{b}_1 + 3t^3\hat{b}_2), \\ &= 2\hat{b}_1 + 9t^2\hat{b}_2 + (6t^3 - 6t)\hat{b}_3, \\ &= 2\hat{b}_1 + 36\hat{b}_2 + 36\hat{b}_3 \text{ m/s}. \end{aligned}$$

***(b) Observer standing in the B-frame*** Note that the position vector and the observer are in the same frame. As such, we can just directly differentiate the position vector to obtain

$$\begin{aligned} {}^N\mathbf{v}^{P/O} &= 2\hat{b}_1 + 9t^2\hat{b}_2, \\ &= 2\hat{b}_1 + 36\hat{b}_2 \text{ m/s}. \end{aligned}$$

**Example 1.4 Velocity of a Particle 3**

Consider a particle $P$ whose position with respect to a fixed point $O$ is given by $\mathbf{r} = 2t\hat{n}_1 + 3t^3\hat{b}_2$ m, where the $\hat{n}_i$ are the unit vectors of a Cartesian inertial frame, $N$, and the $\hat{b}_i$ are the unit vectors of a Cartesian rotating reference, $B$, which rotates with respect to a stationary frame $N$ at a constant angular velocity ${}^N\boldsymbol{\omega}^B = 2\hat{b}_1 + 3\hat{b}_2$ rad/s. Find the velocity of the particle after 2 s as measured by an observer standing in the $N$-frame.

Notice that, in this case, only a part of the position vector is described in the $B$-frame, hence when finding the velocity of the particle as measured by a stationary observer, we need to account *only* for the rotation associated with that component, namely, $3t^3\hat{b}_2$. In other words, we can write

$$\begin{aligned} {}^N\mathbf{v}^{P/O} &= 2\hat{n}_1 + 9t^2\hat{b}_2 + (2\hat{b}_1 + 3\hat{b}_2) \times 3t^3\hat{b}_2 \\ &= 2\hat{n}_1 + 9t^2\hat{b}_2 + 6t^3\hat{b}_3 = 2\hat{n}_1 + 36\hat{b}_2 + 48\hat{b}_3 \text{ m/s}. \end{aligned}$$

**Flipped Classroom Exercise 1.2**

In this exercise, we will walk through the process of finding the derivative of the unit vectors $\frac{\mathrm{d}\hat{b}_i}{\mathrm{d}t}$ for a general three-dimensional rotation. The rotation is carried out through three successive angles $(\theta, \phi, \psi)$. To this end, following Figure 1.2, we will assume that the rotation is a 1-2-3 $(\theta, \phi, \psi)$ rotation. The first rotation goes from the $N$- to the $E$-frame. The second rotation goes from the $E$- to the $C$-frame while the third rotation goes from the $C$- to the $B$-frame.

Using the assumed definition of the successive rotations, you can write

$$\begin{aligned} {}^N\boldsymbol{\omega}^E &= \dot{\theta}\hat{e}_1 = \omega_1\hat{e}_1, \\ {}^E\boldsymbol{\omega}^C &= \dot{\phi}\hat{c}_1 = \omega_2\hat{c}_2, \\ {}^C\boldsymbol{\omega}^B &= \dot{\psi}\hat{b}_3 = \omega_3\hat{b}_2, \end{aligned}$$

Thus, ${}^N\boldsymbol{\omega}^B = {}^N\boldsymbol{\omega}^E + {}^E\boldsymbol{\omega}^C + {}^C\boldsymbol{\omega}^B = \dot{\theta}\hat{e}_1 + \dot{\phi}\hat{c}_2 + \dot{\psi}\hat{b}_3 = \omega_1\hat{e}_1 + \omega_2\hat{c}_2 + \omega_3\hat{b}_3$.

Next, you need to do the following:

1. Express the unit vector **n** in terms of the unit vector **b** using the 1-2-3 rotation derived previously.
2. Assume small angles $(\mathrm{d}\theta, \mathrm{d}\phi, \mathrm{d}\psi)$ and linearize the rotation matrix as well as the angular velocity vector. Note that this assumption is not restrictive, since any general rotation can be described in terms of series of tiny successive rotations. Show that upon linearizing you obtain

$$\begin{bmatrix} \hat{n}_1 \\ \hat{n}_2 \\ \hat{n}_3 \end{bmatrix} = \mathcal{I} \begin{bmatrix} \hat{b}_1 \\ \hat{b}_2 \\ \hat{b}_3 \end{bmatrix} + \begin{pmatrix} 0 & -\mathrm{d}\psi & \mathrm{d}\phi \\ \mathrm{d}\psi & 0 & -\mathrm{d}\theta \\ -\mathrm{d}\phi & \mathrm{d}\theta & 0 \end{pmatrix} \begin{bmatrix} \hat{b}_1 \\ \hat{b}_2 \\ \hat{b}_3 \end{bmatrix}$$

and that for small angles ${}^N\boldsymbol{\omega}^B = \dot{\theta}\hat{b}_1 + \dot{\phi}\hat{b}_2 + \dot{\psi}\hat{b}_3$.

3. Using the previous equations, show that we can write

$$\begin{bmatrix} \mathrm{d}\hat{b}_1 \\ \mathrm{d}\hat{b}_2 \\ \mathrm{d}\hat{b}_3 \end{bmatrix} = \begin{pmatrix} 0 & -\mathrm{d}\psi & \mathrm{d}\phi \\ \mathrm{d}\psi & 0 & -\mathrm{d}\theta \\ -\mathrm{d}\phi & \mathrm{d}\theta & 0 \end{pmatrix} \begin{bmatrix} \hat{b}_1 \\ \hat{b}_2 \\ \hat{b}_3 \end{bmatrix}.$$

4. Divide the previous equation by $\mathrm{d}t$ and show that

$$\begin{bmatrix} \dot{\hat{b}}_1 \\ \dot{\hat{b}}_2 \\ \dot{\hat{b}}_3 \end{bmatrix} = \begin{pmatrix} 0 & -\omega_3 & \omega_2 \\ \omega_2 & 0 & -\omega_1 \\ -\omega_2 & \omega_1 & 0 \end{pmatrix} \begin{bmatrix} \hat{b}_1 \\ \hat{b}_2 \\ \hat{b}_3 \end{bmatrix}.$$

5. Show that the previous equation can be written as

$$\dot{\mathbf{b}} = {}^N\boldsymbol{\omega}^B \times \mathbf{b}.$$

## 1.5 Acceleration of a Particle in a Three-dimensional Space

To find a general expression for the acceleration of a particle in a three-dimensional space with respect to an inertial frame, we differentiate Equation (1.8) once with respect to time, noting that the term on the left-hand side is defined in the stationary $N$-frame while all terms on the right-hand side are described in the rotating $B$ frame. In other words, we need to find ${}^N\frac{\mathrm{d}}{\mathrm{d}t}({}^N\mathbf{v}^{P/O})$, where the superscript on the left-hand side of the derivative is used to denote that the change of the quantity included within the derivative is observed from the inertial frame. Differentiating each term in Equation (1.8) yields

$$\begin{aligned}
{}^N\frac{\mathrm{d}}{\mathrm{d}t}({}^N\mathbf{v}^{P/O}) &= {}^N\mathbf{a}^{P/O}, \\
{}^N\frac{\mathrm{d}}{\mathrm{d}t}({}^B\mathbf{v}^{P/O}) &= {}^B\frac{\mathrm{d}}{\mathrm{d}t}({}^B\mathbf{v}^{P/O}) + {}^N\boldsymbol{\omega}^{\mathbf{B}} \times {}^B\mathbf{v}^{P/O} = {}^B\mathbf{a}^{P/O} + {}^N\boldsymbol{\omega}^{\mathbf{B}} \times {}^B\mathbf{v}^{P/O}, \\
{}^N\frac{\mathrm{d}}{\mathrm{d}t}({}^N\boldsymbol{\omega}^B \times \mathbf{r}) &= {}^B\frac{\mathrm{d}}{\mathrm{d}t}({}^N\boldsymbol{\omega}^B \times \mathbf{r}) + {}^N\boldsymbol{\omega}^B \times ({}^N\boldsymbol{\omega}^B \times \mathbf{r}), \\
&= {}^N\boldsymbol{\alpha}^B \times \mathbf{r} + {}^N\boldsymbol{\omega}^{\mathbf{B}} \times {}^B\mathbf{v}^{P/O} + {}^N\boldsymbol{\omega}^B \times ({}^N\boldsymbol{\omega}^B \times \mathbf{r}),
\end{aligned}$$

where ${}^N\boldsymbol{\alpha}^B = {}^B\frac{\mathrm{d}}{\mathrm{d}t}({}^N\boldsymbol{\omega}^B)$ is the angular acceleration vector. Using the previous equation, we can write

$$\boxed{{}^N\mathbf{a}^{P/O} = {}^B\mathbf{a}^{P/O} + 2{}^N\boldsymbol{\omega}^B \times {}^B\mathbf{v}^{P/O} + {}^N\boldsymbol{\alpha}^B \times \mathbf{r} + {}^N\boldsymbol{\omega}^B \times ({}^N\boldsymbol{\omega}^B \times \mathbf{r})} \qquad (1.9)$$

Equation (1.9) represents the general formula for the acceleration of a particle in space. The first term on the right-hand side represents the acceleration of the particle in the rotating frame. The second term represents the *Coriolis* acceleration, which, as evident, is perpendicular to the velocity of the particle as measured in the rotating frame and the direction of the frame's rotation. The third term is the tangential acceleration, and finally the fourth term is the normal acceleration.

While it is possible to use the acceleration formula as described in the form shown in Equation (1.9), it is absolutely unnecessary to memorize it since one can use the basic understanding utilized throughout its derivation to calculate the acceleration of any particle in space. To demonstrate this, consider the following series of examples:

**Example 1.5 Acceleration of a Particle**

Consider a particle $P$ whose position with respect to a fixed point $O$ is given by $\mathbf{r} = 2t\hat{b}_1 + 3t^3\hat{b}_2$ m, where the $\hat{n}_i$ are the unit vectors of a Cartesian inertial frame, $N$, and the $\hat{b}_i$ are the unit vectors of a Cartesian rotating reference, $B$, which rotates with respect to the stationary frame $N$ at a constant angular velocity ${}^N\boldsymbol{\omega}^B = 2\hat{b}_1 + 3\hat{b}_2$ rad/s. Find the acceleration of the particle after 2 s as measured by an observer standing in the $N$-frame.

In Example 1.3, the velocity of the particle with respect to the inertial observer was found to be

$$ {}^N\mathbf{v}^{P/O} = 2\hat{b}_1 + 9t^2\hat{b}_2 + (6t^3 - 6t)\hat{b}_3 \text{ m/s}. $$

To calculate the acceleration in the reference frame, we differentiate the previous expression, taking into account the rotational component of any term described in a rotating frame. Note also that the angular acceleration ${}^N\boldsymbol{\alpha}^B$ vanishes because the angular velocity is constant. As such, we can write

$$ \begin{aligned} {}^N\mathbf{a}^{P/O} &= 18t\hat{b}_2 + (18t^2 - 6)\hat{b}_3 + (2\hat{b}_1 + 3\hat{b}_2) \times (2\hat{b}_1 + 9t^2\hat{b}_2 + (6t^3 - 6t)\hat{b}_3), \\ &= 18(t^3 - t)\hat{b}_1 + (30t - 12t^3)\hat{b}_2 + (36t^2 - 12)\hat{b}_3, \\ &= 108\hat{b}_1 - 36\hat{b}_2 + 132\hat{b}_3 \text{ m/s}^2. \end{aligned} $$

**Example 1.6 Simple Pendulum Kinematics**

For the simple pendulum of rigid cable length $l$ shown in Figure 1.7, find the velocity and acceleration of particle $P$ with respect to point $O$. Choose your rotating $B$-frame such that ${}^N\boldsymbol{\omega}^B = \dot{\theta}\hat{b}_3$ and express all your answers in the $B$-frame.

We start by defining the position vector from $O$ to $P$ as

$$ \mathbf{OP} = -l\hat{b}_2. $$

It follows that the velocity of point $P$ as observed by an inertial observer at point $O$ is

$$ {}^N\mathbf{v}^{P/O} = \frac{\mathrm{d}\mathbf{OP}}{\mathrm{d}t} = -\dot{l}\hat{b}_2 + {}^N\boldsymbol{\omega}^B \times \mathbf{OP}. $$

Since the pendulum does not change length, $\dot{l} = 0$. Therefore we can write

$$ {}^N\mathbf{v}^{P/O} = \dot{\theta}\hat{b}_3 \times \mathbf{OP} = l\dot{\theta}\hat{b}_1. $$

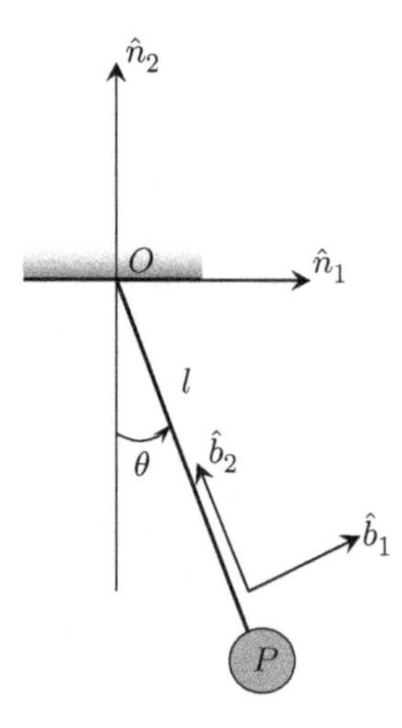


**Figure 1.7** Kinematics of a simple pendulum.

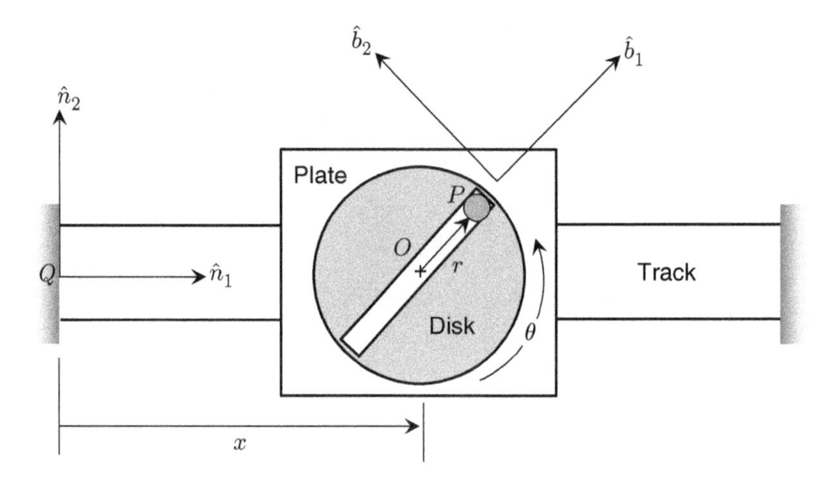


**Figure 1.8** Two-dimensional motion of a particle.

The acceleration of point $P$ with respect to point $O$ can be written as

$$^N\mathbf{a}^{P/O} = \frac{\mathrm{d}^N\mathbf{v}^{P/O}}{\mathrm{d}t} = l\ddot{\theta}\hat{b}_1 + {}^N\boldsymbol{\omega}^B \times {}^N\mathbf{v}^{P/O} = l\ddot{\theta}\hat{b}_1 + l\dot{\theta}^2\hat{b}_2.$$

**Example 1.7 Single-rotation Kinematics**

Consider the system shown in Figure 1.8, which consists of a plate sliding on a track. A disk is mounted on top of the moving plate and is free to rotate through an angle $\theta$ about point $O$. A particle, $P$, is free to move in a groove drilled across the disk as shown. Define two frames: an inertial frame located at point $Q$ and a rotating frame such that $\hat{b}_3 = \hat{n}_3$ and $^N\boldsymbol{\omega}^B = \dot{\theta}\hat{b}_3$ to obtain

- the velocity of the particle $P$ with respect to $O$ as measured by an observer standing in the $B$-frame, ${}^B\mathbf{v}^{P/O}$;
- the velocity of the particle $P$ with respect to $O$ as measured by an observer standing in the $N$-frame, ${}^N\mathbf{v}^{P/O}$;
- the acceleration of the particle $P$ with respect to $Q$ as measured by an observer standing in the $N$-frame, ${}^N\mathbf{a}^{P/Q}$.

Describe your answers in the $B$-coordinate system.

To find ${}^B\mathbf{v}^{P/O}$, we define the position vector **OP** and differentiate it once with respect to time; that is,

$$\mathbf{OP} = r\hat{b}_1, \qquad {}^B\mathbf{v}^{P/O} = \dot{r}\hat{b}_1.$$

To find ${}^N\mathbf{v}^{P/O}$, we carry out the same procedure but account for the relative rotation between the $B$- and $N$-frame; that is,

$$\begin{aligned}\mathbf{OP} &= r\hat{b}_1,\\ {}^N\mathbf{v}^{P/O} &= \dot{r}\hat{b}_1 + \dot{\theta}\hat{b}_3 \times r\hat{b}_1,\\ {}^N\mathbf{v}^{P/O} &= \dot{r}\hat{b}_1 + r\dot{\theta}\hat{b}_2.\end{aligned}$$

To find ${}^N\mathbf{a}^{P/Q}$, we define the position vector **QP** and differentiate it twice with respect to time, accounting for the relative rotation between the $B$- and $N$-frame. This yields

$$\begin{aligned}\mathbf{QP} &= x\hat{n}_1 + r\hat{b}_1,\\ {}^N\mathbf{v}^{P/Q} &= \dot{x}\hat{n}_1 + \dot{r}\hat{b}_1 + \dot{\theta}\hat{b}_3 \times r\hat{b}_1,\\ {}^N\mathbf{v}^{P/Q} &= \dot{x}\hat{n}_1 + \dot{r}\hat{b}_1 + r\dot{\theta}\hat{b}_2,\\ {}^N\mathbf{a}^{P/Q} &= \ddot{x}\hat{n}_1 + (\ddot{r} - r\dot{\theta}^2)\hat{b}_1 + (r\ddot{\theta} + 2\dot{r}\dot{\theta})\hat{b}_2.\end{aligned}$$

To express the previous answer in the $B$-frame, we still need to express $\hat{n}_1$ in terms of the $B$-coordinates. Note that the $B$-frame was formed by carrying out a counter-clockwise rotation around the $\hat{n}_3$ axis. Hence, using the definition of the 3-rotation, we can write $\hat{n}_1 = \cos\theta\hat{b}_1 - \sin\theta\hat{b}_2$. This yields

$${}^N\mathbf{a}^{P/Q} = (\ddot{x}\cos\theta + \ddot{r} - r\dot{\theta}^2)\hat{b}_1 + (r\ddot{\theta} + 2\dot{r}\dot{\theta} - \ddot{x}\sin\theta)\hat{b}_2.$$

**Example 1.8 Radar-tracking Kinematics**

A radar station is tracking a rocket that has just lifted off vertically with a velocity $v_r$ and acceleration $a_r$. The station is located such that the distance between the radar station and the rocket is $R$ (Figure 1.9). Calculate $\dot{R}$, $\ddot{R}$, $\dot{\theta}$, $\ddot{\theta}$ in terms of $v_r$, $a_r$, $R$, and $\theta$ only.
We begin by calculating ${}^N\mathbf{v}^{P/O}$. To this end, we let

$$\mathbf{OP} = R\hat{b}_1, \qquad {}^N\mathbf{v}^{P/O} = \dot{R}\hat{b}_1 + R\dot{\theta}\hat{b}_2.$$

Using $v_r\hat{n}_1 = v_r\sin\theta\hat{b}_1 + v_r\cos\theta\hat{b}_2$, we conclude that $\dot{R} = v_r\sin\theta$ and $\dot{\theta} = v_r\cos\theta/R$.

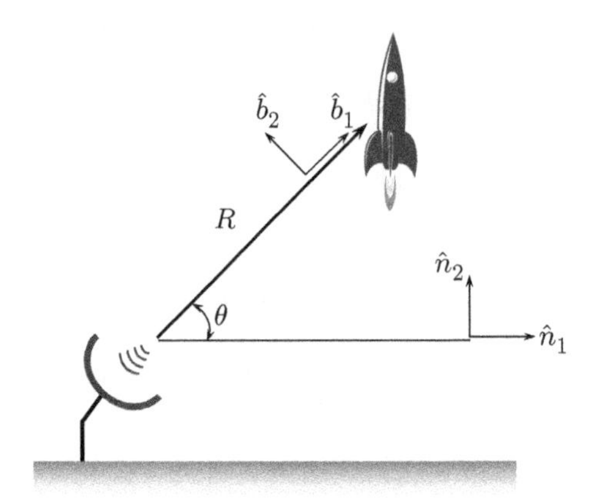


**Figure 1.9** A rocket detected by a radar station.

Next, we calculate the acceleration vector

$$^N\mathbf{a}^{P/O} = (\ddot{R} - R\dot{\theta}^2)\hat{b}_1 + (2\dot{R}\dot{\theta} + R\ddot{\theta})\hat{b}_2.$$

Using $a_r\hat{n}_1 = a_r \sin\theta\hat{b}_1 + a_r \cos\theta\hat{b}_2$, we obtain

$$\ddot{R} = a_r \sin\theta + \frac{v_r^2}{R}\cos^2\theta,$$

$$\ddot{\theta} = \frac{a_r}{R}\cos\theta - \frac{v_r^2}{R^2}\sin 2\theta.$$

### Example 1.9 Cylindrical Coordinates

Obtain the general velocity and acceleration expressions for the motion of a particle as described in a cylindrical coordinate system.

We start by defining a stationary frame, the $N$-frame, as shown in Figure 1.10. Subsequently, we define a rotating frame, the $B$-frame, which is formed through a rotation angle $\phi$, such that $\hat{b}_2 \equiv \hat{n}_2$ is always aligned with the direction of $z$, and $\hat{b}_3$ is always aligned with the direction of $r$. Using this understanding, we can write $^N\boldsymbol{\omega}^B = \dot{\phi}\hat{n}_2 = \dot{\phi}\hat{b}_2$, $^N\boldsymbol{\alpha}^B = \ddot{\phi}\hat{n}_2 = \ddot{\phi}\hat{b}_2$. In the $B$-frame, the position vector from $O$ to $P$ can be written as

$$\mathbf{OP} = z\hat{b}_2 + r\hat{b}_3.$$

The velocity of the particle as measured by a stationary observer at $O$ can be written as

$$^N\mathbf{v}^{P/O} = \dot{z}\hat{b}_2 + \dot{r}\hat{b}_3 + {}^N\boldsymbol{\omega}^B \times (z\hat{b}_2 + r\hat{b}_3),$$

$$^N\mathbf{v}^{P/O} = r\dot{\phi}\hat{b}_1 + \dot{z}\hat{b}_2 + \dot{r}\hat{b}_3.$$

The acceleration of the particle as measured by a stationary observer at $O$ can be obtained by differentiating the previous equation once with respect to time, taking into account the relative

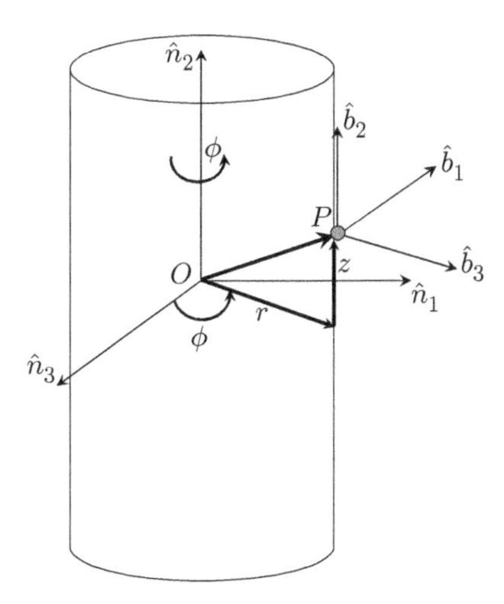


**Figure 1.10** Description of the motion of a particle in a cylindrical coordinate system.

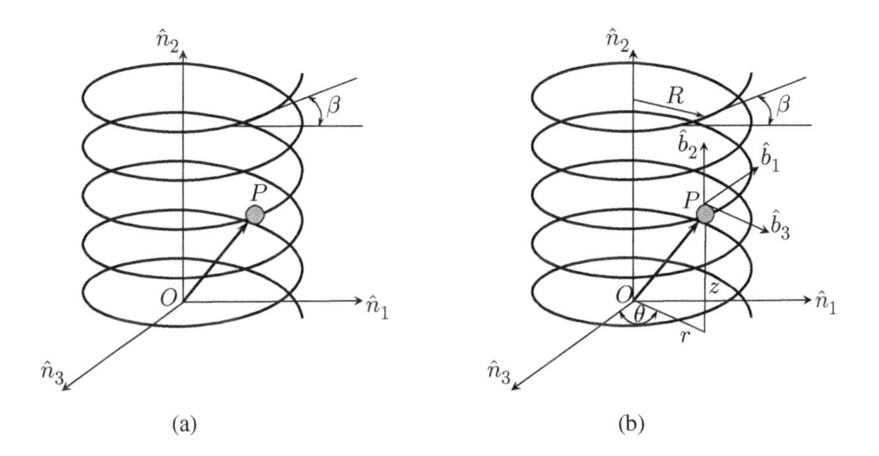


**Figure 1.11** Motion on a helix.

rotations between the two frames. This yields

$$
\begin{aligned}
{}^{N}\mathbf{a}^{P/O} &= (r\ddot{\phi} + \dot{r}\dot{\phi})\hat{b}_1 + \ddot{z}\hat{b}_2 + \ddot{r}\hat{b}_3 + {}^{N}\boldsymbol{\omega}^{B} \times (r\dot{\phi}\hat{b}_1 + \dot{z}\hat{b}_2 + \dot{r}\hat{b}_3) \\
{}^{N}\mathbf{a}^{P/O} &= (r\ddot{\phi} + 2\dot{r}\dot{\phi})\hat{b}_1 + \ddot{z}\hat{b}_2 + (\ddot{r} - r\dot{\phi}^2)\hat{b}_3.
\end{aligned}
$$

### Example 1.10 Helix

Consider the motion of a particle along the helical path shown in Figure 1.11a, where $R$ and $\beta$ are respectively its radius and angle. Find the velocity and acceleration of the particle with respect to point $O$.

To describe the kinematics of the problem, we use cylindrical coordinates and define the rotating frame as shown in Figure 1.11b, where ${}^N\boldsymbol{\omega}^B = \dot{\theta}\hat{b}_2$. We define the position vector **OP** as

$$\mathbf{OP} = r\hat{b}_3 + z\hat{b}_2 = R\hat{b}_3 + R\theta\tan\beta\hat{b}_2.$$

The velocity can then be written as

$${}^N\mathbf{v}^{P/O} = (R\dot{\theta}\tan\beta)\hat{b}_2 + \dot{\theta}\hat{b}_2 \times (R\hat{b}_3 + R\theta\tan\beta\hat{b}_2) = R\dot{\theta}\hat{b}_1 + R\dot{\theta}\tan\beta\hat{b}_2,$$

and the acceleration is

$${}^N\mathbf{a}^{P/O} = R\ddot{\theta}\hat{b}_1 + R\ddot{\theta}\tan\beta\hat{b}_2 - R\dot{\theta}^2\hat{b}_3.$$

 **Flipped Classroom Exercise 1.3**

Point $P$ on a thin disk of radius $r$ rotates about its own axis at point $B$ through an angle $\theta$. The yoke where it is mounted also rotates about the line $OB$ through an angle $\beta$. The entire assembly rotates about the $\hat{n}_2$-axis shown in the figure through angle $\phi$. To fully describe the kinematics of the problem, define three rotating frames such that ${}^N\boldsymbol{\omega}^A = \dot{\phi}\hat{n}_2$, ${}^A\boldsymbol{\omega}^C = -\dot{\beta}\hat{a}_1$, and ${}^C\boldsymbol{\omega}^D = \dot{\theta}\hat{c}_3$, then find the following:

- ${}^C\mathbf{a}^{P/B}$
- ${}^A\mathbf{a}^{P/B}$

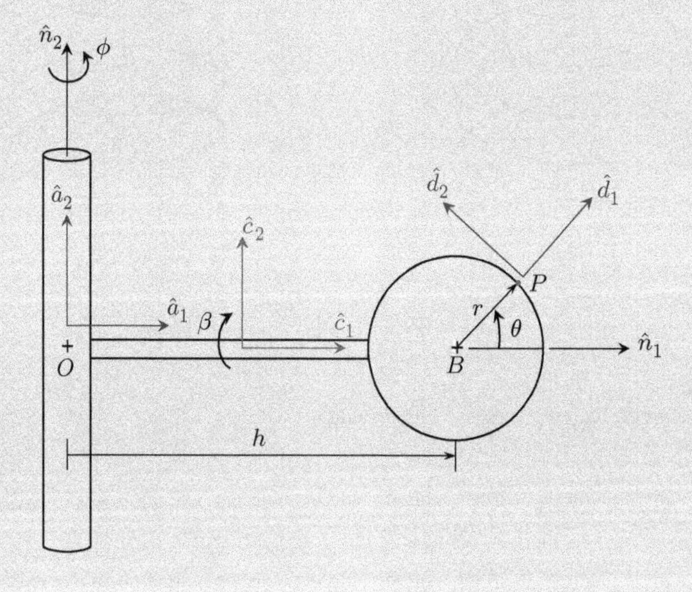

To find ${}^{C}\mathbf{a}^{P/B}$, you need to do the following:

1. Define the position vector from $B$ to $P$ in the easiest possible way. Note that regardless of how this complex system rotates, $r$ will always be in the direction of $\hat{d}_1$.
2. Find the velocity as seen by an observer in the $C$ frame. Note that $\dot{r} = 0$ because the radius of the circle does not change length
3. Show that the acceleration is given by

$$\begin{aligned} {}^{C}\mathbf{a}^{P/B} &= r\ddot{\theta}\hat{d}_2 + {}^{C}\boldsymbol{\omega}^{D} \times r\dot{\theta}\hat{d}_2, \\ &= r\ddot{\theta}\hat{d}_2 - r\dot{\theta}\hat{d}_1. \end{aligned}$$

To find ${}^{A}\mathbf{a}^{P/B}$, you need to do the following:

1. Define the position vector from $B$ to $P$.
2. Show that the velocity of point $P$ with respect to point $B$ as observed in the $A$-frame is given by

$$ {}^{A}\mathbf{v}^{P/B} = {}^{A}\boldsymbol{\omega}^{D} \times r\hat{d}_1 = (-\dot{\beta}\hat{c}_1 + \dot{\theta}\hat{d}_3) \times r\hat{d}_1. $$

   Note that, to carry the cross product between vectors described in two different frames, it is much more convenient to rotate one of them such that they are both described in the same frame.
3. Show that the acceleration of point $P$ with respect to point $B$ as observed in the $A$ frame is given by

$$\begin{aligned} {}^{A}\mathbf{a}^{P/B} &= r\ddot{\theta}\hat{d}_2 - (r\ddot{\beta}\sin\theta + r\dot{\beta}\dot{\theta}\cos\theta)\hat{d}_3 + {}^{C}\boldsymbol{\omega}^{D} \times (r\dot{\theta}\hat{d}_2 - r\dot{\beta}\sin\theta\hat{d}_3), \\ &= -(r\dot{\theta}^2 + r\dot{\beta}^2\sin^2\theta)\hat{d}_1 + (r\ddot{\theta} - r\dot{\beta}^2\cos\theta\sin\theta)\hat{d}_2, \\ &\quad - (r\ddot{\beta}\sin\theta + 2r\dot{\theta}\dot{\beta}\cos\theta)\hat{d}_3. \end{aligned}$$

## Exercises

1.1 Find the rotation matrix necessary to transform a frame $N$ to another frame $B$ by performing a 1-2-1 rotation using angles $(\theta, \phi, \psi)$.

1.2 A position vector from point $O$ to point $P$ is defined as $\mathbf{OP} = 2t\hat{n}_1 + 3\sin t\hat{c}_2 + t^3\hat{b}_1$ m, where ${}^{N}\boldsymbol{\omega}^{C} = 2\hat{c}_2 + 3t\hat{c}_1$ rad/s and ${}^{C}\boldsymbol{\omega}^{B} = 2\sin t\hat{b}_2 + 3\hat{b}_3$ rad/s. Here, $N$ is an inertial frame, while $B$ and $C$ are rotating frames. Find the velocity and acceleration of point $P$ with respect to point $O$ as observed in the inertial frame at $t = \pi$ s.

1.3 A position vector from point $O$ to point $P$ is defined as $\mathbf{OP} = 2t\hat{a}_1 + 3\sin t\hat{a}_2 + t^3\hat{b}_1$ m, where ${}^N\boldsymbol{\omega}^A = 2\hat{a}_2 + 3t\hat{a}_1$ rad/s and ${}^A\boldsymbol{\omega}^B = 3\hat{b}_3$ rad/s. Here, $N$ is an inertial frame, while $B$ and $A$ are rotating frames. Find the velocity and acceleration of point $P$ with respect to point $O$ as observed in the inertial frame at $t = \frac{\pi}{2}$ s.

1.4 Derive the velocity and acceleration of a particle in a spherical coordinate system. Use Figure 1.12 for the definition of the rotation angles.

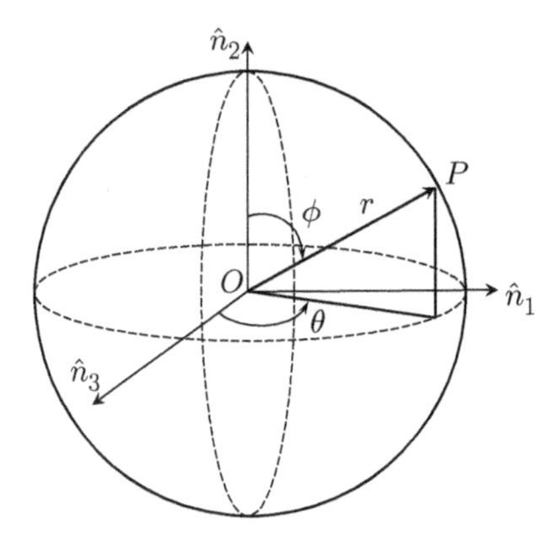


**Figure 1.12** Exercise 1.4.

1.5 In kinematics, the derivative of the acceleration is defined as the Jerk, $\mathbf{J}$. Imagine we live in a world in which the Jerk replaces the acceleration in Newton's second law. Using the general acceleration formula we derived in the chapter, derive a general formula for the Jerk, ${}^N\mathbf{J}^{P/O}$ of a particle $P$ with respect to an inertial point, $O$.

1.6 In the two-dimensional problem shown in Figure 1.13, a particle $P$ of mass $m$ slides in a pipe which rotates with an angle $\phi$ relative to a rod of length $l$ as shown in the figure. The rod, in turn, rotates with an angle $\theta$ relative to the vertical. Define $r$ as the distance from point $Q$ to point $P$ and your rotating coordinates systems such that ${}^N\boldsymbol{\omega}^A = \dot{\theta}\hat{n}_3$ and ${}^A\boldsymbol{\omega}^B = \dot{\phi}\hat{a}_3$, then find:

(a) ${}^A\mathbf{v}^{P/O}$
(b) ${}^N\mathbf{v}^{Q/O}$
(c) ${}^A\mathbf{a}^{P/Q}$
(d) ${}^B\mathbf{a}^{P/Q}$

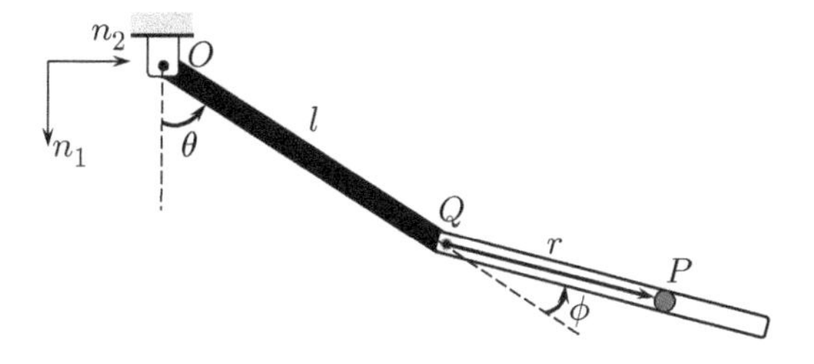


**Figure 1.13** Exercise 1.6.

1.7 For the system shown in Figure 1.14, find ${}^N\mathbf{v}^{P/Q}$, ${}^N\mathbf{a}^{P/Q}$, ${}^N\boldsymbol{\alpha}^B$, and ${}^D\mathbf{a}^{P/O}$. Express all your answers in the $B$-frame and define your frames such that ${}^N\boldsymbol{\omega}^D = \dot{\phi}\hat{n}_2$, ${}^D\boldsymbol{\omega}^B = \dot{\theta}\hat{d}_3$.

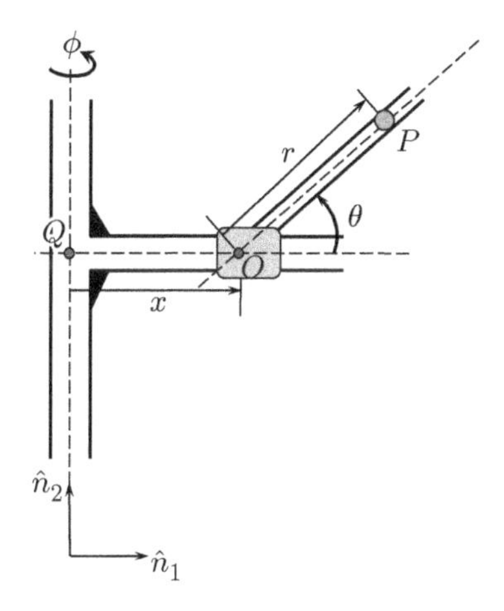


**Figure 1.14** Exercise 1.7.

1.8 For the system shown in Figure 1.15, the column is forced to rotate with constant angular velocity $\dot{\phi}$. A rod of length $d$ is welded onto one end to the column (point $O$) and is pinned at the other end to a rod of length $l$ (point $G$). The rod, of length $l$, is welded to a frictionless pipe at point $S$. A particle $P$ of mass $m$ is free to slide inside the pipe. Define your coordinate system such that ${}^N\boldsymbol{\omega}^A = \dot{\phi}\hat{a}_2$ and ${}^A\boldsymbol{\omega}^B = \dot{\theta}\hat{b}_3$, then find the following:

(a) ${}^A\mathbf{v}^{P/S}$

(b) ${}^A\mathbf{a}^{P/S}$

(c) ${}^A\mathbf{a}^{P/O}$

Express all your answers in the $B$ coordinate system.

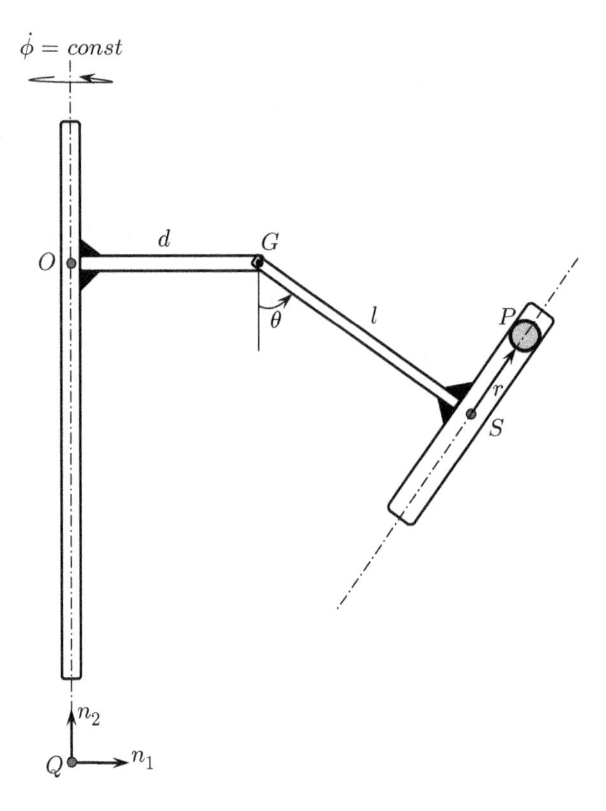


**Figure 1.15** Exercise 1.8.

1.9 The bead shown in Figure 1.16 is constrained to move along the path shown, the radius of which changes with $\theta$ according to $r = \frac{a}{1+\theta}$. Obtain the velocity and acceleration of the particle with respect to point $O$ as observed in the inertial frame. Assuming $\dot{\theta} = 2$ rad/s, what are the velocity and acceleration at $\theta = \pi/2$?

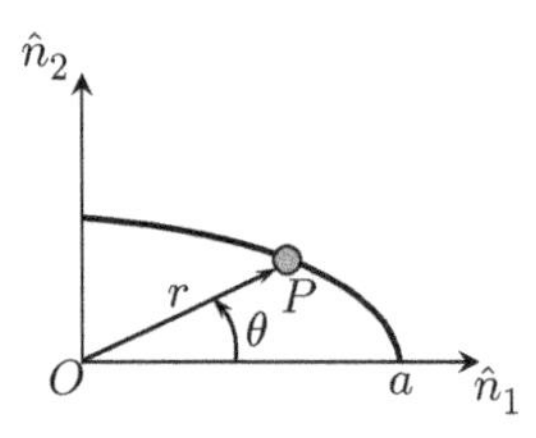


**Figure 1.16** Exercise 1.9.

1.10 Two particles $P_1$ and $P_2$ are rotating with constant angular velocities on the circular paths shown. The angular velocities of $P_1$ and $P_2$ with respect to an inertial frame are $\dot{\theta}$ and $\dot{\phi}$, respectively. Find the velocity and acceleration of $P_1$ with respect to an observer moving with $P_2$.

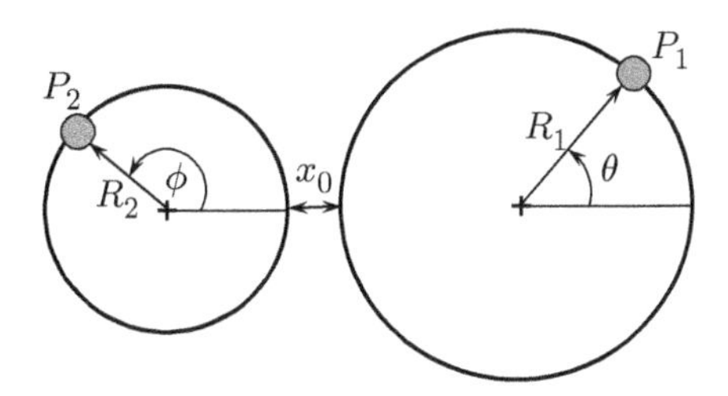


**Figure 1.17** Exercise 1.10.

1.11 A particle moves on the inside surface of a cone of half angle $\alpha$. The axis of the cone is vertical with the vertex pointing downwards. Find the velocity and acceleration of the particle with respect to an inertial point.

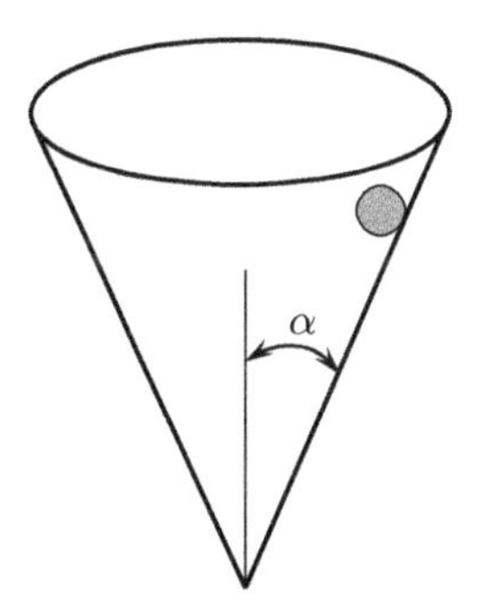


**Figure 1.18** Exercise 1.11.

# 2

# Dynamics of Particles: Vectorial Approach

In Chapter 1, we learned how to find the position, velocity, and acceleration of a particle – its kinematic description – without worrying about the forces causing its motion. While this constitutes an important first step towards learning dynamics, kinematics does not lend itself naturally to obtaining the equations governing the motion of the particle. Such equations are what we are after in this book because, through their solution, we can describe and predict motion of particles. This chapter outlines the most common approach – the vectorial approach – through which the equations governing the motion of particles are derived. To this end, we introduce Newton's second law of dynamics and demonstrate how it can be used to relate the forces acting on a particle to its ensuing motion. Examples involving planar and non-planar motion are discussed.

## 2.1 Newton's Second Law of Dynamics

The most common approach to relate forces acting on a particle to its ensuing motion is by using Newton's second law of dynamics, which is widely known as the *vectorial approach* to mechanics. Newton described his second law for the first time in the *Principia*, where he stated that:

> *The alteration of motion is proportional to the motive force impressed; and is made in the direction of the right line in which that force is impressed.*

In most modern books of classical mechanics, this statement has been rephrased to be more accessible to today's readers. As such, it is more common to see Newton's second law stated as:

> *The acceleration of an object as produced by a net force is directly proportional to the magnitude of the net force, in the same direction as the net force, and inversely proportional to the mass of the object.*

---

*Dynamics of Particles and Rigid Bodies: A Self-Learning Approach*, First Edition. Mohammed F. Daqaq.

Companion website: www.wiley.com/go/daqaq/dynamics

Mathematically, Newton's second law can be written as:

$$\mathbf{F} = m\mathbf{a} \tag{2.1}$$

where **F** is the force vector acting on the particle, $m$ is the mass of the particle, and **a** is its acceleration vector measured with respect to an inertial frame. Newton arrived at this equation using empirical observations of particles in motion. Nevertheless, many historians argue that Euler was the one who first stated Newton's second law mathematically in the vectorial form widely-used today [1].

Since the motion of a free particle in a Cartesian coordinate system is described by three coordinates, which are commonly denoted as $(x, y, z)$, Equation (2.1) holds true along the three different unit vectors $(\hat{n}_1, \hat{n}_2, \hat{n}_3)$ describing the Cartesian coordinate system; that is,

$$F_{\hat{n}_1} = ma_{\hat{n}_1}, \qquad F_{\hat{n}_2} = ma_{\hat{n}_2}, \qquad F_{\hat{n}_3} = ma_{\hat{n}_3} \tag{2.2}$$

When using Newton's second law of dynamics, it is important to keep in mind that Equation (2.2) holds in this simple form *only* when the following two conditions are satisfied:

**Conditions for Applying Newton's Second Law of Dynamics**

1. The mass of the particle does not vary with time. Otherwise, one has to revert to the more general form, which states that the net force acting on a particle is equal to the change in its linear momentum. The case where the mass of the system changes with time will be addressed in more detail in Chapter 7.
2. The acceleration of the particle is measured with respect to an observer standing in an inertial frame.

Assuming that these conditions are satisfied, the process of implementing Newton's second law to obtain the equations governing the motion of a particle is straightforward. This is especially the case when one has mastered the kinematic description of particles as described previously in Chapter 1, and knows how to describe the forces acting on the particle. In what follows, we provide a systematic procedure which can be easily followed to obtain the equations governing the motion of a particle by using Newton's second law of dynamics.

**Procedure for Implementing Newton's Second Law of Dynamics**

1. Define the inertial frame and any rotating frames needed to describe the acceleration of the particle with respect to the inertial frame.
2. Find the acceleration of the particle with respect to the inertial frame.
3. Draw a free-body diagram of the forces acting on the particle. If needed, project these forces onto the frames defined.
4. Equate the force vector in a given direction to the associated mass times acceleration vector in the same direction, as described by Equation (2.2).

**Example 2.1 Simple Pendulum Dynamics**

Consider the simple pendulum shown in Figure 2.1. Obtain the equations governing the motion of the pendulum and the tension in the cable.

To find the equations of motion, we will follow the procedure described above:

1. *Define the inertial frame and any rotating frames needed to describe the acceleration of the particle with respect to the inertial frame.*

   In this example, we use point $O$ as our inertial point, Figure 2.1. It is located in an inertial frame, $N$. We also define a rotating frame, the $B$-frame, such that $\hat{b}_1$ is always normal to the cable and ${}^N\boldsymbol{\omega}^B = \dot{\theta}\hat{b}_3$. The choice of the $B$-frame is made merely to simplify the calculation of the acceleration of point $P$ with respect to point $O$ as observed in the inertial frame.
2. *Find the acceleration of the particle with respect to the inertial frame.*

   To this end, we write

$$\mathbf{OP} = -l\hat{b}_2,$$
$${}^N\mathbf{v}^{P/O} = l\dot{\theta}\hat{b}_1,$$
$${}^N\mathbf{a}^{P/O} = l\ddot{\theta}\hat{b}_1 + l\dot{\theta}^2\hat{b}_2.$$

3. *Draw a free-body diagram of the forces acting on the particle. If needed, project these forces into the frames defined.*

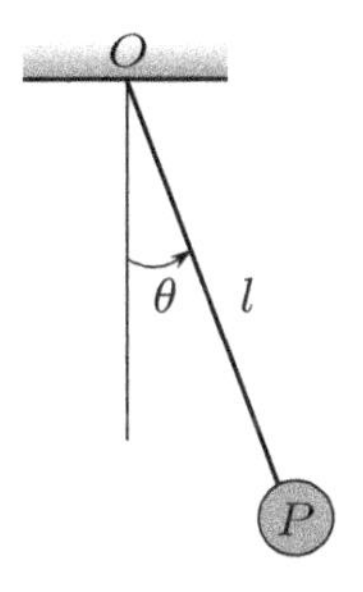


**Figure 2.1** Schematic of a simple pendulum.

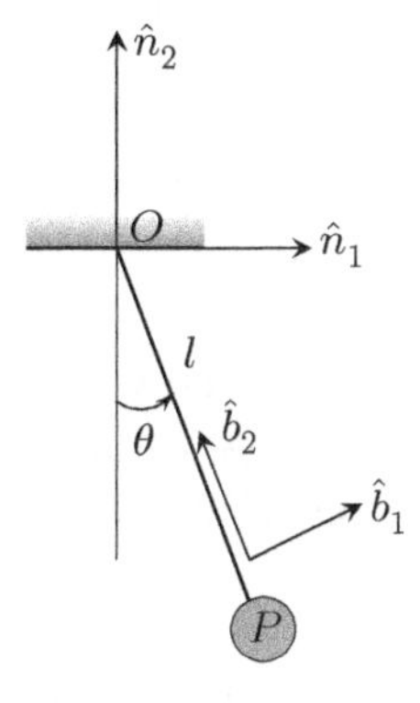


**Figure 2.2** Frames used to describe the motion of the pendulum in Example 2.1.

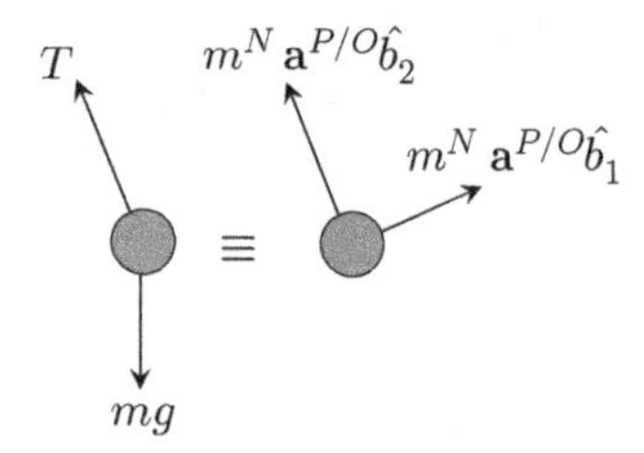


**Figure 2.3** Free-body diagram of the simple pendulum.

There are two forces acting on the particle.

- the tension in the cable acting in the positive $\hat{b}_2$ direction, i.e. $T\ \hat{b}_2$
- the weight of the particle acting in the negative $\hat{n}_2$ direction, i.e. $-mg\ \hat{n}_2$.

Because the acceleration is described in the $B$-frame, it is convenient to describe the weight in the $B$-frame using $-mg\ \hat{n}_2 = -mg\sin\theta\ \hat{b}_1 - mg\cos\theta\ \hat{b}_2$ (see Figure 2.3).

4. *Equate the force vector in a given direction to the associated mass times acceleration vector in the same direction.*

*Direction of* $\hat{b}_1$:

$$\mathbf{F}\hat{b}_1 = m^N\mathbf{a}^{P/O}\hat{b}_1,$$

$$-mg\sin\theta = ml\ddot{\theta},$$

$$\ddot{\theta} + \frac{g}{l}\sin\theta = 0.$$

*Direction of* $\hat{b}_2$:

$$\mathbf{F}\hat{b}_2 = m^N\mathbf{a}^{P/O}\hat{b}_2,$$

$$T - mg\cos\theta = ml\dot{\theta}^2,$$

$$T = ml\dot{\theta}^2 + mg\cos\theta.$$

The equation in the $\hat{b}_1$ direction is known as the equation of motion. Given initial conditions on the angle and angular velocity of the particle, this can be solved to find $\theta(t)$. This requires the solution of a second-order non-linear ordinary differential equation. Once the first equation has been solved, the second equation can be used to find the tension in the cable as a function of the angle $\theta$. Note that the tension is the sum of the static component $mg\cos\theta$ plus a dynamic component $ml\dot{\theta}^2$.

### Example 2.2 Motion of a Particle in a Uniform Gravitational Field

In undergraduate dynamics, you must have learned about the motion of a particle in a uniform gravitational field, also known as the projectile motion. Let us discuss this problem again through this example.

A ball of mass, $m$, is thrown with an initial velocity $v_0$ at an angle $\theta$ from the horizon, as shown in Figure 2.4. Find the path, $y(x)$, that the particle takes as it travels, assuming no frictional losses due to interaction with the atmosphere.

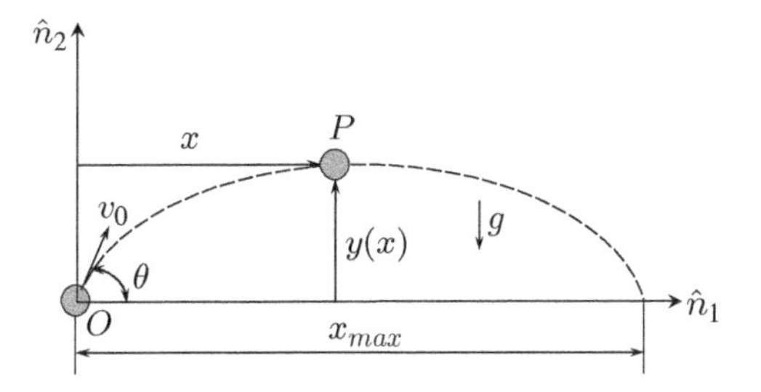


**Figure 2.4** Projectile motion.

To solve this problem, we begin by defining the inertial $N$-frame located at point $O$. Using the $N$-frame, we obtain the acceleration of point $P$ with respect to point $O$ as:

$$^N\mathbf{a}^{P/O} = \ddot{x}\hat{n}_1 + \ddot{y}\hat{n}_2.$$

Next, we draw the free-body diagram shown in Figure 2.5 and apply Newton's second law to obtain

*Direction of* $\hat{n}_1$:

$$\mathbf{F}\hat{n}_1 = m^N\mathbf{a}^{P/O}\hat{n}_1 \qquad \ddot{x} = 0,$$

*Direction of* $\hat{n}_2$:

$$\mathbf{F}\hat{n}_2 = m^N\mathbf{a}^{P/O}\hat{n}_2, \qquad \ddot{y} = -g.$$

To obtain $y(x)$, we need to solve the previous system of equations, which represent two second-order uncoupled linear differential equations. To this end, we specify the initial conditions of the particle at time $t = 0$. We note that at $t = 0$, $x(0) = 0$, $y(0) = 0$, $\dot{x}(0) = v_0 \cos\theta$, and $\dot{y}(0) = v_0 \sin\theta$. Upon integrating these equations using the specified initial conditions, we obtain

$$x(t) = v_0 \cos\theta t, \qquad y(t) = -\frac{g}{2}t^2 + v_0 \sin\theta t.$$

These expressions describe how $x$ and $y$ vary with time. However, we are after the relationship between $y$ and $x$. To obtain this relationship, we solve the first equation for $t$ and substitute in the second equation to obtain

$$y(x) = \frac{-g}{2v_0^2\cos^2\theta}x^2 + \tan\theta x,$$

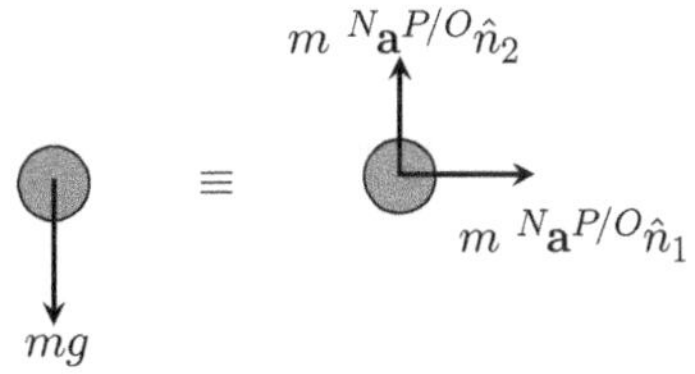

$m\ {}^N\mathbf{a}^{P/O}\hat{n}_2$

$mg \quad \equiv \quad m\ {}^N\mathbf{a}^{P/O}\hat{n}_1$

**Figure 2.5** Free-body diagram of a projectile in motion.

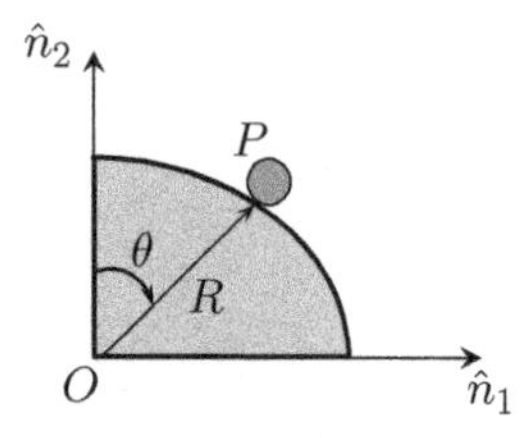


**Figure 2.6** Particle sliding along a smooth surface of radius $R$.

The previous equation can be used to find the value of $x$ at which the projectile hits the ground; in other words, $x_{\text{max}}$. This can be done by setting $y(x) = 0$ and obtaining $x_{\text{max}} = \frac{x_0^2}{g} \sin 2\theta$. Furthermore, to obtain the initial angle of projection at which $x_{\text{max}}$ is maximized, we let $\mathrm{d}x_{\text{max}}/\mathrm{d}\theta = 0$ and obtain $\theta_0 = 45°$.

**Example 2.3 Losing Contact with a Surface**

Figure 2.6 shows a particle of mass $m$ which is displaced slightly from its equilibrium position at the top of the fixed sphere of radius $R$ and slides down the surface. Find the angle at which the particle loses contact with the surface.

To solve this problem, we define the rotating frame, $B$, such that ${}^N\boldsymbol{\omega}^B = -\dot{\theta}\hat{b}_3$ and find the acceleration of the particle $P$ with respect to point $O$:

$$
{}^N\mathbf{a}^{P/O} = R\ddot{\theta}\hat{b}_1 - R\dot{\theta}^2\hat{b}_2.
$$

Next, we draw the free-body diagram shown in Figure 2.7 and use it to apply Newton's second law, thus obtaining the equation of motion. There are two forces acting on the body: the reaction force, $R_N\hat{b}_2$, and the weight, $-mg\hat{n}_2$. Thus, we can write

*Direction of* $\hat{b}_1$ :

$$
\begin{aligned}
\mathbf{F}\hat{b}_1 &= m^N\mathbf{a}^{P/O}\hat{b}_1 \\
\ddot{\theta} - \frac{g}{R}\sin\theta &= 0.
\end{aligned}
$$

*Direction of* $\hat{b}_2$ :

$$
\begin{aligned}
\mathbf{F}\hat{b}_2 &= m\ {}^N\mathbf{a}^{P/O}\hat{b}_2, \\
N - mg\cos\theta &= -mR\dot{\theta}^2.
\end{aligned}
$$

Now that the equations of motion are available, we turn our attention to finding the critical angle, $\theta_{cr}$, at which the particle loses contact with the surface. To this end, we notice that the reaction force, $R_N$, vanishes at $\theta_{cr}$. As such, at $\theta_{cr}$, the previous set of equations can be reduced to

$$
\begin{aligned}
\ddot{\theta} - \frac{g}{R}\sin\theta &= 0, \\
\frac{g}{R}\cos\theta_{cr} &= \dot{\theta}_{cr}^2.
\end{aligned}
$$

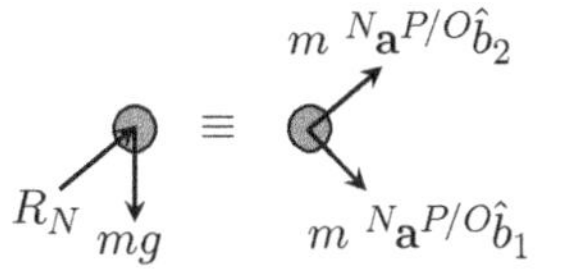

**Figure 2.7** Free-body diagram of a particle sliding along a circular surface.

We note that the first of these is valid for any value of $\theta$, whereas the latter is valid only at $\theta_{cr}$. To find $\theta_{cr}$, we integrate the first equation with respect to $\theta$. To this end, we use the common knowledge that $\ddot{\theta} = \frac{\mathrm{d}\dot{\theta}}{\mathrm{d}\theta}\frac{\mathrm{d}\theta}{\mathrm{d}t}$ or $\ddot{\theta} = \dot{\theta}\frac{\mathrm{d}\dot{\theta}}{\mathrm{d}\theta}$; $\ddot{\theta}d\theta = \dot{\theta}\mathrm{d}\dot{\theta}$. This yields

$$\int_{\dot{\theta}=0}^{\dot{\theta}_{cr}} \dot{\theta}\mathrm{d}\dot{\theta} = \int_{\theta=0}^{\theta_{cr}} \frac{g}{R}\sin\theta \mathrm{d}\theta; \qquad \dot{\theta}_{cr}^2 = \frac{2g}{R}(1 - \cos\theta_{cr}).$$

Next, we eliminate $\dot{\theta}_{cr}$ from the previous equations and solve for $\theta_{cr}$ to obtain

$$\theta_{cr} = \cos^{-1}\frac{2}{3}.$$

It is worth noting that, the critical angle is independent of the radius of the surface and the mass of the particle.

**Example 2.4 Influence of the Earth's Rotation on Projectile Dynamics**

A stone of mass, $m$, is dropped from a parachute at a height, $h = 3000$ m above the ground. Assuming that the stone is dropped from rest vertically downwards and that you are in Abu Dhabi, where the latitude is approximately 24°N, find the location where the stone hits the ground.

We solve this problem by defining the frames shown in Figure 2.8: the $N$-frame is an inertial frame located at the center of the Earth, and the $E$-frame is a rotating frame formed by rotating around the $\hat{n}_1$ axis such that ${}^N\omega^A = \Omega\hat{e}_1$, where $\Omega$ is the frequency of rotation of the Earth. The $B$-frame is formed at a constant angle, $\lambda$, of rotation around the $\hat{e}_3$ axis.

To find the location where the stone hits the ground, we first define its position in space as measured with respect to the center of the Earth.

$$\mathbf{OP} = \mathbf{OQ} + \mathbf{QP} = (R_0 + z)\hat{b}_1 + x\hat{b}_2 + y\hat{b}_3,$$

where $R_0$ is the radius of the earth, $x$ is the distance along the $\hat{b}_2$ axis (south) measured from point $Q$, and $y$ is the distance along the $\hat{b}_3$ axis (east) measured from point $Q$. The velocity and acceleration are given by

$$\begin{aligned}
{}^N\mathbf{v}^{P/O} &= (z - y\Omega\sin\lambda)\hat{b}_1 + (\dot{x} - y\Omega\cos\lambda)\hat{b}_2 \\
&\quad + (\dot{y} + (R_0 + z)\Omega\sin\lambda + x\Omega\cos\lambda)\hat{b}_3, \\
{}^N\mathbf{a}^{P/O} &= (\ddot{z} - 2\Omega\dot{y}\cos\lambda)\hat{b}_1 + (\ddot{x} - 2\Omega\dot{y}\cos\lambda)\hat{b}_2 \\
&\quad + (\ddot{y} + 2\Omega(\dot{x}\sin\lambda + \dot{z}\sin\lambda))\hat{b}_3 + \mathcal{O}(\Omega^2),
\end{aligned}$$

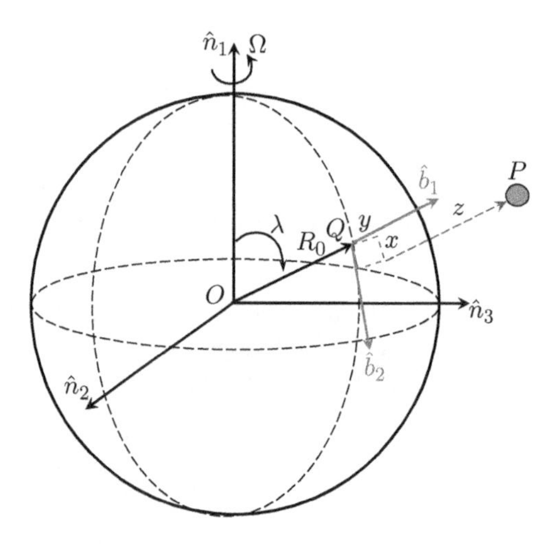


**Figure 2.8** Influence of the Earth's rotation on projectile dynamics.

where $\mathcal{O}(\Omega^2)$ are terms containing $\Omega^2$. These terms can be neglected because the frequency of the Earth's rotation is very small: approximately, $\Omega = 7.29 \times 10^{-5}$ rad/s.

Next, we draw the free-body diagram and implement Newton's law in the $\hat{b}_i$ directions. Note that the only force acting on the body is its weight in the $\hat{b}_1$ direction. This yields:

*Direction of* $\hat{b}_1$ :

$$\mathbf{F}\hat{b}_1 = m^N\mathbf{a}^{P/O}\hat{b}_1, \qquad -g = \ddot{z} - 2\Omega\dot{y}\sin\lambda,$$

*Direction of* $\hat{b}_2$ :

$$\mathbf{F}\hat{b}_2 = m^N\mathbf{a}^{P/O}\hat{b}_2, \qquad \ddot{x} - 2\Omega\dot{y}\cos\lambda = 0,$$

*Direction of* $\hat{b}_3$ :

$$\mathbf{F}\hat{b}_3 = m^N\mathbf{a}^{P/O}\hat{b}_3, \qquad \ddot{y} + 2\Omega(\dot{x}\cos\lambda + z\sin\lambda) = 0.$$

The location at which the stone hits the ground can be found by solving this set of differential equations for which we need to specify the initial conditions. Since the stone was released vertically downward from rest at a height, $h$, the initial conditions can be written as: $x(0) = y(0) = 0$, $\dot{x}(0) = \dot{y}(0) = \dot{z}(0) = 0$, and $z(0) = h$. Integrating the previous equations using the initial conditions yields

$$\dot{z} = -gt + 2\Omega y\sin\lambda, \qquad \dot{x} = 2\Omega y\cos\lambda.$$

Rearranging, we obtain

$$\ddot{y} = 2\Omega gt\sin\lambda,$$

which upon integrating twice with respect to time and using the initial conditions yields

$$y = \frac{1}{3}\Omega gt^3\sin\lambda.$$

Substituting this into the equations governing $\dot{x}$ and $\dot{z}$ while neglecting terms of order $\Omega^2$, yields

$$x = 0, \qquad z = h - \frac{1}{2}gt^2.$$

The time required for the stone to reach the ground, $t_{cr}$ can be found by setting $z = 0$ in the previous equation and obtaining $t_{cr} = \sqrt{\frac{2h}{g}}$. Substituting $t_{cr}$ back into the equation for $y$, we obtain

$$y = \frac{2}{3}\Omega h\sqrt{\frac{2h}{g}}\sin\lambda \qquad \text{(east)}.$$

Using the information given in the problem statement: $h = 3000$ m, $\lambda = 90° - 24° = 66°$, we obtain $y = 32.9$ cm (east). Therefore, the stone will hit the ground 32.9 cm east of where it was released. The time required to reach the ground is approximately 24.7 **s**.

 **Flipped Classroom Exercise 2.1**

A bead is constrained to slide on a wire that is bent in the shape of a cycloid having the parametric equation $x = a(\theta - \sin\theta), y = a(1 - \cos\theta)$. Find the differential equation governing the motion of the bead.

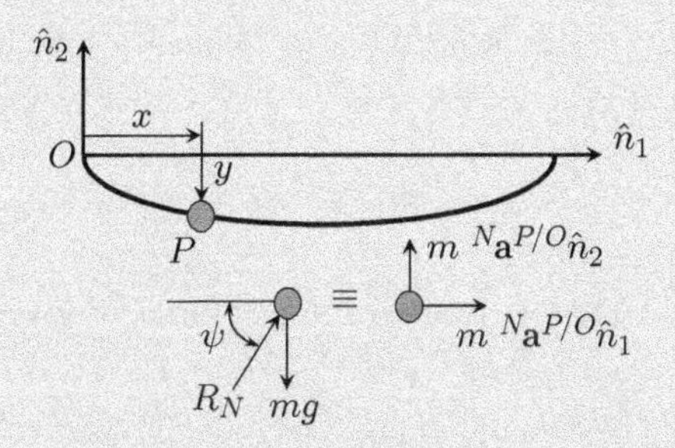


1. Using the $N$-frame as a reference frame, show that the acceleration of point $P$ with respect to point $O$ is given by

$${}^N\mathbf{a}^{P/O} = a(\ddot{\theta} - \cos\theta\ddot{\theta} + \dot{\theta}^2\sin\theta)\hat{n}_1 - a(\ddot{\theta}\sin\theta + \dot{\theta}^2\cos\theta)\hat{n}_2.$$

2. Draw a free-body diagram of the system. Note that the reaction on the bead is always normal to the cycloid.
3. Using Newton's second law, write the two equations that govern the two unknowns in the system. What are these unknowns?
4. Rearrange the resulting equations and reduce them into one equation governing the dynamics of the angle $\theta$. Show that this equation can be written as

$$\ddot{\theta} + \frac{\sin\theta}{2(1-\cos\theta)}\dot{\theta}^2 - g\frac{\sin\theta}{2a(1-\cos\theta)} = 0.$$

### Flipped Classroom Exercise 2.2

A cable is attached to a particle of mass, $m$, and is wound completely around a disk of radius $R$, as shown in the figure. The disk rotates with a constant angular velocity $\Omega$. Find the change of length, $l(t)$ and the tension in the cable as function of time assuming planar motion in the horizontal plane.

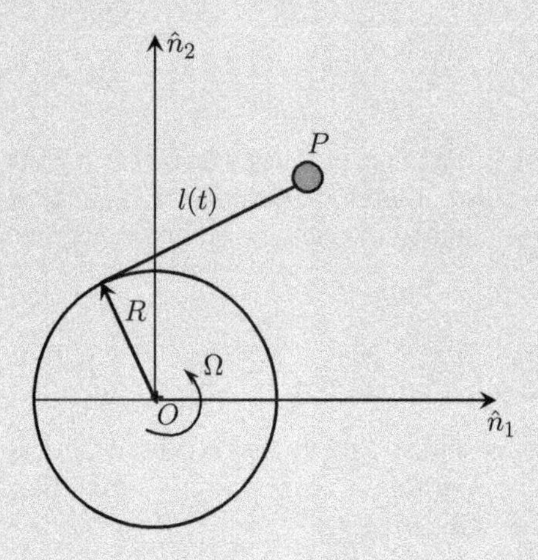


1. Define the rotating $B$-frame such that $\hat{b}_1$ aligns with the the direction of the cable. Show that

$$^N\boldsymbol{\omega}^B = \left(\Omega + \frac{\dot{l}}{R}\right)\hat{b}_3.$$

2. Show that the acceleration from point $O$ to point $P$ is given by

$$^N\mathbf{a}^{P/O} = -l\left(\Omega + \frac{\dot{l}}{R}\right)^2 \hat{b}_1 + \left(\frac{l\ddot{l}}{R} + \frac{\dot{l}^2}{R} - R\Omega^2\right)\hat{b}_2.$$

3. Apply Newton's second law along the $\hat{b}_i$ directions and show that

$$T = ml\left(\Omega + \frac{\dot{l}}{R}\right)^2$$

$$l\ddot{l} + \dot{l}^2 - R^2\Omega^2 = 0.$$

4. Using the initial condition $l(0) = 0$, $\dot{l}(0) = R\Omega$, show that

$$l(t) = R\Omega t,$$

$$T = 4mR\Omega^3 t.$$

## 2.2 Stiffness and Viscous Damping

Two elements that are widely used in dynamics and vibrations are the spring and damper. The first is used to represent a structural element that stores potential energy due to its elasticity; the second is used to represent energy dissipation that is linearly proportional to the velocity of motion, also known as viscous or *Rayleigh* damping.

When an elastic element is deflected a distance $x$, as shown in Figure 2.9, it pulls back with a force, $F_s$. Assuming linear elastic behavior, the force in the spring is equal to $F_s = kx$, where $k$ is the stiffness coefficient of the elastic element. Similarly, assuming a linear viscous damping behavior, when one end of a viscous damper moves with a velocity $\dot{x}$, the damper pulls back with a force equal to $F_d = c\dot{x}$, where $c$ is the damping coefficient. Note that, in general, elastic and dissipation mechanisms can exhibit non-linear dependence on the deflection and velocity respectively. For instance, under large deformations, stiffness elements can exhibit a hardening/softening behavior in which the stiffness coefficient becomes a function of the displacement itself; in other words, $k = k(x)$. Similarly, drag forces acting on an immersed body in motion are generally proportional to the square of the body's velocity.

**Example 2.5 Spring Force**

Consider the spherical pendulum in Figure 2.10. The pendulum consists of a mass $m$ and an elastic cable with a spring of constant $k$ of initial unstretched length, $l_0$. Find the equations governing the motion of the system.

As shown in Figure 2.11, we define two rotating coordinate systems such that ${}^N\boldsymbol{\omega}^C = \dot{\phi}\hat{n}_3$ and ${}^C\boldsymbol{\omega}^B = -\dot{\theta}\hat{c}_3$. Hence, in the $B$-frame, the angular velocity vector can be expressed as ${}^N\boldsymbol{\omega}^B = \dot{\phi}\sin\theta\hat{b}_1 - \dot{\theta}\hat{b}_2 + \dot{\phi}\cos\theta\hat{b}_3$. The position vector from point $O$ to point $P$ can be written as $\mathbf{OP} = -(l_0 + r)\hat{b}_3$, where $r$ is the change of length in the cable due to its elasticity. The velocity and acceleration can be written as

$$
\begin{aligned}
{}^N\mathbf{v}^{P/O} &= (l_0 + r)\dot{\theta}\hat{b}_1 + (l_0 + r)\dot{\phi}\sin\theta\hat{b}_2 - \dot{r}\hat{b}_3, \\
{}^N\mathbf{a}^{P/O} &= \hat{b}_1[(l_0 + r)\ddot{\theta} + 2\dot{r}\dot{\theta} - (l_0 + r)\dot{\phi}^2\sin\theta\cos\theta] + \\
&\quad \hat{b}_2[2(l_0 + r)\dot{\theta}\dot{\phi}\cos\theta + (l_0 + r)\ddot{\phi}\sin\theta + 2\dot{r}\dot{\phi}\sin\theta] + \\
&\quad \hat{b}_3[-\ddot{r} + (l_0 + r)\dot{\theta}^2 + (l_0 + r)\dot{\phi}^2\sin^2\theta].
\end{aligned}
$$

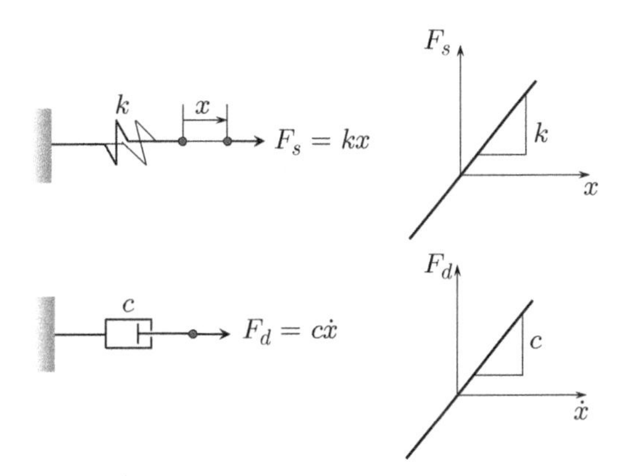


**Figure 2.9** Graphical representation of stiffness (spring) and viscous damping (dashpot).

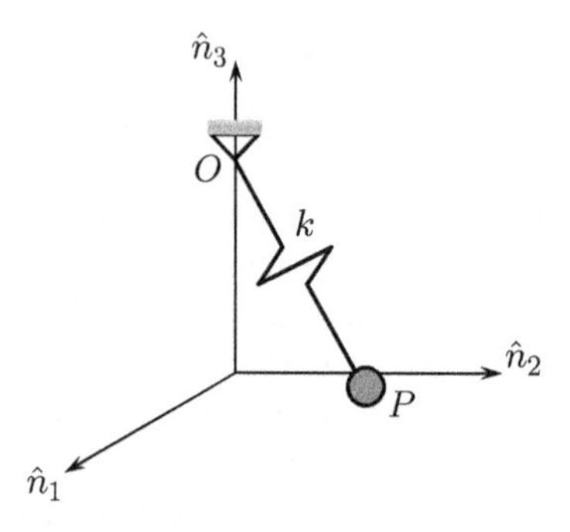


**Figure 2.10** A spherical pendulum with an elastic cable.

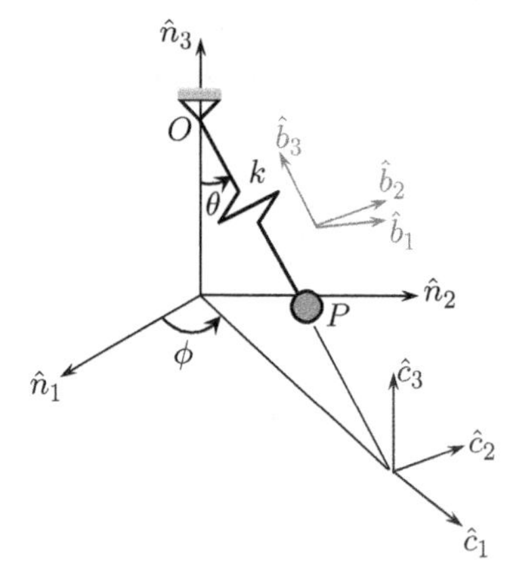


**Figure 2.11** Coordinate system used to describe the kinematics/dynamics of the spherical pendulum.

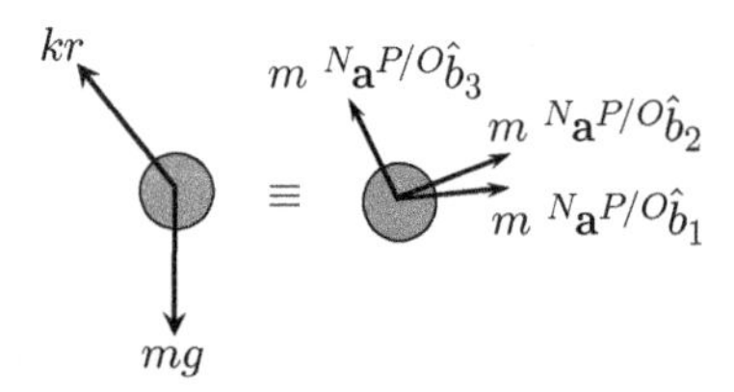


**Figure 2.12** Free-body diagram of a spherical pendulum.

As shown in Figure 2.12, there are two forces acting on the body: the force in the spring $kr\ \hat{b}_3$ and the weight $-mg\hat{c}_3$. Implementing Newton's second law along the three directions of the rotating frame, we obtain:

*Direction of* $\hat{b}_1$ :

$$\mathbf{F}\hat{b}_1 = m^N\mathbf{a}^{P/O}\hat{b}_1,$$

$$(l_0 + r)\ddot{\theta} + g\sin\theta + 2\dot{r}\dot{\theta} = (l_0 + r)\dot{\phi}^2\sin\theta\cos\theta,$$

*Direction of* $\hat{b}_2$ :

$$\mathbf{F}\hat{b}_2 = m^N\mathbf{a}^{P/O}\hat{b}_2,$$
$$2(l_0 + r)\dot{\theta}\dot{\phi}\cos\theta + (l_0 + r)\ddot{\phi}\sin\theta + 2\dot{r}\dot{\phi}\sin\theta = 0,$$

*Direction of* $\hat{b}_3$ :

$$\mathbf{F}\hat{b}_3 = m^N\mathbf{a}^{P/O}\hat{b}_3,$$
$$\ddot{r} + \frac{k}{m}r + (l_0 + r)\dot{\theta}^2 = (l_0 + r)\dot{\phi}^2\sin^2\theta + mg\cos\theta.$$

**Flipped Classroom Exercise 2.3**

The particle shown in the figure, which has a mass, $m$, is constrained to move in the plane shown by four springs. Each of the springs $s_i$ has a stiffness $k$ and an unstretched length $l$ in the position shown in the figure. Obtain the equations governing the motion of the particle when it is arbitrarily displaced. Motion occurs in the horizontal plane and, as a result, there is no need to include gravitational forces.

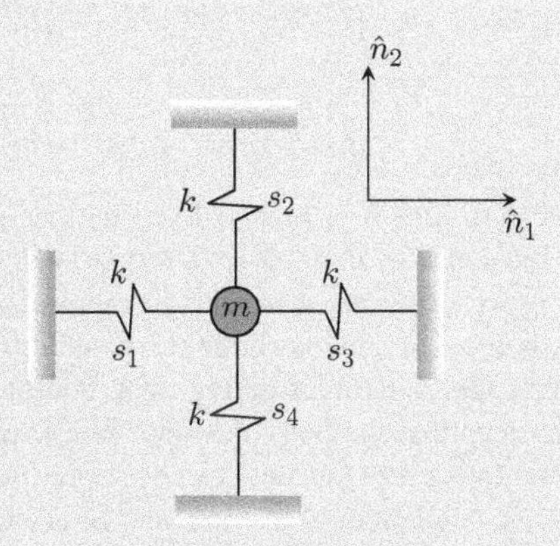


1. Define your inertial, $N$-frame, at the center. Displace the particle a distance $x$ along the $\hat{n}_1$-axis and a distance $y$ along the $\hat{n}_2$-axis.
2. Using the resulting geometry, show that the force in each spring can be described as

$$\mathbf{F}_1 = k\left[(l + x)\left(\frac{l}{\sqrt{(l + x)^2 + y^2}} - 1\right)\hat{n}_1 + y\left(\frac{l}{\sqrt{(l + x)^2 + y^2}} - 1\right)\hat{n}_2\right],$$

$$\mathbf{F}_2 = k\left[x\left(\frac{l}{\sqrt{x^2 + (l - y)^2}} - 1\right)\hat{n}_1 + (l - y)\left(\frac{l}{\sqrt{x^2 + (l - y)^2}} - 1\right)\hat{n}_2\right],$$

$$\mathbf{F}_3 = k\left[(x - l)\left(\frac{l}{\sqrt{(l - x)^2 + y^2}} - 1\right)\hat{n}_1 + y\left(\frac{l}{\sqrt{(l - x)^2 + y^2}} - 1\right)\hat{n}_2\right],$$

$$\mathbf{F}_4 = k\left[x\left(\frac{l}{\sqrt{x^2+(l+y)^2}}-1\right)\hat{n}_1+(l+y)\left(\frac{l}{\sqrt{x^2+(l+y)^2}}-1\right)\hat{n}_2\right].$$

3. Implement Newton's equations along the $\hat{n}_1$ and $\hat{n}_2$ directions to show that

$$\ddot{x} = \sqrt{\frac{k}{m}}\left[(l+x)\left(\frac{l}{\sqrt{(l+x)^2+y^2}}-1\right)+x\left(\frac{l}{\sqrt{x^2+(l-y)^2}}-1\right)\right.$$
$$\left.+(x-l)\left(\frac{l}{\sqrt{(l-x)^2+y^2}}-1\right)+x\left(\frac{l}{\sqrt{x^2+(l+y)^2}}-1\right)\right]$$
$$\ddot{y} = \sqrt{\frac{k}{m}}\left[y\left(\frac{l}{\sqrt{(l+x)^2+y^2}}-1\right)+(l-y)\left(\frac{l}{\sqrt{x^2+(l-y)^2}}-1\right)\right.$$
$$\left.+y\left(\frac{l}{\sqrt{(l-x)^2+y^2}}-1\right)+(l+y)\left(\frac{l}{\sqrt{x^2+(l+y)^2}}-1\right)\right].$$

## 2.3 Dry Friction

When two bodies are forced to slide with respect to each other, micro-scale asperities on their surface interact causing local elastic and plastic deformations that result in a force known as dry friction. One of the most widely used models to represent energy losses due to friction between moving bodies is the dry friction model or Coulomb's law of friction. The model is simple and states that, if the net forces acting on a body are less than what is required to overcome frictional forces, then the body does not move. In such cases, frictional forces preventing the body from moving are static and can be approximated by $F_s = \mu_s N$, where $\mu_s$ is the non-dimensional static coefficient of friction, and $N$ is the normal force exerted by each surface on the other, directed perpendicular to the surface.

A common approach to obtaining the empirical value of the static friction coefficient, $\mu_s$, between two surfaces, is by putting a block made of one material on the surface of a ramp made from the other material as shown in Figure 2.13. The sliding angle is then increased incrementally up to the point where sliding is initiated. At this point, one can carry out a simple static equilibrium analysis and show that $\tan\theta_{cr} = \mu_s$.

Once the forces acting on the body overcome the static frictional forces and the body starts moving, dry friction forces drop in magnitude and are approximated by $F_k = \mu_k N$, where $\mu_k$ is the non-dimensional kinetic coefficient of friction. The kinetic friction coefficient is smaller than the static one and usually ranges between zero and slightly above one. Since the dynamic friction force always opposes the direction of the velocity, it can be represented in the following form:

$$F_k = -\mu_k N \text{sign}(\dot{\mathbf{x}}),$$

where $\text{sign}(\dot{\mathbf{x}})$ refers to the sign of the velocity.

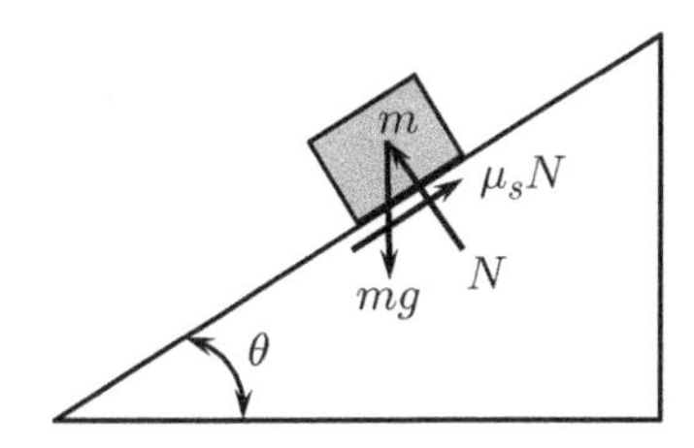


**Figure 2.13** A method to obtain the static friction coefficient, $\mu_s$.

### Example 2.6 Dry Friction

For the system shown in Figure 2.14, the vertical column is forced to rotate with a constant angular velocity $\Omega\,\hat{n}_2$. A pipe is welded to the vertical column, making a constant angle $\theta_0$ with the $\hat{n}_2$ direction. A particle $P$ of mass $m$ slides inside the pipe. The kinetic coefficient of friction between the mass and the internal surface of the pipe is $\mu_k$. Find the equations governing the motion of the particle.

In addition to the inertial frame, we use two rotating frames to describe the motion of the particle. The first coordinate system is the $A$-frame, which is formed such that ${}^N\boldsymbol{\omega}^A = \Omega\,\hat{n}_2 = \Omega\,\hat{a}_2$. The second coordinate frame is formed by performing a constant angle rotation of $\theta_0$ around the $-\hat{a}_3$ axis.

With these coordinates, we can write the position, velocity, and acceleration of point $P$ with respect to point $O$ measured in the inertial frame as follows:

$$\begin{aligned}
\mathbf{OP} &= r\hat{e}_2, \\
{}^N\mathbf{v}^{P/O} &= \dot{r}\hat{e}_2 - r\Omega\sin\theta_0\hat{e}_3, \\
{}^N\mathbf{a}^{P/O} &= -r\Omega^2\sin\theta_0\cos\theta_0\hat{e}_1 + (\ddot{r} - r\Omega^2\sin^2\theta_0)\hat{e}_2 - 2\dot{r}\Omega\sin\theta_0\hat{e}_3.
\end{aligned}$$

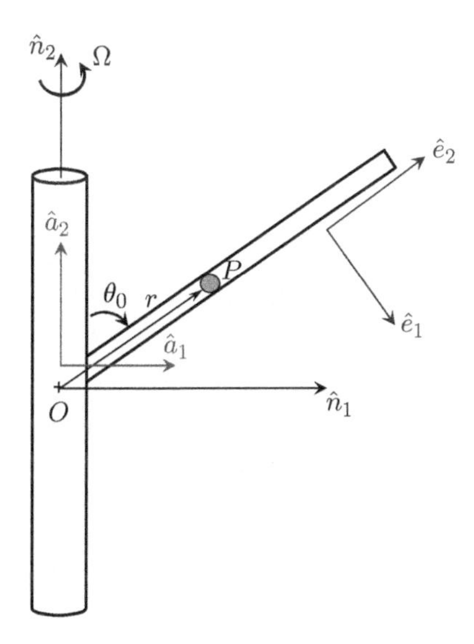


**Figure 2.14** Motion of a particle in a rotating pipe with friction.

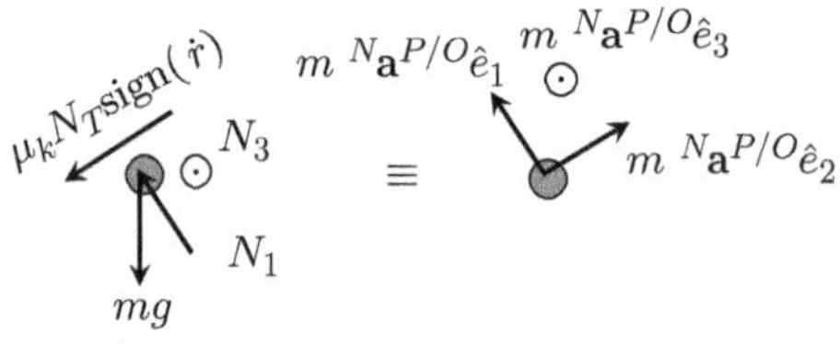

**Figure 2.15** Free-body diagram of a particle moving inside a pipe.

To implement Newton's second law, we construct the free-body diagram shown in Figure 2.15 and note that the reaction force is decomposed into two components, $N_1$ and $N_3$, such that $N_T = \sqrt{N_1^2 + N_3^2}$. This yields:

*Direction of* $\hat{e}_1$ :

$$\mathbf{F}\hat{e}_1 = m^N\mathbf{a}^{P/O}\hat{e}_1,$$

$$N_1 = m\sin\theta_0(g - r\Omega^2\cos\theta_0).$$

*Direction of* $\hat{e}_2$ :

$$\mathbf{F}\hat{e}_2 = m^N\mathbf{a}^{P/O}\hat{e}_2,$$

$$\ddot{r} - r\Omega^2\sin^2\theta_0 + \frac{\mu_k}{m}N_T\mathrm{sgn}(\dot{r}) + g\cos\theta_0 = 0.$$

*Direction of* $\hat{e}_3$ :

$$\mathbf{F}\hat{e}_3 = m^N\mathbf{a}^{P/O}\hat{e}_3,$$

$$N_3 = -2m\Omega\dot{r}\sin\theta_0.$$

Manipulating these equations, we can write:

$$N_T = m\sin\theta_0\sqrt{4\Omega^2\dot{r}^2 + (g - r\Omega^2\cos\theta_0)^2},$$

$$\ddot{r} - r\Omega^2\sin^2\theta_0 + \mu_k\sin\theta_0\sqrt{4\Omega^2\dot{r}^2 + (g - r\Omega^2\cos\theta_0)^2}\mathrm{sgn}(\dot{r}) + g\cos\theta_0 = 0.$$

### Flipped Classroom Exercise 2.4

In the system shown in the figure, particle $m_1$ slides without friction on the surface of mass $m_2$. As a result, mass $m_2$ slides on the horizontal surface a distance $x$. The kinetic friction coefficient between mass $m_2$ and the horizontal surface is $\mu_k$. Find the equations that govern the motion of the two masses.

1. There are four unknowns in this problem. What are they?
2. Using the coordinate system shown in the figure, show that the accelerations of mass $m_1$ and $m_2$ is given by

$$^N\mathbf{a}^{P_1/O} = (\ddot{x} - \ddot{y}\cos\theta)\hat{n}_1 - (\ddot{y}\sin\theta)\hat{n}_2,$$

$$^N\mathbf{a}^{P_2/O} = \ddot{x}\hat{n}_1.$$

3. Draw a free-body diagram for each mass, then use Newton's equations to show that:

$$m_2\ddot{x} + m_1\ddot{y}\sin\theta\tan\theta + \mu_k(m_1 + m_2)g = \mu_k m_1 \sin\theta\ddot{y} + m_1 g\tan\theta,$$

$$\ddot{y}\sin\theta\tan\theta - g\tan\theta = \ddot{x} - \ddot{y}\cos\theta.$$

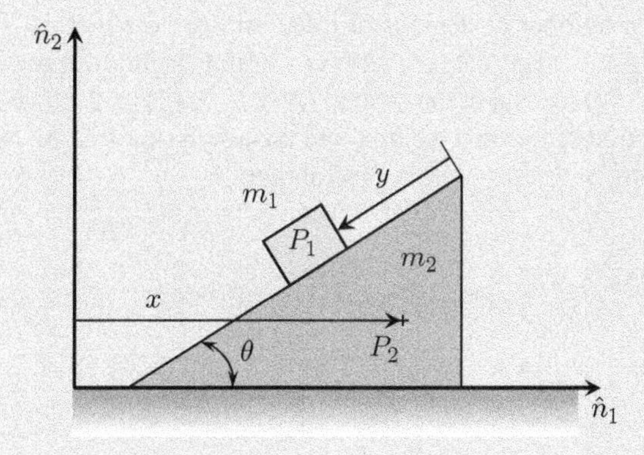


## 2.4 Dynamics of a System of Particles

The motion of a free particle in a free space can be described by three coordinates. Therefore, to define the motion of $\mathcal{N}$ particles in space, we need to use $3 \times N$ coordinates. Nonetheless, in many instances, the motion of the particle is not completely free and is constrained kinematically to a certain path. We call such constraints kinematic constraints. For instance, a particle that is constrained to move along the $x$-axis of a Cartesian frame is subjected to two constraints, $y = 0$ and $z = 0$. As such, the particle is only free to move in one direction and is said to have one degree of freedom. For one particle, the number of degrees of freedom is equal to three minus the number of constraints. For $\mathcal{N}$ particles, it is given by:

$$\boxed{DOF = 3\mathcal{N} - \mathcal{C}} \tag{2.4}$$

where $DOF$ is the number of degrees of freedom, $\mathcal{N}$ is the number of particles, and $\mathcal{C}$ is the number of constraints.

A pendulum of a fixed cable length, $l$, moving in a two-dimensional space has one degree of freedom because its motion involves one particle subject to two constraints: $z = 0$ and $x^2 + y^2 = l^2$. The number of degrees of freedom of a given system of particles decides the minimum number of equations needed to fully describe the motion of that system. Therefore, for the pendulum, only one equation defined in terms of the rotation angle $\theta$ (as derived in Example 2.1) is necessary to describe the motion of the pendulum.

To implement Newton's second law on a system of particles, we implement it on each particle separately, taking into account the influence of the internal forces between the particles. Per Newton's third law, which states that for every action there is a reaction equal in magnitude and opposite in direction, the internal forces between two particles are always equal but opposite in direction.

**Example 2.7 Gantry Crane**

Consider the motion of the gantry crane shown in Figure 2.16. The payload has a mass, $m_1$, while the trolley has mass, $m_2$. Find the equations that govern the motion of the system.

To begin, let us define our unknowns and our degrees of freedom. In the problem, there are four unknowns: the displacement, $x$, the angle $\theta$, the tension in the cable $T$, and the vertical reaction at the pin $Q_y$.

To determine the number of degrees of freedom, we note that $m_1$ is subjected to two constraints: $z = 0$ and the fixed cable length constraint. Similarly, mass $m_2$ is subject to two constraints, $y = z = 0$. As such, the system has $3 \times 2 - 4 = 2\ DOF$.

To find the equation of motion, we first find the acceleration of $m_1, m_2$ with respect to point $O$ as measured in the inertial $N$-frame. We obtain:

*Mass* $m_1$:

$$^N\mathbf{a}^{P/O} = (\ddot{x} + l\ddot{\theta}\cos\theta - l\dot{\theta}^2\sin\theta)\hat{n}_1 + (l\ddot{\theta}\sin\theta + l\dot{\theta}^2\cos\theta)\hat{n}_2,$$

*Mass* $m_2$:

$$^N\mathbf{a}^{P/O} = \ddot{x}\hat{n}_1.$$

Next, we draw the free-body diagram for each mass, and implement Newton's second law on each mass separately along the possible coordinates (Figure 2.17).

*Mass* $m_1$:

$$\mathbf{F}\hat{n}_1 = m_1\ {}^N\mathbf{a}^{P/O}\hat{n}_1,$$

$$T\cos\theta - m_1 g = m_1(l\ddot{\theta}\sin\theta + l\dot{\theta}^2\cos\theta),$$

$$\mathbf{F}\hat{n}_2 = m_1\ {}^N\mathbf{a}^{P/O}\hat{n}_2,$$

$$-T\sin\theta = m_1(\ddot{x} + l\ddot{\theta}\cos\theta - l\dot{\theta}^2\sin\theta),$$

*Mass* $m_2$:

$$\mathbf{F}\hat{n}_1 = m_2\ {}^N\mathbf{a}^{P/O}\hat{n}_1,$$

$$m_2\ddot{x} = T\sin\theta,$$

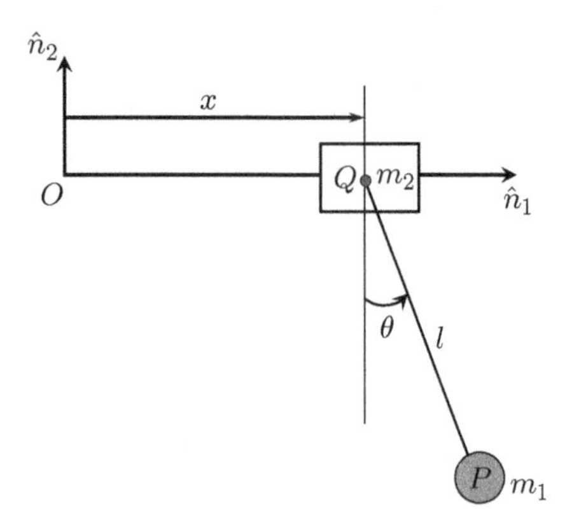


**Figure 2.16** A schematic of a gantry crane.

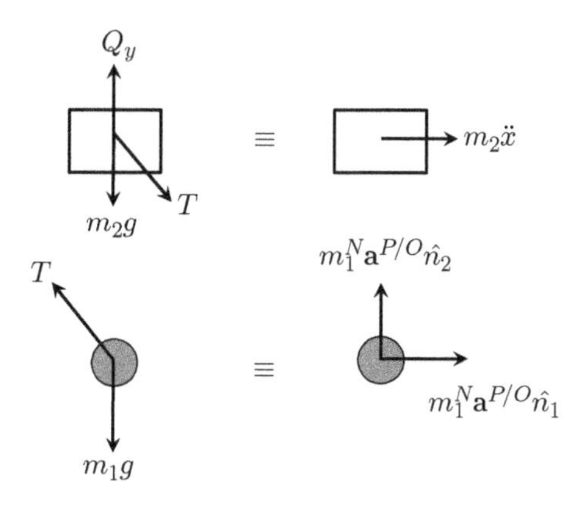


**Figure 2.17** Free-body diagram of a gantry crane.

$$\mathbf{F}\hat{n}_2 = m_2\,{}^N\mathbf{a}^{P/O}\hat{n}_2,$$

$$Q_y = m_2 g + T\cos\theta.$$

Note that the tension force, $T$, is an internal force, so it appears on each body (equal in magnitude and opposite in direction). The preceding four equations can be solved together for the four unknowns.

### Flipped Classroom Exercise 2.5

For a system of $\mathcal{N}$ particles, each of mass $m_i$ and located at $\mathbf{r}_i$, show that

$$\mathbf{F} = m\ddot{\mathbf{r}}_G,$$

where $\mathbf{F} = \sum_{i=1}^{\mathcal{N}} \mathbf{F}_i$ is the net external force acting on the system of particles, $\mathbf{F}_i$ is the net external force acting on the $i$th particle, $m = \sum_{i=1}^{\mathcal{N}} \mathbf{m}_i$, $\mathbf{r}_G = \sum_{i=1}^{\mathcal{N}} m_i\mathbf{r}_i / \sum_{i=1}^{\mathcal{N}} m_i$ is the center of mass of the particles, and $\mathbf{f}_{ij}$ is the internal force exerted by the $j$th particle on the $i$th particle such that $\mathbf{f}_{ij} = -\mathbf{f}_{ji}$, $\mathbf{f}_{ii} = 0$.

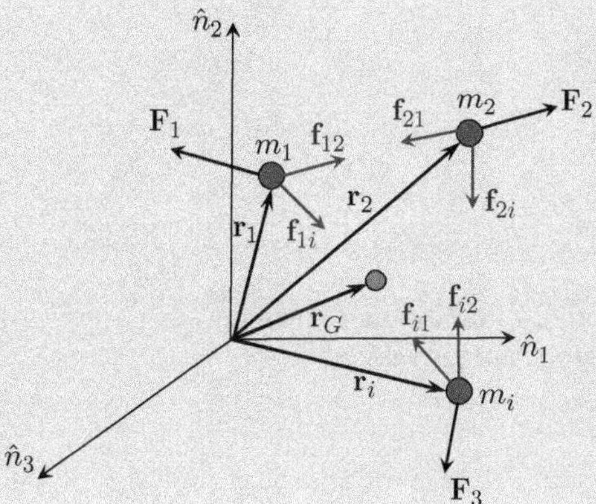

To show that $\mathbf{F} = m\ddot{\mathbf{r}}_G$, carry out the following steps:

1. Implement Newton's law on each particle and show it is equal to

$$m_i\ddot{\mathbf{r}}_i = \mathbf{F}_i + \sum_{j=1}^{\mathcal{N}} \mathbf{f}_{ji}, \qquad i = 1, 2, 3, \ldots, \mathcal{N}.$$

2. Add all the previous equations together from $i = 1$ to $\mathcal{N}$ and show that

$$\sum_{i=1}^{\mathcal{N}} m_i\ddot{\mathbf{r}}_i = \sum_{i=1}^{\mathcal{N}} \mathbf{F}_i.$$

3. Using the definition of the center of mass $\mathbf{r}_G$, show that

$$\mathbf{F} = m\ddot{\mathbf{r}}_G.$$

Note that the acceleration of the center of mass does not depend on the internal forces.

**Flipped Classroom Exercise 2.6**

For the system of two particles shown in the figure, particle $m_1$ is allowed to slide a distance $r$ on the frictionless massless rod of length, $l$. On one end, the mass is attached to a spring of stiffness $k$, and, on the other end, it is attached to a viscous damper of damping coefficient $c$. The second mass $m_2$ is fixed to the end of the massless rod as shown. Assume that the initial unstretched length of the spring is $r_0$ and find the equations governing the motion of the system.

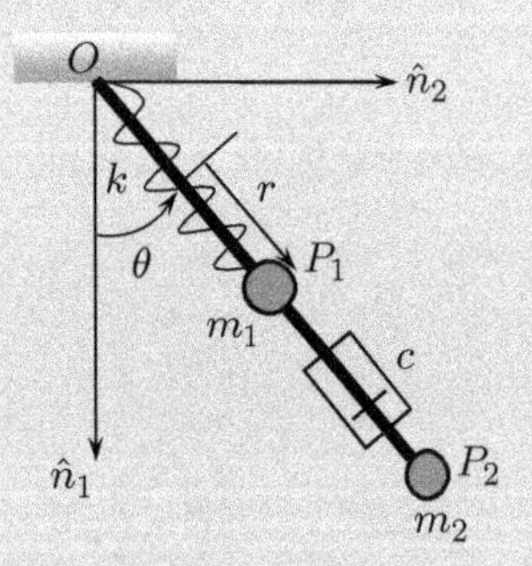


1. Define the rotating frame $B$-frame such that ${}^N\boldsymbol{\omega}^B = \dot{\theta}\hat{b}_3$ and show that the acceleration of $m_1$ and $m_2$ are respectively:

*Mass* $m_1$ :

$${}^N\mathbf{a}^{P_1/O} = (\ddot{r} - (r_0 + r)\dot{\theta}^2)\hat{b}_1 + (2\dot{r}\dot{\theta} + (r_0 + r)\ddot{\theta})\hat{b}_2,$$

*Mass* $m_2$ :

$$^N\mathbf{a}^{P_2/O} = -l\dot{\theta}^2\hat{b}_1 + l\ddot{\theta}\hat{b}_2.$$

2. Draw a free-body diagram for each mass. Note there are four forces acting on $m_1$ and four forces acting on $m_2$. Note also that there are four unknowns in this problem. What are they? (Hint: do not forget about the shear force acting on $m_2$.)
3. Apply Newton's second law on each mass independently and show that

$$\ddot{r} + \frac{k}{m_1}r - (r_0 + r)\dot{\theta}^2 + \frac{c}{m_1}\dot{r} = g\cos\theta,$$

$$\ddot{\theta} + \frac{g}{l}\sin\theta = \frac{m_1}{m_2}(2\dot{r}\dot{\theta} + (r_0 + r)\ddot{\theta}) + \frac{m_1}{m_2}\frac{g}{l}\sin\theta.$$

## 2.5 Newton's Law of Gravitation

Newton's law of gravitation states that a particle attracts every other particle in the universe via a force that is directly proportional to the product of their masses and inversely proportional to the square of the distance between their centers. The law was derived by Newton based on empirical observations. Mathematically, the force of gravitation, $\mathbf{F}_{12}$ can be expressed as:

$$\mathbf{F}_{12} = G\frac{m_1 m_2}{\mathbf{r}^2}\hat{\mathbf{r}} = G\frac{m_1 m_2}{\mathbf{r}^3}\mathbf{r} \tag{2.5}$$

where, as shown in Figure 2.18, $m_1$ and $m_2$ are the masses of the interacting particles, $G$ is the gravitational constant ($6.674 \times 10^{-11}$ N(m/kg)$^2$), $\mathbf{r}$ is the position vector connecting the two masses, $\mathbf{r}^2$ is the square of the distance between the interacting particles, and $\hat{\mathbf{r}}$ is a unit vector in the direction of $\mathbf{r}$.

**Example 2.8 Gravitational Attraction**

In Figure 2.19, a particle of mass $m_1$ is attached to a cable of length $l$ and is allowed to oscillate. The other particle, which is of mass $m_2$, is constrained to slide without friction along the horizontal rod. Derive the equations that govern the motion of this system of particles, accounting for the mutual attraction due to their own gravitational fields.

We first obtain the acceleration of both particles as observed in the reference, $N$, frame; that is:

*Mass* $m_1$:

$$^N\mathbf{a}^{P/O} = (l\ddot{\theta}\cos\theta - l\dot{\theta}^2\sin\theta)\hat{n}_1 + (l\ddot{\theta}\sin\theta + l\dot{\theta}^2\cos\theta)\hat{n}_2,$$

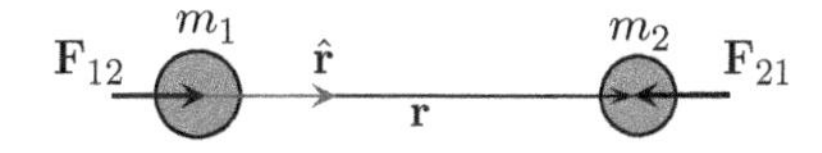


**Figure 2.18** Two particles interacting due to Newton's law of gravitation.

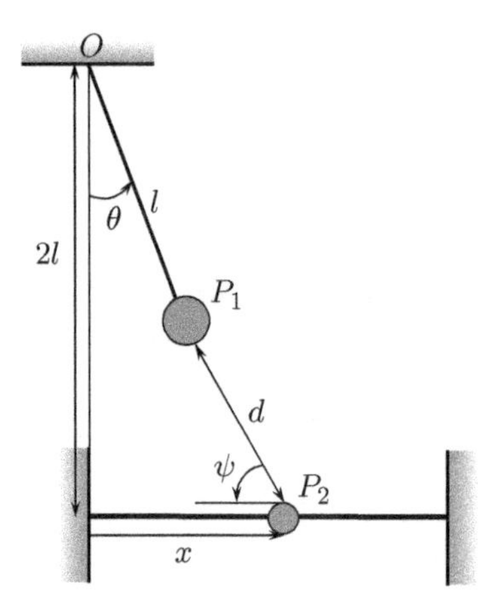


**Figure 2.19** Dynamics of two particles with mutual gravitational interactions.

*Mass* $m_2$:

$$^N\mathbf{a}^{P/O} = \ddot{x}\hat{n}_1.$$

To find the equations governing the motion of each particle, we refer to Figure 2.20 and implement Newton's second law to obtain:

*Mass* $m_1$:

$$\mathbf{F}\hat{n}_1 = m_1\ {}^N\mathbf{a}^{P_1/O}\hat{n}_1,$$

$$-T\sin\theta + F_A\cos\psi = m_1(l\ddot{\theta}\cos\theta - l\dot{\theta}^2\sin\theta),$$

$$\mathbf{F}\hat{n}_2 = m_1\ {}^N\mathbf{a}^{P_1/O}\hat{n}_2,$$

$$T\cos\theta - F_A\sin\psi - m_1 g = m_1(l\ddot{\theta}\sin\theta + l\dot{\theta}^2\cos\theta),$$

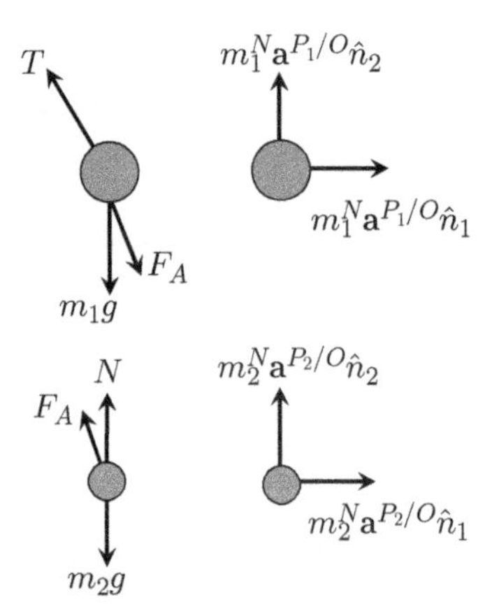


**Figure 2.20** Free-body diagram of two particles with mutual gravitational interactions.

*Mass* $m_2$:

$$\mathbf{F}\hat{n}_1 = m_2\ {}^N\mathbf{a}^{P_2/O}\hat{n}_1,$$

$$-F_A\cos\psi = m_2\ddot{x},$$

where $F_A = Gm_1m_2/d^2$ is the attraction force between the two masses,

$$d = \sqrt{(x - l\sin\theta)^2 + l^2(2 - \cos\theta)^2},$$

is the distance between the two masses, and $\psi$ is the angle that $F_A$ makes with the $\hat{n}_1$ axis where:

$$\cos\psi = \frac{x - l\sin\theta}{d}, \qquad \sin\psi = \frac{2l - l\cos\theta}{d}.$$

**Flipped Classroom Exercise 2.7**

At one point, physicists and transportation entrepreneurs discussed the idea of speedy travel between two points on the Earth's surface by drilling a straight tunnel between them (they were not necessarily diametrically opposite each other). If, as shown, an object of mass $m$ is dropped into the tube, how long does it take to reach the other end? Ignore friction, and assume (erroneously) that the density of the Earth is constant.

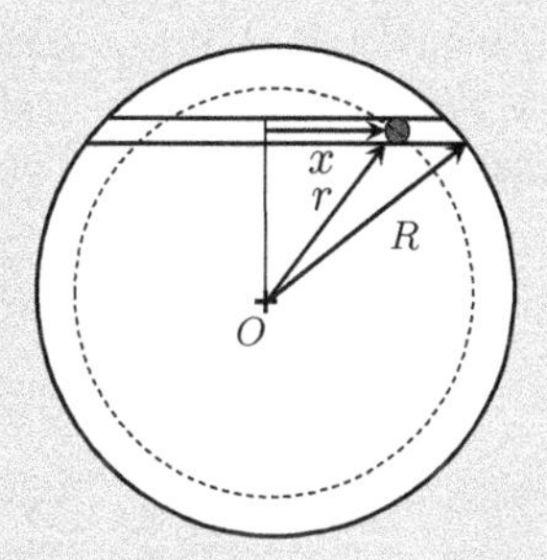


1. Since we are interested in the $x$ dynamics, set the inertial coordinate at point $O$ and find the acceleration of the mass $m$.
2. Draw a free-body diagram for the mass. Note that there is only one force acting on the mass, which is that due to the interaction between the mass of the earth and the dropped object, $m$ (hint: only the mass of of the Earth within the dashed circle contributes to the interaction force).
3. Apply Newton's second law on the mass $m$ and show that

$$\ddot{x} + \frac{GM}{R^3}x = 0.$$

4. The last equation is that of an undamped harmonic oscillator, meaning that when the object is dropped from rest, it will continue to go back and forth between the two sides of the tube. Knowing the properties of undamped oscillators, show that the time required to go from one end of the tube to the other is 42 min, regardless of the object's weight and the location of the tunnel entrance and exit points. This is very fast!

## Exercises

2.1 Consider a bead that slides without friction on a hoop, as shown in Figure 2.21. The hoop is constrained to rotate at a constant angular velocity, $\Omega$. Derive the equation governing the motion for the bead.

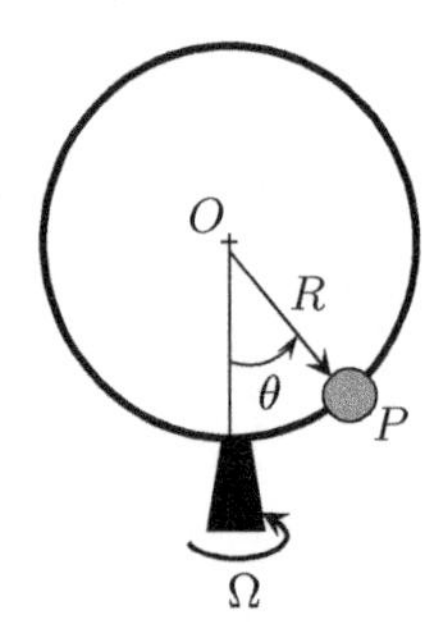


**Figure 2.21** Exercise 2.1.

2.2 In the two-dimensional problem shown in Figure 2.22, a particle $P$ of mass $m_1$ slides without friction on a circular surface of radius, $R$. The circular surface is forced to move to the right over the horizontal surface with an acceleration of $\alpha g$, where $\alpha$ is a constant and $g$ is the gravitational acceleration. If the small particle is given a small nudge to the right at $\theta = 0$, find the critical angle, $\theta_{cr}$, at which it loses contact with the circular surface.

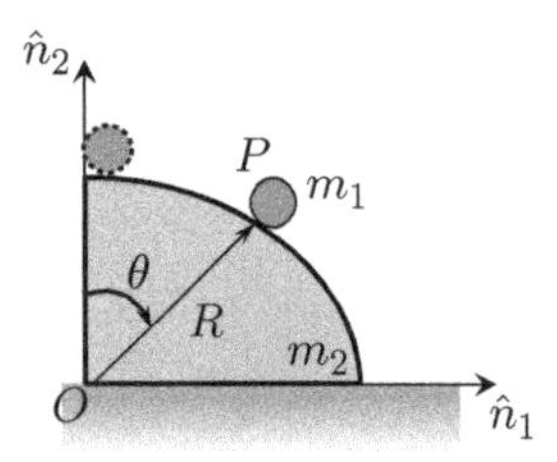


**Figure 2.22** Exercise 2.2.

2.3 A rock is thrown straight up with an initial velocity $v_0$ by someone standing on the equator. Find a formula to calculate where the rock lands (distance and direction). Your answer should be in terms of $v_0$, $g$, and $\Omega$, where $\Omega$ is the frequency of the Earth's rotation.

2.4 A basketball is thrown upward with initial velocity $v_0$. Assume that the drag force from the air is $F = m\alpha v$. What is the speed of the ball, $v_f$, when it hits the ground? (An implicit equation is sufficient.) Does the ball spend more time or less time in the air than it would if it were thrown in a vacuum?

2.5 Figure 2.23 depicts two masses $m_1$ and $m_2$, which are attached to each other with an elastic cable that has an unstretched length $l_0$ and stiffness $k$. The masses are free to slide inside a frictionless massless elbow-pipe as shown in the figure. The pipe is free to rotate with an angle $\theta$. Derive the equations that govern the motion of the system.

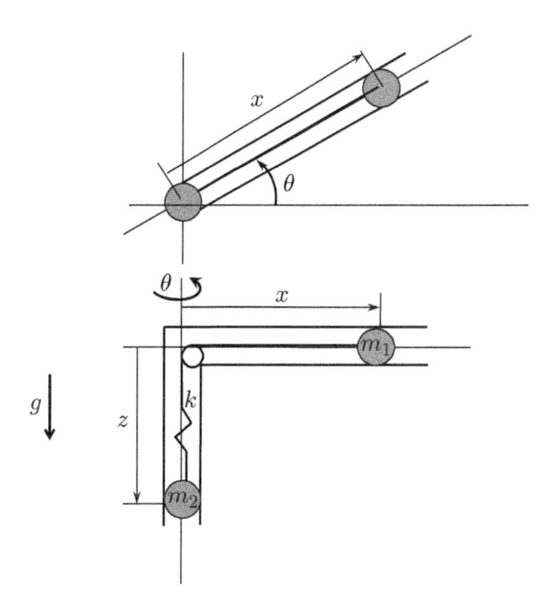


**Figure 2.23** Exercise 2.5.

2.6 For the double pendulum system shown in Figure 2.24, find the equations governing the motion of the system. The external force, $F(t)$, is always perpendicular to the point mass.

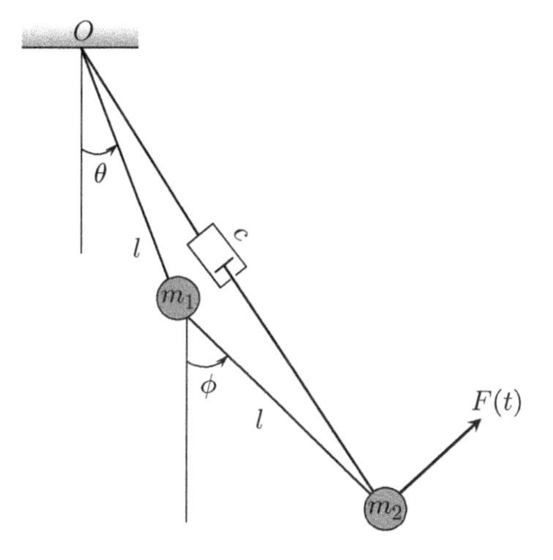


**Figure 2.24** Exercise 2.6.

2.7 The massless disk shown in Figure 2.25 rotates freely in the horizontal plane about a vertical axis through $O$. A block of mass $m$ is attached to $O$ through a spring of stiffness $k$ and a dashpot of a damping coefficient $c$. Use Newton's equations to obtain the equations of motion in terms of the coordinate $r, \theta$.

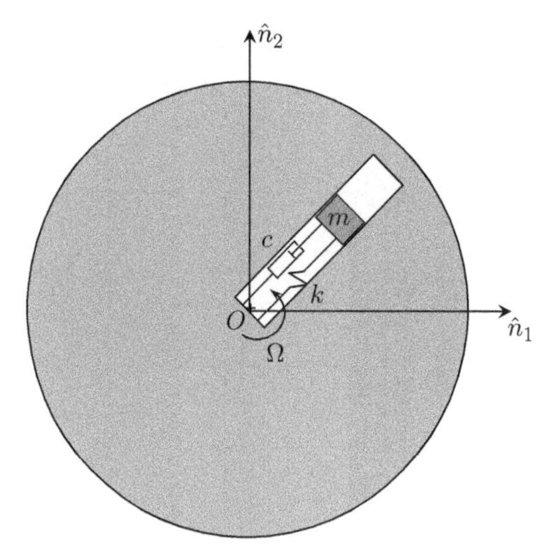


**Figure 2.25** Exercise 2.7.

2.8 Figure 2.26 depicts a spring attached at one end to a disk of radius $R$ and on the other to a particle of mass, $m$. The disk rotates with a constant angular velocity $\Omega$. Find the equations that govern the motion of the system. Assume the spring was initially unstretched.

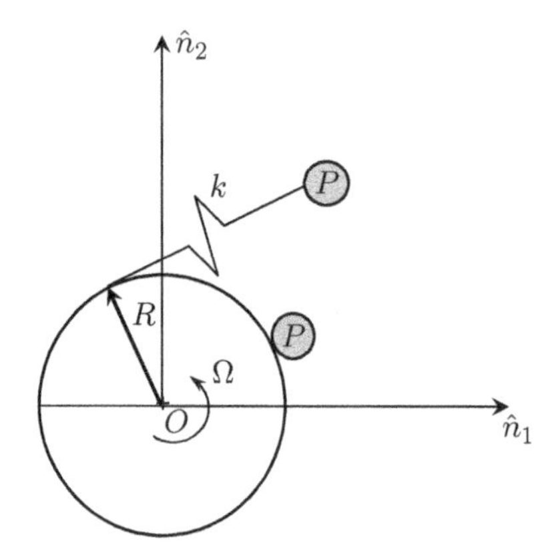


**Figure 2.26** Exercise 2.8.

2.9 For the system shown in Figure 2.27 find the equations of motion and the tension in the cable.

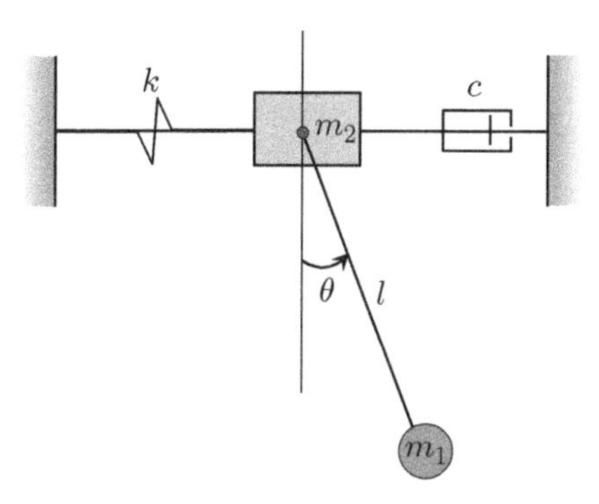


**Figure 2.27** Exercise 2.9.

2.10 A particle of mass $m$ is given an initial velocity $v_0$ at the top of the fixed circular surface of radius $R$, as shown in Figure 2.28. The particle is connected to a spring of stiffness $k$ of unstretched length $0.5R$ and is allowed to slide around the smooth surface. Find the maximum allowable initial velocity $v_0$ such that the block does not lose contact with the surface when it passes through the lowest point.

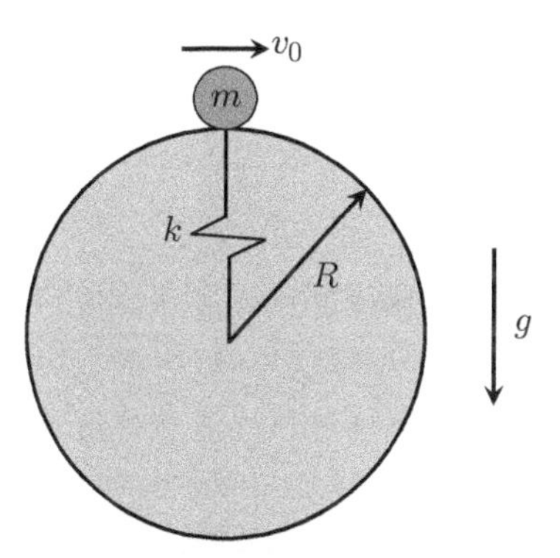


**Figure 2.28** Exercise 2.10.

2.11 A block of mass $m$ slides without friction inside the massless pipe shown in Figure 2.29. The pipe rotates with a constant angular velocity $\Omega$ about point $O$. Find the equation of motion of the block, and the reaction exerted by the pipe on the rod.

2.12 An individual is throwing balls from the window of his third-floor apartment. The window is at a height $h$ above level ground. At what inclination angle should the balls be thrown so that they travel the maximum possible horizontal distance?

2.13 Repeat Exercise 2.12, but now find the inclination angle that maximizes/minimizes the travel time.

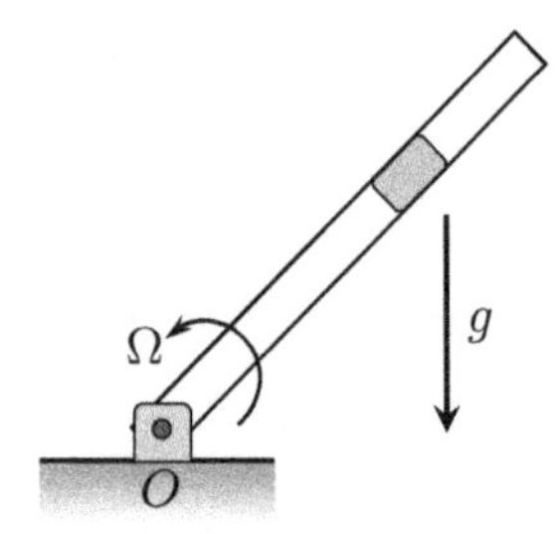


**Figure 2.29** Exercise 2.11.

2.14 Consider the motion of two pendula attached through a rigid string of length $l$, as shown in Figure 2.30. The left-hand pendulum is given an initial condition and allowed to oscillate; write the equations governing the motion of the system and find the tension in the cable.

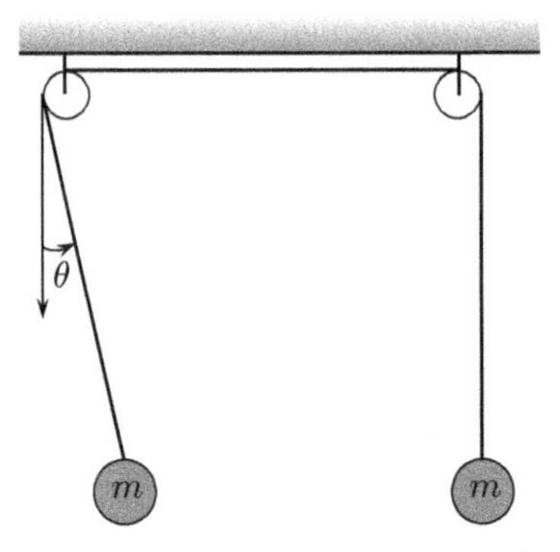


**Figure 2.30** Exercise 2.14.

## Reference

1. L. Euler (1750) "Discovery of a new principle of mechanics", *Memoires de L'Academie Royale des Sciences,* **Par. XXIII,** 196.

# 3

# Dynamics of Rigid Bodies: Vectorial Approach

In this chapter, we extend the vectorial approach based on Newton's second law of dynamics to rigid bodies. To this end, we introduce the concepts of the center of mass, mass moment of inertia, and principal axis of rotation. We used these concepts to model planar and non-planar motion of rigid bodies.

## 3.1 Center of Mass

In dynamics, the center of mass of a given mass distribution is the point at which the weighted relative position of the distributed mass sums to zero. In other words, as shown in Figure 3.1, if we have a collection of $\mathcal{N}$ particles each of mass, $m_i$, randomly distributed in space such that their position vectors measured with respect to the same fixed point are given by $\mathbf{r}_i$, then the center of mass $\mathbf{r}_G$ is the location of the point that satisfies:

$$\sum_{i=1}^{\mathcal{N}} m_i(\mathbf{r}_i - \mathbf{r}_G) = 0,$$

which upon solving for $\mathbf{r}_G$ gives

$$\boxed{\mathbf{r}_G = \frac{\sum_{i=1}^{\mathcal{N}} m_i \mathbf{r}_i}{\sum_{i=1}^{\mathcal{N}} m_i}} \tag{3.1}$$

The center of mass was first defined by Archimedes, who demonstrated that the torque exerted by a number of weights on the suspension point of a lever arm is equal to that exerted by their sum when the sum was placed at a unique point along the lever known as the center of mass.

*Dynamics of Particles and Rigid Bodies: A Self-Learning Approach*, First Edition. Mohammed F. Daqaq.

Companion website: www.wiley.com/go/daqaq/dynamics

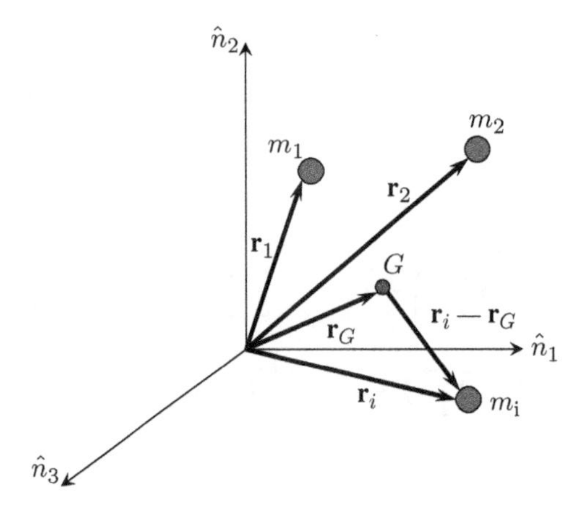


**Figure 3.1** Center of mass, $G$, of $\mathcal{N}$ particles.

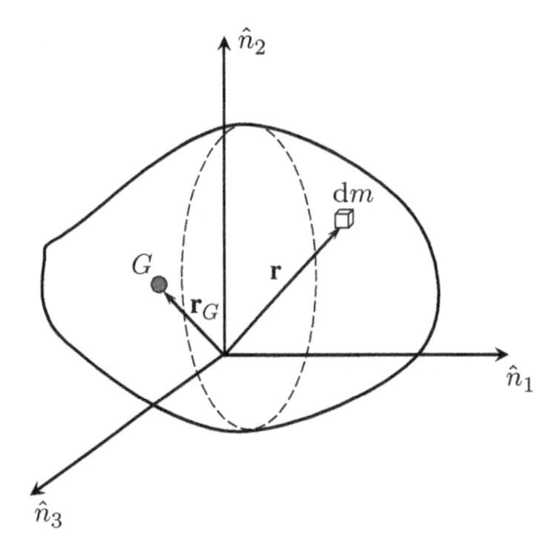


**Figure 3.2** Center of mass, $G$, of a continuous volume.

When the discrete masses are replaced by a continuous volume, $\mathcal{V}$, of a uniform density, $\rho$, as shown in Figure 3.2, the volume can be discretized into differential masses, $\mathrm{d}m = \rho \mathrm{d}\mathcal{V}$, and the sum in Equation (3.1) can be replaced by an integral over the whole volume to obtain:

$$\int_{\mathcal{V}} \rho(\mathbf{r} - \mathbf{r}_G)\mathrm{d}\mathcal{V} = 0,$$

which upon solving for $\mathbf{r}_G$ yields

$$\boxed{\mathbf{r}_G = \frac{\int_{\mathcal{V}} \mathbf{r}\mathrm{d}\mathcal{V}}{\mathcal{V}}} \tag{3.2}$$

## 3.2 Mass Moment of Inertia

The mass moment of inertia, also known as the second moment of mass, is a quantity used in dynamics to measure the resistance of a body to rotation about a given axis. For one particle of mass, $m_i$, the moment of inertia about a given axis passing through an arbitrary point is defined as $I = m_i r_i^2$, where $r_i$ is the distance between the axis of interest and a parallel axis located at the particle. The concept of the moment of inertia was first introduced by Christiaan Huygens in his study of compound pendulum dynamics [1].

As shown in Figure 3.3, the moment of inertia of a particle about the set of axes whose origin is located at point $O$, can be written as

$$
\begin{aligned}
I_{11} &= m_i(y_i^2 + z_i^2) && \text{Moment of inertia of the particle about the } \hat{n}_1\text{-axis,}\\
I_{22} &= m_i(x_i^2 + z_i^2) && \text{Moment of inertia of the particle about the } \hat{n}_2\text{-axis,}\\
I_{33} &= m_i(x_i^2 + y_i^2) && \text{Moment of inertia of the particle about the } \hat{n}_3\text{-axis.}
\end{aligned}
$$

Here, it is worth emphasizing how the distance $r_i$ is measured for each moment of inertia term. For $I_{11}$, $r_i$ is the distance between the $\hat{n}_1$ axis and a parallel axis passing through the particle. In other words, $r_i^2 = y_i^2 + z_i^2$. For $I_{22}$, $r_i$ is the distance between the $\hat{n}_2$ axis and a parallel axis located at the particle. In other words, $r_i^2 = x_i^2 + z_i^2$.

In addition to the previous terms, which are called the *principal moment of inertia* terms, there are six more terms called the *product of inertia* terms, which are defined as

$$
\begin{aligned}
I_{12} = I_{21} &= -m_i x_i y_i,\\
I_{13} = I_{31} &= -m_i x_i z_i,\\
I_{23} = I_{32} &= -m_i y_i z_i.
\end{aligned}
$$

The product of inertia terms do not represent resistance to rotation about a given axis. Nevertheless, they have a physical interpretation. Note that the product of inertia terms have a negative sign, which means they represent tendency rather than resistance. Indeed, given an

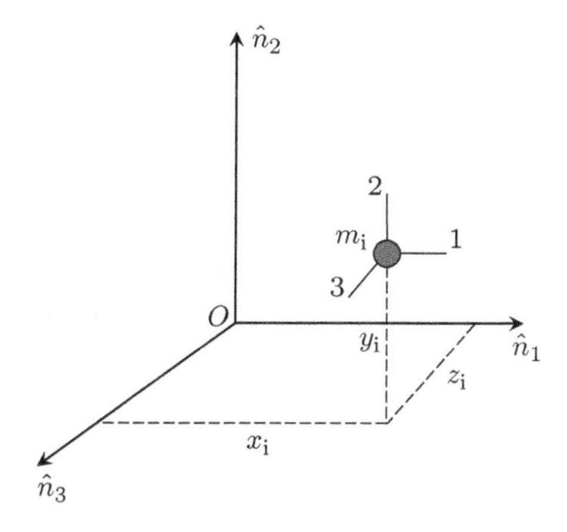


**Figure 3.3** Moment of inertia of one particle about a set of axes whose origin is located at $O$.

initial angular velocity about one axis, the body cannot sustain its rotation about that axis if the product of inertia terms are not zero. In other words, the body tends to start rotating about an axis different from the one it started rotating about.

The nine inertia terms are commonly expressed in terms of an inertia matrix, in the following form:

$$I = \begin{pmatrix} I_{11} & I_{12} & I_{13} \\ I_{21} & I_{22} & I_{23} \\ I_{31} & I_{32} & I_{33} \end{pmatrix}.$$

For $\mathcal{N}$ particles, the moment of inertia about an axis is the sum of the moments of inertia of each particle about that axis; that is

$$I = \begin{pmatrix} \sum_{i=1}^{\mathcal{N}} m_i(y_i^2 + z_i^2) & -\sum_{i=1}^{\mathcal{N}} m_i x_i y_i & -\sum_{i=1}^{\mathcal{N}} m_i x_i z_i \\ -\sum_{i=1}^{\mathcal{N}} m_i x_i y_i & \sum_{i=1}^{\mathcal{N}} m_i(x_i^2 + z_i^2) & -\sum_{i=1}^{\mathcal{N}} m_i y_i z_i \\ -\sum_{i=1}^{\mathcal{N}} m_i x_i z_i & -\sum_{i=1}^{\mathcal{N}} m_i y_i z_i & \sum_{i=1}^{\mathcal{N}} m_i(x_i^2 + y_i^2) \end{pmatrix} \tag{3.3}$$

When considering a continuous volume, $\mathcal{V}$, with a uniform density, $\rho$, the summation over particles is transformed into an integral over the volume by considering a differential mass, $\mathrm{d}m = \rho \mathrm{d}\mathcal{V}$. This yields

$$I = \begin{pmatrix} \int_{\mathcal{V}} \rho(y^2 + z^2)\mathrm{d}\mathcal{V} & -\int_{\mathcal{V}} \rho xy \mathrm{d}\mathcal{V} & -\int_{\mathcal{V}} \rho xz \mathrm{d}\mathcal{V} \\ -\int_{\mathcal{V}} \rho xy \mathrm{d}\mathcal{V} & \int_{\mathcal{V}} \rho(x^2 + z^2)\mathrm{d}\mathcal{V} & -\int_{\mathcal{V}} \rho yz \mathrm{d}\mathcal{V} \\ -\int_{\mathcal{V}} \rho xz \mathrm{d}\mathcal{V} & -\int_{\mathcal{V}} \rho yz \mathrm{d}\mathcal{V} & \int_{\mathcal{V}} \rho(x^2 + y^2)\mathrm{d}\mathcal{V} \end{pmatrix} \tag{3.4}$$

### Example 3.1 Moment of Inertia of a Collection of Particles

Four masses $(m, 2m, 3m, 4m)$ are located respectively at $(2a, 0, 0)$, $(0, 2a, a)$, $(a, 0, 3a)$, $(3a, a, 2a)$ from the origin, as shown in Figure 3.4. Find their center of mass and the inertia matrix about the origin.

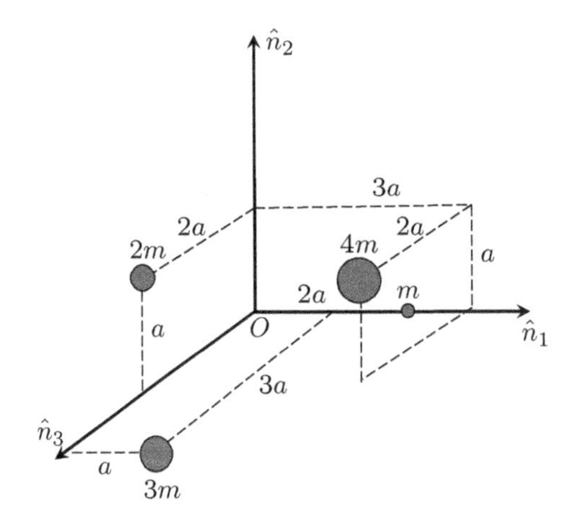


**Figure 3.4** Example 3.1.

To find the center of mass, we use the relation

$$\mathbf{r}_G = \frac{\sum_{i=1}^{\mathcal{N}} m_i \mathbf{r}_i}{\sum_{i=1}^{\mathcal{N}} m_i},$$

which yields

$$x_G = \frac{\sum_{i=1}^{\mathcal{N}} m_i x_i}{\sum_{i=1}^{\mathcal{N}} m_i} = \frac{2ma + 3ma + 12ma}{10m} = 1.7a,$$

$$y_G = \frac{\sum_{i=1}^{\mathcal{N}} m_i y_i}{\sum_{i=1}^{\mathcal{N}} m_i} = \frac{4ma + 4ma}{10m} = 0.8a,$$

$$z_G = \frac{\sum_{i=1}^{\mathcal{N}} m_i y_i}{\sum_{i=1}^{\mathcal{N}} m_i} = \frac{2ma + 9ma + 8ma}{10m} = 1.9a.$$

Thus, $\mathbf{r}_G = 1.7a\,\hat{n}_1 + 0.8a\,\hat{n}_2 + 1.9a\,\hat{n}_3$.

To find the moment of inertia, we use Equation (3.3), where

$$\begin{aligned}
I_{11} &= 2m((2a)^2 + a^2) + 3m(3a)^2 + 4m(a^2 + 4a^2) = 57ma^2,\\
I_{22} &= m(2a)^2 + 2m(a)^2 + 3m(a^2 + 9a^2) + 4m(9a^2 + 4a^2) = 88ma^2,\\
I_{33} &= m(2a)^2 + 2m(2a)^2 + 3ma^2 + 4m(9a^2 + a^2) = 55ma^2,\\
I_{12} = I_{21} &= -4m(3a)(a) = -12ma^2,\\
I_{13} = I_{31} &= -(3m(3a)a + 4m(6a)) = -33ma^2,\\
I_{23} = I_{32} &= -2m(2a)a - 4m(2a)(a) = -12ma^2.
\end{aligned}$$

This yields,

$$I = ma^2 \begin{pmatrix} 57 & -12 & -33 \\ -12 & 88 & -12 \\ -33 & -12 & 55 \end{pmatrix}.$$

**Example 3.2 Moment of Inertia of a Thin Plate**

The thin triangular plate shown in Figure 3.5 has a uniform area density $\rho$. Find its center of mass and moment of inertia matrix about the set of axes passing through point $O$.

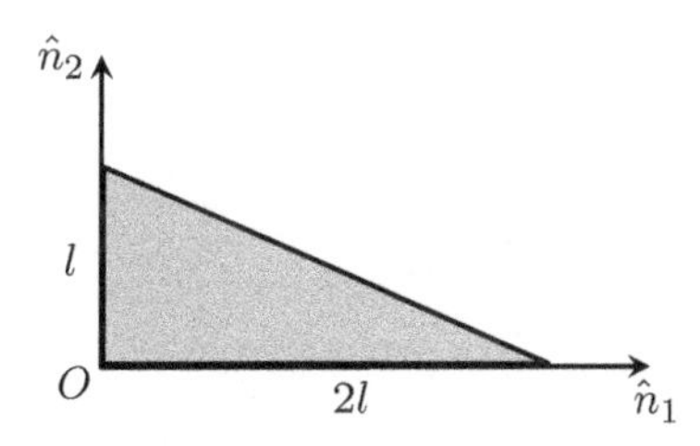


**Figure 3.5** Thin plate of area density $\rho$.

To find the center of mass, we use Equation (3.2). We note that the plate is thin, and so the integral over the volume is transformed into an integral over the area. The distance along the $\hat{n}_1$-axis is defined as $x$ while that along the $\hat{n}_2$-axis is defined as $y$. It follows that

$$x_G = \frac{\int_0^{2l}\int_0^{y=l-\frac{x}{2}} \rho x \mathrm{d}y\mathrm{d}x}{\int_0^{2l}\int_0^{y=l-\frac{x}{2}} \rho \mathrm{d}y\mathrm{d}x} = \frac{2l}{3},$$

$$y_G = \frac{\int_0^{2l}\int_0^{y=l-\frac{x}{2}} \rho y \mathrm{d}y\mathrm{d}x}{\int_0^{2l}\int_0^{y=l-\frac{x}{2}} \rho \mathrm{d}y\mathrm{d}x} = \frac{l}{3}.$$

The limits on the integrals above are as follows: when $x$ goes from 0 to $2l$, $y$ goes from 0 to its maximum value at each value of $x$, as represented by the equation of the line $y = l - \frac{x}{2}$.

To find the mass moment of inertia about the set of axes shown, we use Equation (3.4), and obtain:

$$I_{11} = \int_0^{2l}\int_0^{l-\frac{x}{2}} \rho y^2 \mathrm{d}y\mathrm{d}x = \frac{\rho l^4}{6},$$

$$I_{22} = \int_0^{2l}\int_0^{l-\frac{x}{2}} \rho x^2 \mathrm{d}y\mathrm{d}x = \frac{2\rho l^4}{3},$$

$$I_{33} = \frac{5\rho l^4}{6}, \qquad I_{33} = I_{11} + I_{22} \text{ for two-dimensional bodies,}$$

$$I_{12} = I_{21} = -\int_0^{2l}\int_0^{l-\frac{x}{2}} \rho xy \mathrm{d}y\mathrm{d}x = -\frac{\rho l^4}{6},$$

$$I_{13} = I_{31} = 0,$$

$$I_{23} = I_{32} = 0.$$

This yields the following inertia matrix:

$$I = \rho l^4 \begin{pmatrix} \frac{1}{6} & -\frac{1}{6} & 0 \\ -\frac{1}{6} & \frac{2}{3} & 0 \\ 0 & 0 & \frac{5}{6} \end{pmatrix}.$$

 **Flipped Classroom Exercise 3.1**

The thin plate shown in the figure has a uniform area density, $\rho$, and a mass, $m$. Find the center of gravity and the mass moment of inertia matrix about the axis passing through point $O$. Express your answers in terms of the mass $m$.

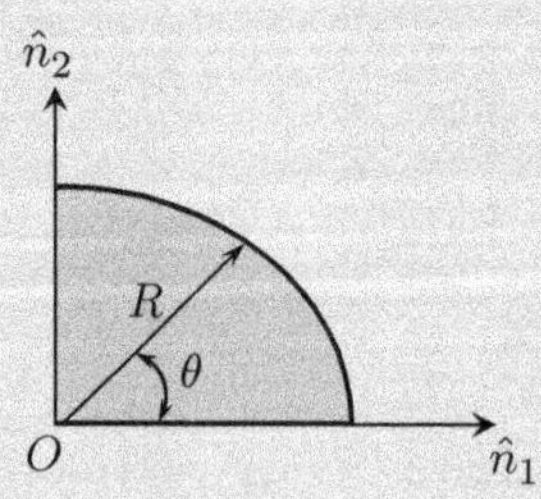

1. Instead of using the Cartesian coordinates, use polar coordinates, such that $x = r\cos\theta, y = r\sin\theta$. Show that $\mathrm{d}A = r\mathrm{d}r\mathrm{d}\theta$.
2. Use Equation (3.2) to show that $r_G = \frac{2}{3}R$ and $\theta_G = \frac{\pi}{4}$.
3. Use Equation (3.4) to show that

$$I = mR^2 \begin{pmatrix} \frac{1}{4} & -\frac{1}{2\pi} & 0 \\ -\frac{1}{2\pi} & \frac{1}{4} & 0 \\ 0 & 0 & \frac{1}{2} \end{pmatrix}.$$

Hint: $I_{11} = \int_0^{\frac{\pi}{2}} \int_0^R \rho r^3 \sin^2\theta \mathrm{d}r\mathrm{d}\theta$.

## 3.3 Parallel Axis Theorem

This section illustrates that by knowing the moment of inertia matrix about a set of axes whose origin is located at a point, say $O$, then one can infer the moment of inertia about another parallel coordinate system passing through another point, say $Q$. To this end, consider Figure 3.6, where we want to find the moment of inertia of an arbitrarily-shaped body about a set of axes whose origin is located at point $O$. However, rather than finding the moment of inertia about point $O$ directly, we define another intermediate set of axes that is parallel to the original system and whose origin is located at point $Q$.

To find the moment of inertia of the body about point $O$, we define a differential mass $\mathrm{d}m = \rho\mathrm{d}\mathcal{V}$ and use Equation (3.4). It follows that

$$I_{11} = \int_{\mathcal{V}} \rho(y^2 + z^2)\mathrm{d}\mathcal{V},$$

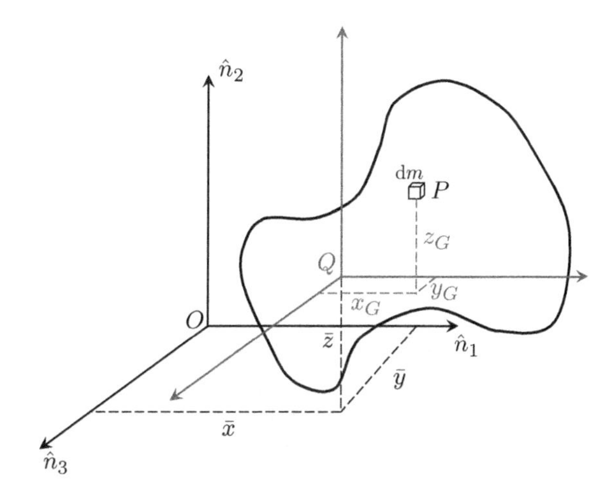


**Figure 3.6** Parallel axis theorem.

where $y$ and $z$ are the distances from point $O$ to point $P$ along the $\hat{n}_2$ and $\hat{n}_3$ axes, respectively; that is

$$x = \bar{x} + x_G, \qquad y = \bar{y} + y_G.$$

Upon substituting the previous equation into the integral for $I_{11}$ and rearranging, we obtain

$$I_{11} = \rho \int_{\mathcal{V}} (\bar{y}^2 + \bar{z}^2)\mathrm{d}\mathcal{V} + \rho \int_{\mathcal{V}} (y_G^2 + z_G^2)\mathrm{d}\mathcal{V} + 2\rho \int_{\mathcal{V}} \bar{y}y_G \mathrm{d}\mathcal{V} + 2\rho \int_{\mathcal{V}} \bar{z}z_G \mathrm{d}\mathcal{V}.$$

Since $\bar{y}$ and $\bar{z}$ are independent of the choice of the location of the differential mass, they can be taken outside the integrals to give:

$$I_{11} = \rho\mathcal{V}(\bar{y}^2 + \bar{z}^2) + \rho \int_{\mathcal{V}} (y_G^2 + z_G^2)\mathrm{d}\mathcal{V} + 2\rho\bar{y} \int_{\mathcal{V}} y_G \mathrm{d}\mathcal{V} + 2\rho\bar{z} \int_{\mathcal{V}} z_G \mathrm{d}\mathcal{V}.$$

This can be further simplified by choosing $Q$ to coincide with the center of mass of the body, $G$. The last two terms then vanish because at the center of mass, $\int_{\mathcal{V}} z_G \mathrm{d}\mathcal{V} = \int_{\mathcal{V}} y_G \mathrm{d}\mathcal{V} = 0$. This yields

$$I_{11} = m(\bar{y}^2 + \bar{z}^2) + I_{11G}.$$

Carrying out the same procedure for all of the other moment of inertia terms, we obtain

$$\boxed{\begin{aligned} I_{11} &= m(\bar{y}^2 + \bar{z}^2) + I_{11G} \\ I_{22} &= m(\bar{x}^2 + \bar{z}^2) + I_{22G} \\ I_{33} &= m(\bar{x}^2 + \bar{y}^2) + I_{33G} \\ I_{12} &= I_{21} = -m\bar{x}\bar{y} + I_{12G} \\ I_{13} &= I_{31} = -m\bar{x}\bar{z} + I_{13G} \\ I_{23} &= I_{32} = -m\bar{y}\bar{z} + I_{23G} \end{aligned}} \tag{3.5}$$

These relations are known as the *parallel axis theorem* or the *Huygens–Steiner theorem* after, respectively, the Dutch and Swiss mathematicians Chrsitiaan Hugens and Jakob Steiner who contributed to its derivation. The parallel axis theorem can be used to find the moment of inertia matrix about any point in space knowing the moment of inertia about an axis passing through the center of mass, $G$.

**Parallel Axis Theorem**

You cannot use the parallel axis theorem given by Equation (3.5) to find the moment of inertia matrix about an arbitrary point, say $O$, by directly using the known moment of inertia matrix about another arbitrary point, say $Q$. The center of mass $G$ should always be an intermediate step. In other words, first use the parallel axis theorem to find the inertia matrix about $G$ using the known inertia matrix about $Q$. Subsequently, use the inertia matrix about $G$ to find the inertia matrix about $O$.

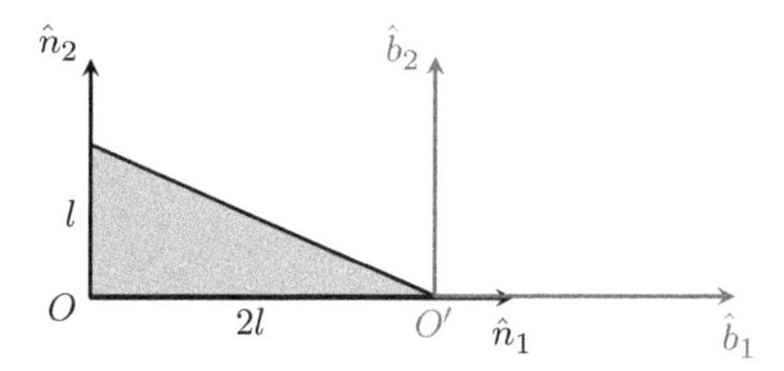


**Figure 3.7** Thin plate of area density $\rho$.

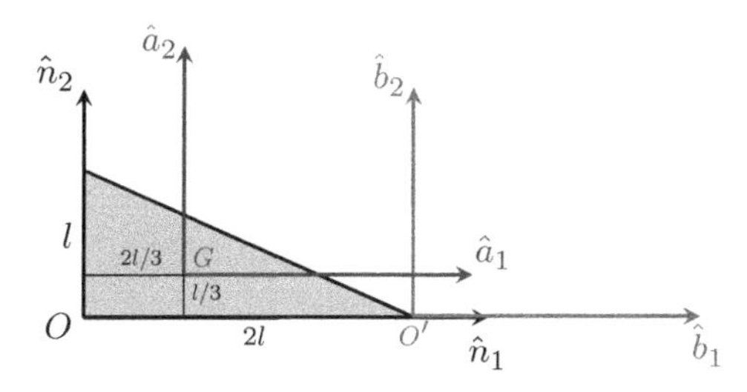


**Figure 3.8** Parallel axis theorem implemented on the thin plate of area density $\rho$.

### Example 3.3 Parallel Axis Theorem

Starting with the inertia matrix from Example 3.2, use the parallel axis theorem to find the inertia matrix around the coordinate system whose origin is located at $O'$ in Figure 3.7.

First, note that we cannot use the parallel axis theorem directly between point $O$ and point $O'$. As such, we construct an intermediate set of axes located at the center of mass $G$, whose position with respect to point $O$ is given by $\mathbf{r}_G = 2l/3\,\hat{n}_1 + l/3\,\hat{n}_2$, as shown in Figure 3.8. Next, we use Equation (3.5) to find the moment of inertia about a coordinate system whose origin is located at point $G$. This yields

$$
\begin{aligned}
I_{11G} &= I_{11} - m\bar{y}^2 = \frac{\rho l^4}{6} - \frac{\rho l^4}{9} = \frac{\rho l^4}{18},\\
I_{22G} &= I_{22} - m\bar{x}^2 = \frac{2\rho l^4}{3} - \frac{4\rho l^4}{9} = \frac{2\rho l^4}{9},\\
I_{33G} &= I_{33} - m(\bar{x}^2 + \bar{y})^2 = \frac{5\rho l^4}{18},\\
I_{12G} &= I_{21G} = I_{12} + m\bar{x}\bar{y} = -\frac{1\rho l^4}{6} + \frac{4\rho l^4}{9} = \frac{1}{18}\rho l^4,\\
I_{13G} &= I_{31G} = I_{23G} = I_{32G} = 0.
\end{aligned}
$$

Using the parallel axis theorem one more time to move from point $G$ to point $O'$ yields

$$
I'_{11} = I_{11G} + m\bar{y}^2 = \frac{\rho l^4}{18} + \frac{\rho l^4}{9} = \frac{\rho l^4}{6},
$$

$$I'_{22} = I_{22G} + m\bar{x}^2 = \frac{2\rho l^4}{9} + \frac{16\rho l^4}{9} = 2\rho l^4,$$

$$I'_{33} = I'_{11} + I'_{22} = \frac{13\rho l^4}{6},$$

$$I'_{12} = I'_{21} = I_{12G} - m\bar{x}\bar{y} = \frac{\rho l^4}{18} - \rho l^2 \left(\frac{4l}{3}\right)\left(\frac{-l}{3}\right) = \frac{1}{2}\rho l^4,$$

while the rest of the inertia terms are zero. The inertia matrix about $O'$ can therefore be written as:

$$I' = \frac{\rho l^4}{6} \begin{pmatrix} 1 & 3 & 0 \\ 3 & 12 & 0 \\ 0 & 0 & 13 \end{pmatrix}.$$

**Flipped Classroom Exercise 3.2**

Starting with the results of Flipped Classroom Exercise 3.1, find the moment of inertia about the set of axes shown in the figure, whose origin is located at point $O'$. Express your answers in terms of the mass $m$.

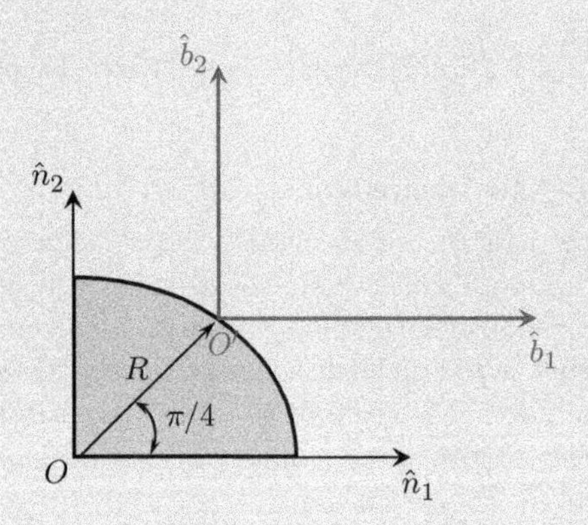


1. Instead of using Cartesian coordinates, use polar coordinates, such that $x = r\cos\theta$, $y = r\sin\theta$. Show that $\mathrm{d}A = r\mathrm{d}r\mathrm{d}\theta$.
2. Use Equation (3.2) to show that $r_G = \frac{2}{3}R$ and $\theta_G = \frac{\pi}{4}$.
3. Use the parallel axis theorem to show that

$$I_G = mR^2 \begin{pmatrix} \frac{1}{36} & -\frac{4\pi-9}{18\pi} & 0 \\ -\frac{4\pi-9}{18\pi} & \frac{1}{36} & 0 \\ 0 & 0 & \frac{1}{18} \end{pmatrix}.$$

4. Use the parallel axis theorem again to show that

$$I_{O'} = mR^2 \begin{pmatrix} \frac{1}{12} & \frac{3\pi-9}{18\pi} & 0 \\ \frac{3\pi-9}{18\pi} & \frac{1}{12} & 0 \\ 0 & 0 & \frac{1}{6} \end{pmatrix}.$$

## 3.4 Rotation of the Inertia Matrix

If the inertia matrix about a set of axes whose origin is located at a given point is known, it is also possible to calculate the moment of inertia about another set of axes formed by rotating the original set about an arbitrary axis. To find the relationship between the original and rotated coordinate systems, we borrow the concept of the rotational kinetic energy of rigid bodies. This will be discussed in further detail in Chapter 6.

For now, all you need to know is that, when a rigid body is rotating with an angular velocity vector $\boldsymbol{\omega}$ about a fixed point $O$, its kinetic energy can be written as:

$$\mathcal{T} = \frac{1}{2}\boldsymbol{\omega}^T I_O \boldsymbol{\omega} \tag{3.6}$$

where $\boldsymbol{\omega}^T$ is the transpose of $\boldsymbol{\omega}$, and $I_O$ is the moment of inertia matrix about a set of axes whose origin is located at point $O$. As discussed in Chapter 1, when the coordinate system is rotated around an arbitrary axis, its new angular velocity vector $\boldsymbol{\omega}^*$ can be related to the original angular velocity vector $\boldsymbol{\omega}$ using a transformation matrix $L$, such that

$$\boldsymbol{\omega}^* = L\boldsymbol{\omega}. \tag{3.7}$$

Note that the kinetic energy of a rotating body does not change by rotating the coordinate system used to describe it; hence we can write

$$\mathcal{T} = \frac{1}{2}\boldsymbol{\omega}^{*T} I_O^* \boldsymbol{\omega}^* = \frac{1}{2}\boldsymbol{\omega}^T I_O \boldsymbol{\omega}. \tag{3.8}$$

Substituting Equation (3.7) into Equation (3.8) yields

$$I_O = L^T I_O^* L, \tag{3.9}$$

or

$$\boxed{I_O^* = L\, I_O\, L^T} \tag{3.10}$$

Equation (3.10) can be used to obtain the moment of inertia matrix about a coordinate system formed using a general rotation $L$ about the original coordinate system, provided that both systems share the same origin.

**Example 3.4 Rotation of the Inertia Matrix**

Starting with the inertia matrix obtained in Example 3.2, find the inertia matrix about a coordinate system formed by a rotation $\theta$ about the $\hat{n}_3$ axis, as shown on Figure 3.9. Find the angle $\theta$ that diagonalizes the inertia matrix.

To find the inertia matrix about the rotated frame, we must define the transformation matrix, $L$. For a third or 3-rotation, the transformation matrix is given by

$$L = \begin{pmatrix} \cos\theta & \sin\theta & 0 \\ -\sin\theta & \cos\theta & 0 \\ 0 & 0 & 1 \end{pmatrix}$$

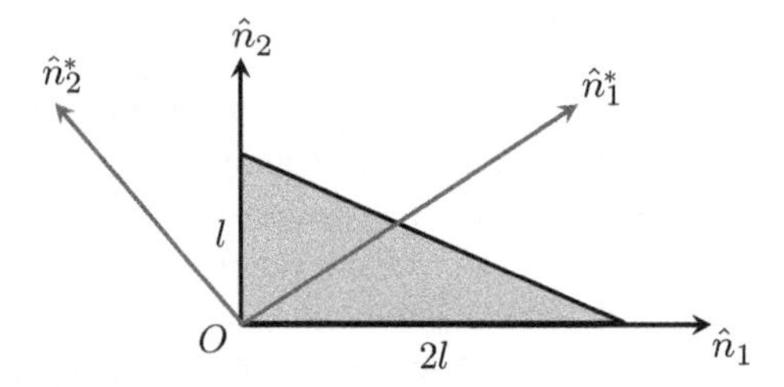


**Figure 3.9** Rotation of the inertia matrix for a triangular shape.

Using Equation (3.10), we can write

$$I_O^* = \rho l^4 \begin{pmatrix} \cos\theta & \sin\theta & 0 \\ -\sin\theta & \cos\theta & 0 \\ 0 & 0 & 1 \end{pmatrix} \begin{pmatrix} \frac{1}{6} & -\frac{1}{6} & 0 \\ -\frac{1}{6} & \frac{2}{3} & 0 \\ 0 & 0 & \frac{5}{6} \end{pmatrix} \begin{pmatrix} \cos\theta & -\sin\theta & 0 \\ \sin\theta & \cos\theta & 0 \\ 0 & 0 & 1 \end{pmatrix},$$

$$= \frac{\rho l^4}{6} \begin{pmatrix} \cos^2\theta - 2\sin\theta\cos\theta + 4\sin^2\theta & 3\sin\theta\cos\theta + 4\sin^2\theta - \cos^2\theta & 0 \\ 3\sin\theta\cos\theta + 4\sin^2\theta - \cos^2\theta & \sin^2\theta + 2\sin\theta\cos\theta + 4\cos\theta & 0 \\ 0 & 0 & 5 \end{pmatrix}.$$

To find the angle $\theta$ that diagonalizes the inertia matrix we set the off-diagonal terms to zero; that is,

$$3\sin\theta\cos\theta + 4\sin^2\theta - \cos^2\theta = 0,$$

$$\tan 2\theta = \frac{2}{3}, \qquad \theta = 16.85^\circ.$$

### *3.4.1 The Principal Axes*

The principal axes are the set of axes about which the inertia matrix is diagonal; that is, about which all product of the inertia terms are zero ($I_{12} = I_{21} = I_{31} = I_{13} = I_{23} = I_{32} = 0$). In the previous example, the principal axes were realized by rotating the original coordinate system through an angle $\theta = 16.85^\circ$ about the $\hat{n}_3$ axis. Nevertheless, the process of finding the principal axes is not always easy, especially when they cannot be realized by performing a simple rotation about one of the three unit vectors forming the original coordinate system.

To generalize the process of diagonalizing inertia matrices and finding the principal axes, we recast the problem in terms of the original transformation:

$$I_O^* = L\, I_O\, L^T.$$

Since the inertia matrix $I_O$ is a real symmetric matrix, and $L^{-1} = L^T$ because a rotation matrix is an orthogonal matrix, then according to matrix theory, there must exist a matrix $L$ that diagonalizes $I_O$ such that the diagonal matrix $I_O^*$ is given by $I_O^* = L\, I_O\, L^T$. Here, the rotation matrix $L$ is constructed such that its rows represent the three mutually orthogonal right eigenvectors of the matrix $I_O$ (see Section I.2.1.5). Furthermore, if each eigenvector were to be normalized to have unity length, then the diagonal elements, also known as the *principal moments of inertia*, are the real eigenvalues of the matrix $I_O$.

**Example 3.5 Principal Axes of Rotation**

For the inertia matrix obtained in Example 3.2, find the rotation matrix $L$ that diagonalizes the inertia matrix. What are the principal moments of inertia?

Our goal is to diagonalize the matrix $I_O$ given by

$$I_O = \frac{\rho l^4}{6}\begin{pmatrix} 1 & -1 & 0 \\ -1 & 4 & 0 \\ 0 & 0 & 5 \end{pmatrix}.$$

Since $I_O$ is a real symmetric matrix, then according to matrix theory, there must exist a matrix $L$ that diagonalizes $I_O$. To find this matrix, we first find the eigenvalues of the matrix $I_O$ using $|I_O - \lambda\mathcal{I}| = 0$, where $\mathcal{I}$ is the identity matrix and $\lambda$ represents the eigenvalues of $I_O$. This yields the following eigenvalues:

$$\lambda_1 = \frac{5-\sqrt{13}}{2}, \qquad \lambda_2 = \frac{5+\sqrt{13}}{2}, \qquad \lambda_3 = 5.$$

The eigenvectors are then obtained using $(I_O - \lambda_i\mathcal{I})\mathbf{e}_i = 0$, where $\mathbf{e}_i$ are the right eigenvectors associated with the $\lambda_i$. This yields

$$\mathbf{e}_1 = \begin{bmatrix} 0.957 \\ 0.2898 \\ 0 \end{bmatrix}, \qquad \mathbf{e}_2 = \begin{bmatrix} 0.2898 \\ -0.9571 \\ 0 \end{bmatrix}, \qquad \mathbf{e}_3 = \begin{bmatrix} 0 \\ 0 \\ 1 \end{bmatrix}.$$

Here, $\mathbf{e}_i$ are normalized such that $\| \mathbf{e}_i \| = 1$. Hence, $L$ can be constructed such that

$$L = \begin{bmatrix} \mathbf{e}_1 \\ \mathbf{e}_2 \\ \mathbf{e}_3 \end{bmatrix} = \begin{pmatrix} 0.9571 & 0.2898 & 0 \\ 0.2899 & -0.9571 & 0 \\ 0 & 0 & 1 \end{pmatrix}.$$

Note that the previous matrix does not resemble any of the general rotations obtained in Chapter 1 for a rotation about a single axis. Nevertheless, when multiplying the middle row with a negative one, which means we are reversing the direction of the second vector, then the previous matrix becomes a rotation matrix about the third axis (3-rotation) with $\sin\theta = 0.2898$, and $\theta = 16.85°$.

**Example 3.6 General Formula**

Show that when one of the principal axes is aligned with the unit vector in the direction of $\hat{n}_3$, then the other two principal axes can be obtained by performing a counterclock-wise rotation $\alpha$ around the $\hat{n}_3$ axis such that

$$\tan(2\alpha) = \frac{2I_{12}}{I_{11} - I_{22}}.$$

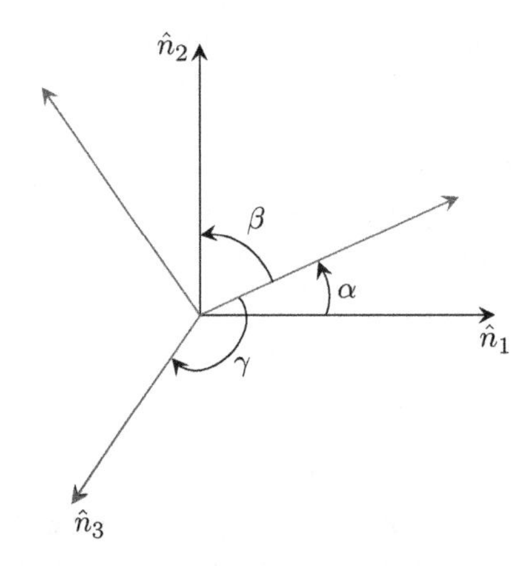


**Figure 3.10** General formula for a single rotation.

Starting with the fact that the principal axes of rotation satisfies the following equation:

$$\begin{pmatrix} I_{11}-\lambda & I_{12} & I_{13} \\ I_{21} & I_{22}-\lambda & I_{23} \\ I_{31} & I_{32} & I_{33}-\lambda \end{pmatrix} \begin{bmatrix} e_{11} \\ e_{22} \\ e_{33} \end{bmatrix} = 0,$$

and noting that the eigenvector $\mathbf{e}_1$ represents the direction cosine vector with the three original axes, we can use Figure 3.10 to write

$$\begin{bmatrix} e_{11} \\ e_{22} \\ e_{33} \end{bmatrix} = \begin{bmatrix} \cos\alpha \\ \cos\beta \\ \cos\gamma \end{bmatrix}.$$

For a rotation about the third axis, the direction cosine vector satisfies the following: $\gamma = 90^0$, $\cos\gamma = 0$, $\alpha + \beta = 90°$, meaning that $\sin\alpha = \cos\beta$ and $\cos^2\alpha + \cos^2\beta = 1$. This yields:

$$(I_{11} - \lambda)\cos\alpha + I_{12}\sin\alpha = 0,$$

$$(I_{12}\cos\alpha + (I_{22} - \lambda)\sin\alpha = 0.$$

Eliminating $\lambda$ from the above equations yields

$$I_{12}(1 - \tan^2\alpha) = (I_{11} - I_{22})\tan\alpha,$$

$$\tan 2\alpha = \frac{2I_{12}}{I_{11} - I_{22}}. \tag{3.11}$$

 **Flipped Classroom Exercise 3.3**

The inertia matrix of a rigid body is given by

$$I = \frac{1}{3}\begin{pmatrix} 4 & 2 & -1 \\ 2 & 7 & -2 \\ -1 & -2 & 4 \end{pmatrix}. \tag{3.12}$$

Obtain the rotation matrix that can be used to rotate the coordinate system to the principal axes.

1. Show that the eigenvalues of $I$ are $\lambda_1 = \lambda_2 = 1, \lambda_3 = 3$.
2. Find the eigenvectors associated with these eigenvalues and normalize them. Note that three orthogonal eigenvectors can be found despite the repeated eigenvalues (algebraic multiplicity). This can be realized by forcing the third eigenvector to be orthogonal to the other two. Show that the eigenvectors can be written as

$$\mathbf{e}_1 = \frac{1}{\sqrt{5}}\begin{bmatrix} -2 \\ 1 \\ 0 \end{bmatrix}, \quad \mathbf{e}_2 = \frac{1}{\sqrt{30}}\begin{bmatrix} 1 \\ 2 \\ 5 \end{bmatrix}, \quad \mathbf{e}_3 = \frac{1}{\sqrt{6}}\begin{bmatrix} -1 \\ -2 \\ 1 \end{bmatrix}.$$

3. Construct the rotation matrix $L$ by using the previous eigenvectors as the rows of $L$.

**Principal Axes**

The principal axes are the set of axes about which the moment of inertia matrix is diagonal. To find the transformation matrix that aligns the coordinate system with the principal axes, you need to take the following steps:

1. Find the eigenvalues of the inertia matrix.
2. Find the right eigenvectors associated with those eigenvalues. Note that three mutually orthogonal right eigenvectors can always be found even in the presence of repeated eigenvalues (algebraic multiplicity).
3. Use the eigenvectors to construct the transformation matrix $L$ such that its rows represent the eigenvectors of the inertia matrix.

Note that if the eigenvectors were normalized to have unity length, then the diagonal elements of the diagonalized inertia matrix, also known as the principal moments of inertia, will be the real eigenvalues of the inertia matrix.

## 3.5 Planar Motion of Rigid Bodies

When using Newton's second law to analyze the dynamics of particles, it is inherently assumed that the moment of inertia of the particle about its center of mass is zero. In other words, we assume that the particle is too small to have any resistance to rotational motion about any axis of rotation located at its center of mass. As such, the free motion of the particle in free space is sufficiently described by three coordinate systems involving translational motions along the three axes of the Cartesian frame (three degrees of freedom).

On the other hand, a rigid body has its mass distributed over a bigger volume such that it can resist rotation along one or more axes of rotation located at its center of mass. As such, the full dynamics of a rigid body involved in translational and rotational motions is described

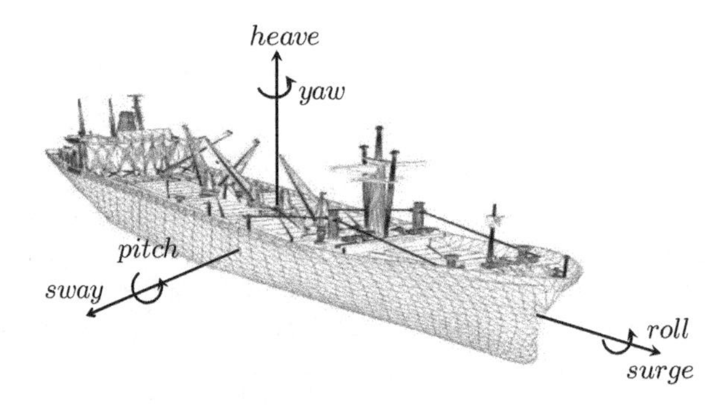


**Figure 3.11** Six degrees of freedom of a rigid body clearly illustrated using ship motion.

by six degrees of freedom: translation along the three coordinates and rotation around the three axes. As shown in Figure 3.11, for ship and flight dynamics, the three translations are commonly referred to as *surge*, *sway*, and *heave*, and the three rotations are described as *pitch*, *roll*, and *yaw*.

A simplified version of rigid-body dynamics is obtained when the body involves planar motion; in other word, translational motions in a plane formed by two coordinates and a rotation about an axis normal to them. In such a scenario, the dynamics can be fully described using three degrees of freedom and the motion is commonly referred to as planar rigid-body motion. In what follows, we derive the equations that can be used to study planar rigid-body dynamics.

Consider the arbitrarily-shaped rigid body shown in Figure 3.12. Without loss of generality, consider the translational motion to occur in the directions of $(\hat{n}_1, \hat{n}_2)$ and the rotation to occur about the third axis $\hat{n}_3$. To describe the motion of the rigid body, we use two different frames. The first is the inertial $N$-frame located at point $Q$, while the second is a rotating $B$-frame located at point $O$. The body is subject to a multitude of external $\mathbf{F}_i$ and internal, $\mathbf{f}'$, forces, as shown.

We first obtain the acceleration of the differential mass, $\mathrm{d}m$ with respect to the inertial point $Q$, that is, ${}^N\mathbf{a}^{P/Q}$. To this end, we write

$$\mathbf{QP} = \mathbf{QO} + \mathbf{r}.$$

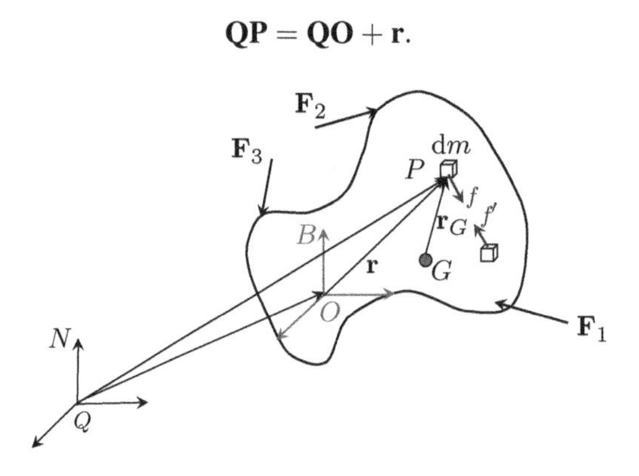


**Figure 3.12** Planar rigid-body dynamics. (*See color plate section for the color representation of this figure.*)

Note that **QO** is described in the inertial frame, whereas **r** is described in the rotating frame. As such, the velocity can be written as

$$ {}^{N}\mathbf{v}^{P/Q} = {}^{N}\mathbf{v}^{O/Q} + \dot{\mathbf{r}} + {}^{N}\boldsymbol{\omega}^{B} \times \mathbf{r}. $$

Since the body considered is rigid (inelastic), the vector **r** does not change length. Therefore, $\dot{\mathbf{r}}$ vanishes. Thus, the previous equation reduces to

$$ {}^{N}\mathbf{v}^{P/Q} = {}^{N}\mathbf{v}^{O/Q} + {}^{N}\boldsymbol{\omega}^{B} \times \mathbf{r}. $$

Using the velocity expression, the acceleration can be written as

$$ {}^{N}\mathbf{a}^{P/Q} = {}^{N}\mathbf{a}^{O/Q} + {}^{N}\boldsymbol{\alpha}^{B} \times \mathbf{r} + {}^{N}\boldsymbol{\omega}^{B} \times ({}^{N}\boldsymbol{\omega}^{B} \times \mathbf{r}), \tag{3.13} $$

where ${}^{N}\boldsymbol{\alpha}^{B}$ is the angular acceleration vector described in the $B$-frame.

According to Newton's second law of dynamics, a differential mass $\mathrm{d}m$ placed at point $P$ and acted upon by a differential force $\mathrm{d}\mathbf{F}$ will accelerate according to

$$ \mathrm{d}\mathbf{F} = {}^{N}\mathbf{a}^{P/Q}\mathrm{d}m. \tag{3.14} $$

Integrating the previous equation over the volume, we obtain

$$ \mathbf{F} = \int_{\mathcal{V}} {}^{N}\mathbf{a}^{P/Q}\mathrm{d}m, \tag{3.15} $$

where **F** is the vectorial sum of all external forces. Note that the effect of internal forces disappears, since they are equal in magnitude and opposite in direction. Substituting Equation (3.13) into Equation (3.15), yields

$$ \mathbf{F} = \int_{\mathcal{V}} ({}^{N}\mathbf{a}^{O/Q} + {}^{N}\boldsymbol{\alpha}^{B} \times \mathbf{r} + {}^{N}\boldsymbol{\omega}^{B} \times ({}^{N}\boldsymbol{\omega}^{B} \times \mathbf{r}))\mathrm{d}m. \tag{3.16} $$

Since $\mathbf{a}^{O/Q}$, ${}^{N}\boldsymbol{\omega}^{B}$, and ${}^{N}\alpha^{B}$ are independent of our choice of $\mathrm{d}m$, we can rewrite Equation (3.16) as:

$$ \mathbf{F} = m\mathbf{a}^{O/Q} + {}^{N}\boldsymbol{\alpha}^{B} \times \int_{\mathcal{V}} \mathbf{r}\mathrm{d}m + {}^{N}\boldsymbol{\omega}^{B} \times ({}^{N}\boldsymbol{\omega}^{B} \times \int_{\mathcal{V}} \mathbf{r}\mathrm{d}m). $$

When choosing point $O$ to coincide with the center of mass, $G$, of the rigid body, we obtain

$$ \mathbf{F} = m{}^{N}\mathbf{a}^{G/Q} + {}^{N}\boldsymbol{\alpha}^{B} \times \int_{\mathcal{V}} \mathbf{r}_{G}\mathrm{d}m + {}^{N}\boldsymbol{\omega}^{B} \times ({}^{N}\boldsymbol{\omega}^{B} \times \int_{\mathcal{V}} \mathbf{r}_{G}\mathrm{d}m). $$

Note that $\int_{\mathcal{V}} \mathbf{r}_{G}\mathrm{d}m = 0$ by the definition of the center of mass. The previous equation then reduces to

$$ \boxed{\mathbf{F} = m{}^{N}\mathbf{a}^{G/Q}} \tag{3.17} $$

Equation (3.17) states that the sum of external forces acting on a rigid body is equal to the mass of the rigid body times the acceleration of its center of mass with respect to an inertial point. Equation (3.17) holds in the vectorial sense; that is, following the planar motion assumption, Equation (3.17) represents two equations applied along the two coordinates $(\hat{n}_1, \hat{n}_2)$ forming the plane.

**Applying Newton's Second Law to Rigid Bodies**

When applying Newton's second law to rigid bodies, keep in mind that the acceleration is that of the center of mass with respect to an inertial frame. Equation (3.17) does not hold in this simple form for an arbitrary point on the body.

### 3.5.1 *Moment about an Inertial Point*

Equation (3.17) is not sufficient to describe the planar motion of rigid bodies since free motion involves three degrees of freedom. As such, we need one more equation. To obtain the third equation, we use the cross product of the vector **QP** with Equation (3.14). This yields

$$\mathbf{QP} \times \mathrm{d}\mathbf{F} = \mathbf{QP} \times {}^N\mathbf{a}^{P/Q} \mathrm{d}m. \tag{3.18}$$

Integrating Equation (3.18) over the volume, we obtain

$$\mathbf{M}_Q = \int_{\mathcal{V}} \mathbf{QP} \times {}^N\mathbf{a}^{P/Q} \mathrm{d}m, \tag{3.19}$$

where $\mathbf{M}_Q$ is the moment exerted by all external forces about point $Q$. Using $\mathbf{QP} = \mathbf{QO} + \mathbf{r}$ in conjunction with Equation (3.13) yields

$$\mathbf{M}_Q = \int_{\mathcal{V}} (\mathbf{QO} + \mathbf{r}) \times ({}^N\mathbf{a}^{O/Q} + {}^N\boldsymbol{\alpha}^B \times \mathbf{r} + {}^N\boldsymbol{\omega}^B \times ({}^N\boldsymbol{\omega}^B \times \mathbf{r}))\mathrm{d}m. \tag{3.20}$$

Noting that $\mathbf{QO}$, ${}^N\mathbf{a}^{O/Q}, {}^N\boldsymbol{\omega}^B$, and ${}^N\boldsymbol{\alpha}^B$ are independent of $\mathrm{d}m$, we can rewrite Equation (3.20) as

$$\begin{aligned}\mathbf{M}_Q = {} & m\,\mathbf{QO} \times {}^N\mathbf{a}^{O/Q} + \mathbf{QO} \times \left({}^N\boldsymbol{\omega}^B \times \left({}^N\boldsymbol{\omega}^B \times \int_{\mathcal{V}} \mathbf{r}\mathrm{d}m\right)\right) \\ & + \mathbf{QO} \times \left({}^N\alpha^B \times \int_{\mathcal{V}} \mathbf{r}\mathrm{d}m\right) + \int_{\mathcal{V}} \mathbf{r}\mathrm{d}m \times {}^N\mathbf{a}^{O/Q} + \int_{\mathcal{V}} \mathbf{r} \times ({}^N\boldsymbol{\alpha}^B \times \mathbf{r})\mathrm{d}m \\ & + \int_{\mathcal{V}} \mathbf{r} \times ({}^N\boldsymbol{\omega}^B \times ({}^N\boldsymbol{\omega}^B \times \mathbf{r}))\mathrm{d}m. \end{aligned} \tag{3.21}$$

Equation (3.21) can be simplified further by choosing point $O$ to coincide with the center of mass $G$ of the body, in which case $\int_{\mathcal{V}} \mathbf{r}_G \mathrm{d}m$ vanishes. This yields

$$\mathbf{M}_Q = m\,\mathbf{QG} \times {}^N\mathbf{a}^{G/Q} + \int_{\mathcal{V}} \mathbf{r}_G \times ({}^N\boldsymbol{\alpha}^B \times \mathbf{r}_G)\mathrm{d}m + \int_{\mathcal{V}} \mathbf{r}_G \times ({}^N\boldsymbol{\omega}^B \times ({}^N\boldsymbol{\omega}^B \times \mathbf{r}_G))\mathrm{d}m. \tag{3.22}$$

Since for planar rigid body motion, ${}^N\boldsymbol{\omega}^B$ involves a rotation about one axis, the last term vanishes because, for planar motion, $\mathbf{r}_G$ is always parallel to $({}^N\boldsymbol{\omega}^B \times ({}^N\boldsymbol{\omega}^B \times \mathbf{r}_G))$. This leads to

$$\mathbf{M}_Q = m\,\mathbf{QG} \times {}^N\mathbf{a}^{G/Q} + \int_{\mathcal{V}} \mathbf{r}_G \times ({}^N\boldsymbol{\alpha}^B \times \mathbf{r}_G)\mathrm{d}m. \tag{3.23}$$

the last term in Equation (3.23) can be further simplified to $I_G\,{}^N\boldsymbol{\alpha}^B$, where $I_G$ is the moment of inertia about the axis of rotation which passes through the center of mass, $G$. This leads to

$$\boxed{\mathbf{M}_Q = m\,\mathbf{QG} \times {}^N\mathbf{a}^{G/Q} + I_G\,{}^N\boldsymbol{\alpha}^B} \tag{3.24}$$

Equation (3.24) is valid for planar rigid-body rotation when summing moments about an inertial point, $Q$.

### 3.5.2 Moment about a Moving Point on the Body

In many instances, the point around which we sum moments is not necessarily an inertial point. In fact, it is sometimes convenient to sum moments about a point on the body that is accelerating in space. To this end, we re-derive the equations by summing moments about point $O$ instead of point $Q$. This yields

$$\mathbf{M}_O = \int_{\mathcal{V}} \mathbf{r}\mathrm{d}m \times {}^N\mathbf{a}^{O/Q} + \int_{\mathcal{V}} \mathbf{r} \times ({}^N\boldsymbol{\alpha}^B \times \mathbf{r})\mathrm{d}m + \int_{\mathcal{V}} \mathbf{r} \times ({}^N\boldsymbol{\omega}^B \times ({}^N\boldsymbol{\omega}^B \times \mathbf{r}))\mathrm{d}m. \tag{3.25}$$

Using Figure 3.12, we let $\mathbf{r} = \mathbf{OG} + \mathbf{r}_G$, and restrict the motion to a plane to obtain

$$\begin{aligned}\mathbf{M}_O = &\int_{\mathcal{V}} \mathbf{OG}\mathrm{d}m \times {}^N\mathbf{a}^{O/Q} + \int_{\mathcal{V}} \mathbf{r}_G\mathrm{d}m \times {}^N\mathbf{a}^{O/Q} \\ &+ \int_{\mathcal{V}} (\mathbf{OG} + \mathbf{r}_G) \times ({}^N\boldsymbol{\alpha}^B \times (\mathbf{OG} + \mathbf{r}_G))\mathrm{d}m.\end{aligned} \tag{3.26}$$

The second term in the previous equation vanishes since $\int_{\mathcal{V}} \mathbf{r}_G \mathrm{d}m = 0$, and the last term can be shown to be $I_O\,{}^N\boldsymbol{\alpha}^B$, where $I_O$ is the moment of inertia about an axis passing through point $O$. This yields

$$\boxed{\mathbf{M}_O = m\,\mathbf{OG} \times {}^N\mathbf{a}^{O/Q} + I_O\,{}^N\boldsymbol{\alpha}^B} \tag{3.27}$$

Equation (3.27) is valid for planar rigid-body motion when summing moments about a moving point on the body.

### 3.5.3 Moment about the Center of Mass or a Fixed Point on the Body

Equation (3.27) can be reduced further when point $O$ is chosen to coincide with point $G$. In this case the first term disappears and the equation becomes

$$\boxed{\mathbf{M}_G = I_G\,{}^N\boldsymbol{\alpha}^B} \tag{3.28}$$

Equation (3.28) is valid only when we sum moments about the center of mass.

Equation (3.27) can also be reduced when point $O$ is chosen to be a fixed point, $S$, on the body. In this case, ${}^N\mathbf{a}^{S/Q}$ vanishes and we obtain

$$\mathbf{M}_S = I_S\, {}^N\boldsymbol{\alpha}^B \tag{3.29}$$

Equation (3.29) is valid only when summing moments about a fixed point located on the body.

**Summary of Equations for Planar Rigid-body Motion**

Free planar rigid-body motion constitutes three degrees of freedom that are sufficiently described by three independent equations. Two of these equations result directly from Newton's second law of dynamics, which, for rigid bodies, can be written as

$$\mathbf{F} = m{}^N\mathbf{a}^{G/Q},$$

where $\mathbf{F}$ is the net external force vector acting on the body, $m$ is the mass of the rigid body, and ${}^N\mathbf{a}^{G/Q}$ is the acceleration of the center of mass $G$ of the rigid body with respect to an inertial point $Q$.

The third equation can be obtained by summing moments about a point on or away from the body. The following points are very common:

1. Point $Q$ is an inertial point in space. In this case, the moment $\mathbf{M}_Q$ exerted by the external forces on point $Q$ is described by

$$\mathbf{M}_Q = m\,\mathbf{QG} \times {}^N\mathbf{a}^{G/Q} + I_G\, {}^N\boldsymbol{\alpha}^B.$$

2. Point $O$ is an accelerating point on the body. In this case, the moment $\mathbf{M}_O$ exerted by the external forces on point $O$ is described by

$$\mathbf{M}_O = m\,\mathbf{OG} \times {}^N\mathbf{a}^{O/Q} + I_O\, {}^N\boldsymbol{\alpha}^B.$$

3. Point $G$ is the center of mass of the rigid body. In this case, the moment $\mathbf{M}_G$ exerted by the external forces on point $G$ is described by

$$\mathbf{M}_G = I_G\, {}^N\boldsymbol{\alpha}^B.$$

4. Point $S$ is a fixed point on the rigid body. In this case, the moment $\mathbf{M}_S$ exerted by the external forces on point $S$ is described by

$$\mathbf{M}_S = I_S\, {}^N\boldsymbol{\alpha}^B.$$

**Procedure for Obtaining the Equations of Motion for Planar Rigid-body Motion**

1. Identify the unknowns in the problem. This includes the degrees of freedom, and the external and internal forces. To determine the unknown forces, draw a free-body diagram.

2. Depending on which unknowns you are trying to obtain, decide whether you need to use Newton's translational equations. If you need these equations, you must obtain the acceleration of the center of mass of the rigid body with respect to an inertial point before implementing Newton's second law of dynamics.
3. Decide which point is the best to sum moments about. Is it an inertial point? Is it the center of gravity, or some other point? Your choice will determine which moment equation you must use. Note that a smart choice can always reduce the number of unknowns appearing in the equations.

**Example 3.7 Internal Forces**

For the compound pendulum of length $l$ and mass $m$ shown in Figure 3.13, find:

(a) the equation of motion
(b) the shear force at $l/2$ as function of the angle $\theta$
(c) the axial force at $l/2$ as function of the angle $\theta$.
For (c), assume that the rod is released from rest when $\theta = \frac{\pi}{2}$.

**(a)** ***The equations of motion*** There are three unknowns in this problem. The motion of the coordinate $\theta$, and the reactions at the pin, $O_x$ and $O_y$, as shown in Figure 3.14. To find these unknowns, there exist three equations: two translational (Newton's second law applied along the $\hat{n}_1$ and $\hat{n}_2$ directions) and one rotational (summing moments about a point).

The simplest way to find the equation of motion for this system is by summing moments about the fixed point $O$, using

$$\mathbf{M}_O = I_O \, {}^N\boldsymbol{\alpha}^B.$$

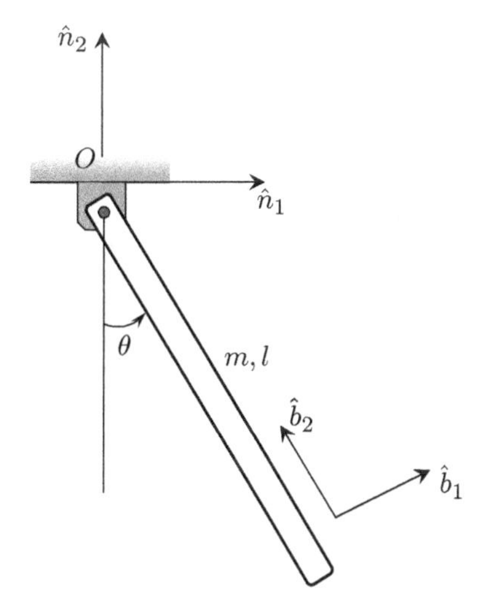


**Figure 3.13** Planar motion of a compound pendulum.

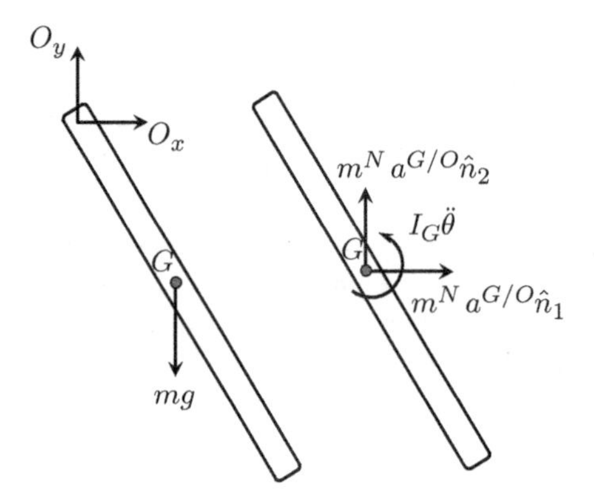


**Figure 3.14** Free-body diagram of a compound pendulum.

As shown in Figure 3.14, only the component of the weight $mg\cos\theta$ results in a moment $\mathbf{M}_O = -mg\frac{l}{2}\cos\theta\,\hat{n}_3$ about point $O$. As such, we can write

$$-mg\frac{l}{2}\cos\theta\,\hat{n}_3 = I_O\ddot{\theta}\,\hat{n}_3,$$

where $I_O = \frac{1}{3}ml^2$ is the moment of inertia of the rod about the $\hat{n}_3$ axis. It follows that

$$\ddot{\theta} + \frac{3}{2}\frac{g}{l}\sin\theta = 0.$$

Note that, in many instances, summing moments about a suitable point, as in this case, alleviates the need to use Newton's translational equations.

**(b)** ***Shear force*** To find the shear force at $\frac{l}{2}$, we must first observe the effect of that force on the dynamics. As such, we take a section at $\frac{l}{2}$ and release all the internal forces, as shown in Figure 3.15. By doing so, the problem involving the dynamics of one rigid body with three unknowns ($O_x$, $O_y$, $\theta$) becomes a problem involving two rigid bodies with six unknowns ($O_x$, $O_y$, $\theta$, $R_N$, $V$, $M_c$). Here, $R_N$, and $V$ represent, respectively, the axial and shear forces acting on the section while $M_c$ is the bending moment.

Next, we need to decide which equations are best to determine the shear force $V$. Since the equation governing the $\theta$ dynamics has already been determined in (a), one can obtain an equation involving only the unknown shear force $V$ by summing forces in the $\hat{b}_1$ direction on the lower section of the rod. Therefore, it makes perfect sense to apply Newton's second law in the $\hat{b}_1$ direction on the lower section of the rod. To this end, we first determine the acceleration of $G_2$ with respect to the inertial point $O$, which can be written as

$${}^N\mathbf{a}^{G_2/O} = \frac{3l}{4}\ddot{\theta}\,\hat{b}_1 + \frac{3l}{4}\dot{\theta}^2\,\hat{b}_2.$$

By applying Newton's second law in the $\hat{b}_1$ direction, we obtain

$$\mathbf{F}\hat{b}_1 = \frac{m}{2}\,{}^N\mathbf{a}^{G_2/O}\hat{b}_1, \qquad V = \frac{m}{2}\left(g\sin\theta + \frac{3}{4}l\ddot{\theta}\right).$$

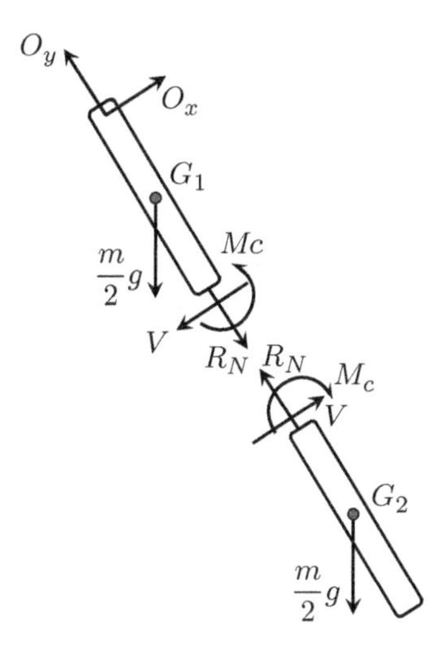


**Figure 3.15** Free-body diagram showing the internal forces.

Note that we are using $\frac{m}{2}$ in the previous equation because we are applying Newton's equation on the lower half rod. We can further eliminate $\ddot{\theta}$ from the previous equation by using $\ddot{\theta} = -\frac{3}{2}\frac{g}{l}\sin\theta$. This yields

$$V = -\frac{1}{16}mg\sin\theta.$$

It follows that the shear force opposes the direction of motion and takes its maximum value of $\frac{1}{16}mg$ when $\theta = \frac{\pi}{2}$.

**(c)** ***Axial force*** To obtain the axial force $R_N$, we sum forces in the $\hat{b}_2$ direction on the lower half of the rod. This yields

$$\mathbf{F}\hat{b}_2 = \frac{m}{2}\,{}^N\mathbf{a}^{G_2/O}\hat{b}_2, \qquad R_N = \frac{m}{2}\left(g\cos\theta + \frac{3}{4}l\dot{\theta}^2\right).$$

It is evident that $R_N$ depends on $\dot{\theta}$, which in turn depends on the initial conditions. To eliminate $\dot{\theta}$ from the equation, we use the provided initial conditions $\theta(0) = \frac{\pi}{2}, \dot{\theta}(0) = 0$, and write

$$\int_{\pi/2}^{\theta}\ddot{\theta}\mathrm{d}\theta = \int_{0}^{\dot{\theta}}\dot{\theta}\mathrm{d}\dot{\theta}.$$

Upon substituting $\ddot{\theta} = -\frac{3}{2}\frac{g}{l}\sin\theta$ into the previous equation and integrating, we obtain

$$\dot{\theta}^2 = \frac{3g}{l}\cos\theta,$$

which upon substitution into the equation for $R_N$ yields

$$R_N = \frac{13}{8}mg\cos\theta.$$

Note that for the initial conditions given, $R_N$ is zero when $\theta = \frac{\pi}{2}$ and takes a maximum value of $\frac{13}{8}mg$ when $\theta = 0$.

**Example 3.8 Slip versus No Slip**

The ladder shown in Figure 3.16 is released from rest when $\theta = 0°$. Assuming that, for some range of the angle $\theta$ slipping does not occur, determine as a function of $\theta$ only (no $\dot{\theta}$ or $\ddot{\theta}$) the normal and frictional forces that are exerted on the ladder by the ground as it falls. At what angle does the ladder start slipping if the static coefficient of friction is $\mu_s = 0.5$. Treat the ladder as a rod of length, $l$, and mass $m$.

Since slipping does not occur, the equation of motion is obtained by summing moments about point $O$ assuming it is a fixed point. This assumption is valid *only* provided the ladder does not lift off the ground before slipping. It follows from Figure 3.17 that

$$\mathbf{M}_O = I_O\ddot{\theta}, \qquad \ddot{\theta} = \frac{3}{2}\frac{g}{l}\sin\theta.$$

To obtain the frictional and normal forces, we apply Newton's second law in the $\hat{n}_1$ and $\hat{n}_2$ directions. To this end, we first find the acceleration of the center of mass $G$ with respect to point $O$, which can be written as

$${}^N\mathbf{a}^{G/O} = \left(\frac{l}{2}\ddot{\theta}\cos\theta - \frac{l}{2}\dot{\theta}^2\sin\theta\right)\hat{n}_1 - \left(\frac{l}{2}\ddot{\theta}\sin\theta + \frac{l}{2}\dot{\theta}^2\cos\theta\right)\hat{n}_2.$$

Summing forces in the $\hat{n}_1$ and $\hat{n}_2$ directions yields

$$\mathbf{F}\hat{n}_1 = m^N\mathbf{a}^{G/O}\hat{n}_1, \qquad F_f = m\left(\frac{l}{2}\ddot{\theta}\cos\theta - \frac{l}{2}\dot{\theta}^2\sin\theta\right).$$

$$\mathbf{F}\hat{n}_2 = m^N\mathbf{a}^{G/O}\hat{n}_2, \qquad R_N - mg = -m\left(\frac{l}{2}\ddot{\theta}\sin\theta + \frac{l}{2}\dot{\theta}^2\cos\theta\right).$$

Since the rod is released from rest at $\theta = 0$, we can write

$$\int_0^{\theta}\ddot{\theta}\mathrm{d}\theta = \int_0^{\dot{\theta}}\dot{\theta}\mathrm{d}\dot{\theta}.$$

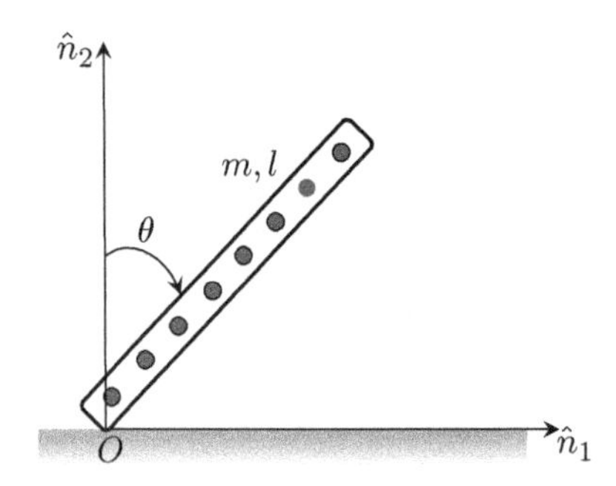


**Figure 3.16** Schematic of a falling ladder released from rest at $\theta = 0$.

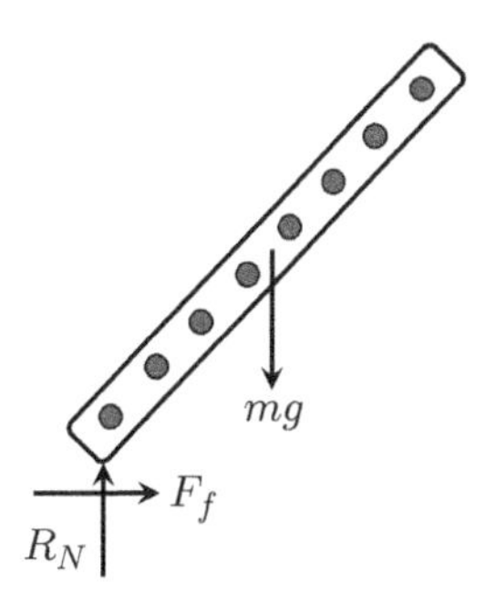


**Figure 3.17** Free-body diagram of a falling ladder.

Using $\ddot{\theta} = \frac{3}{2}\frac{g}{l}\sin\theta$, we obtain

$$\dot{\theta}^2 = \frac{3g}{l}(1 - \cos\theta).$$

Substituting the previous equation back into the expressions for $F_f$ and $R_N$, we obtain

$$F_f = \frac{3mg}{2}\sin\theta\left(\frac{3}{2}\cos\theta - 1\right).$$

$$R_N = \frac{mg}{4}(1 - 3\cos\theta)^2.$$

Just before the ladder starts slipping, the acceleration of point $O$ is zero and the coefficient of friction is $\mu_s$. As such, to find the angle at which the ladder slips, we let $F_f = \mu_s R_N$. This yields

$$\mu_s(1 - 3\cos\theta_{cr})^2 = 6\sin\theta_{cr}\left(\frac{3}{2}\cos\theta_{cr} - 1\right),$$

which, upon solving the previous equation for $\mu_s = 0.5$, yields $\theta_{cr} \approx 37.8°$.

**Example 3.9 Losing Contact**

A hollow sphere of mass $m$ and radius $r$ rolls without slipping on the surface of radius $R$. The motion takes place in a single plane. Find the angle $\theta$ at which the sphere loses contact with the surface if it starts from rest with a tiny nudge.

Define the inertial $N$-frame and the rotating $B$-frame, as shown in Figure 3.18. It follows that ${}^N\boldsymbol{\omega}^B = -\dot{\theta}\hat{b}_3$ and that the acceleration of the center of mass, $G$, of the rolling sphere with respect to point $O$ is

$${}^N\mathbf{a}^{G/O} = (R+r)\ddot{\theta}\hat{b}_1 - (R+r)\dot{\theta}^2\hat{b}_2.$$

Applying Newton's second law along the directions of $\hat{b}_1$ and $\hat{b}_2$, yields

$$\mathbf{F}\hat{b}_1 = m\,{}^N\mathbf{a}^{G/O}\hat{b}_1, \qquad F_f = mg\sin\theta - m(R+r)\ddot{\theta}.$$

$$\mathbf{F}\hat{b}_2 = m{}^N\mathbf{a}^{G/O}\hat{b}_2, \qquad R_N - mg\cos\theta = -m(R+r)\dot{\theta}^2.$$

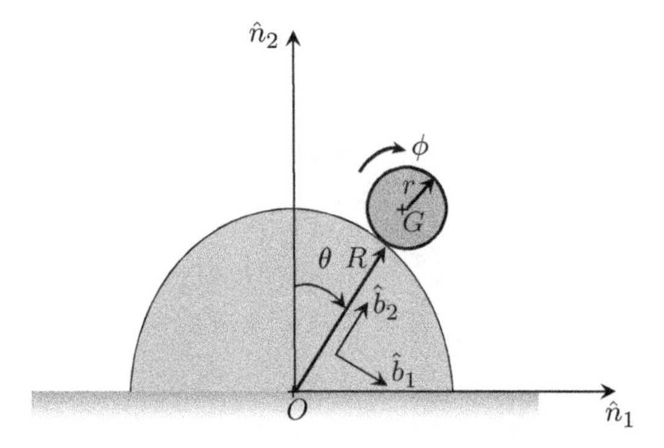


**Figure 3.18** A sphere rolling on a spherical surface.

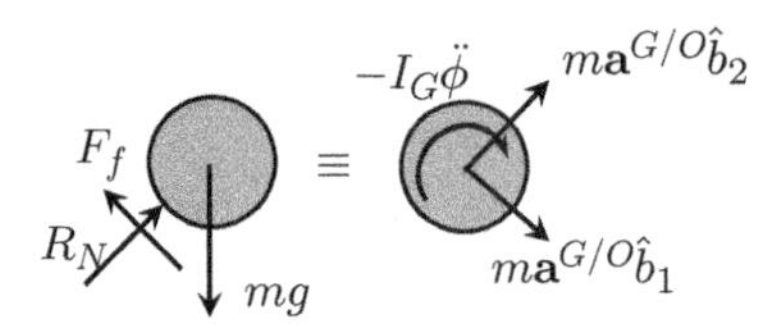


**Figure 3.19** Free-body diagram of the rolling sphere.

By referring to Figure 3.19, we sum moments about the center of mass, $G$, of the sphere, and obtain

$$\mathbf{M}_G = -I_G\ddot{\phi}, \qquad F_f r = I_G\ddot{\phi},$$

where $I_G = \frac{2}{3}mr^2$. Note that $\phi$ and $\theta$ can be related via $(R+r)\phi = r\theta$. Substituting this relation and the expression we obtained for $F_f$ back into the previous equation and simplifying, we obtain

$$\ddot{\theta} = \frac{3}{5(R+r)}g\sin\theta.$$

When the sphere loses contact with the surface, $R_N$ vanishes. Based on the sum of forces in the $\hat{b}_2$ direction, this results in the following equation:

$$\dot{\theta}^2 = \frac{g}{(R+r)}\cos\theta.$$

Since the sphere is released from rest at $\theta = 0$, we can write

$$\int_0^{\theta}\ddot{\theta}\mathrm{d}\theta = \int_0^{\dot{\theta}}\dot{\theta}\mathrm{d}\dot{\theta}.$$

Upon substituting $\ddot{\theta} = \frac{3}{5(R+r)}g\sin\theta$ into the previous equation and integrating, we obtain

$$\dot{\theta}^2 = \frac{6}{5}\frac{g}{R+r}(1-\cos\theta).$$

Equating the two equations describing $\dot{\theta}^2$ yields $\cos\theta = \frac{6}{11}$ or $\theta = \cos^{-1}\frac{6}{11}$. It is quite interesting to note that the angle at which the sphere loses contact does not depend on any of the design parameters.

 **Flipped Classroom Exercise 3.4**

A point mass, $3m$, is welded to a slender rod of mass $m$ and length $l$ at a distance $d$ from point $O$, which rests in a corner. The body is released from rest with $\theta = 30°$.

(a) Find the distance $d$ such that the rod has the greatest angular acceleration as it falls.
(b) Use the result that you obtained for $d$ in (a) to find the critical angle $\theta_{cr}$ when the rod first loses contact with the wall.

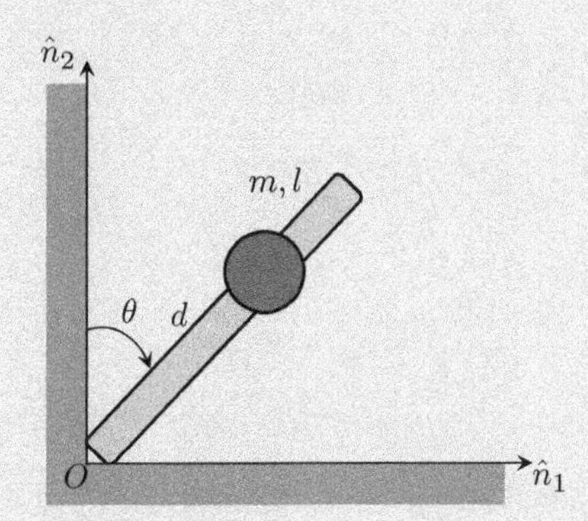


**(a)** ***Distance for greatest angular acceleration***

1. Define your rotating frame $B$ such that ${}^N\omega^B = -\dot{\theta}\hat{b}_3$.
2. Find the location of the center of mass, $G$.
3. Find the moment of inertia, $I_O$, about point $O$.
4. Assume that point $O$ is stationary (valid up until the point loses contact with either the wall or the ground) and sum moments about $O$ to obtain the following equation:

$$\ddot{\theta} = \frac{4mgr_G}{I_O}\sin\theta.$$

   where $r_G$ is the distance between point $O$ and point $G$.
5. Maximize $\ddot{\theta}$ for any value of $\theta$. Show that

$$d_{\text{max}} = \frac{\sqrt{5}-1}{6}l.$$

**(b)** ***Critical angle***

1. Find the acceleration of point $G$ with respect to the inertial point, $O$.
2. Apply Newton's second law along the $\hat{b}_1$ and $\hat{b}_2$ directions.
3. The rod loses contact with the wall when the reaction at the wall vanishes. Use this information to show that $\theta_{cr} = \cos^{-1}\frac{\sqrt{3}}{3}$.

### Flipped Classroom Exercise 3.5

A chimney of boards, of total length $l$ and mass $m$, which initially stands upright, is given a tiny kick, and it topples over. At what height $h$ is the chimney most likely to break due to tensile forces in the rods? Assume that the chimney consists of boards stacked on top of each other, and that each board is attached to the two adjacent ones with tiny rods at each end, as shown below. The goal is to determine which rod in the chimney has the maximum tension. Assume that the width of the chimney, $2r$, is very small compared to its length, $l$, and that no slipping occurs at the corner.

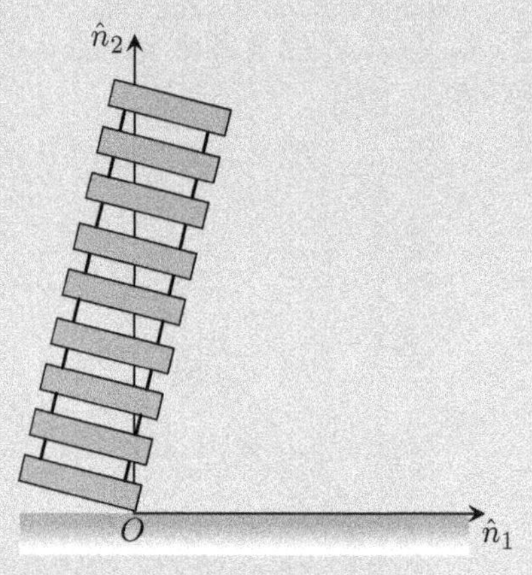


To solve this problem, you need to do the following:

1. Define your inertial frame to be at point $O$ and define the rotating frame such that ${}^N\omega^B = -\dot{\theta}\hat{b}_3$.
2. Find the equation of motion of the chimney by summing moments about point $O$ and show that

$$\ddot{\theta} = \frac{3}{2}\frac{g}{l}\sin\theta.$$

3. Take a section at an arbitrary point of height $h$ and release the internal forces in the rods. Note that you have two tensile forces and a shear force. Draw the free-body diagram.

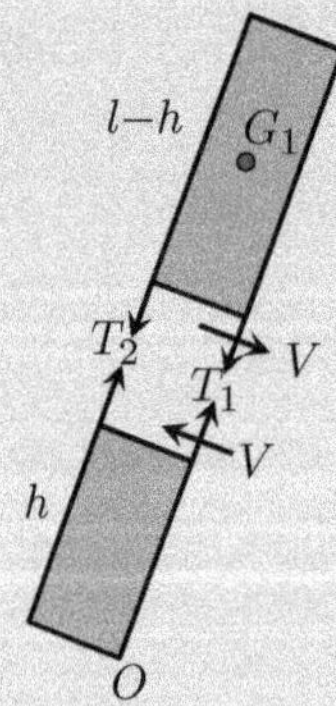

4. It is best to work with the upper section because it simplifies the analysis significantly. Implement Newton's equations on the upper section and use the assumption that $r/l \ll 1$ to show that

$$T_1 - T_2 = mg(1-h^*)\cos\theta + m\frac{l}{2}(1-h^*)(1+h^*)\dot{\theta}^2,$$

$$V = m\frac{l}{2}(1-h^*)(1+h^*)\ddot{\theta} - m(1-h^*)\sin\theta,$$

where $h^* = h/l$.

5. Sum moments about the center of gravity of the upper part and show that

$$(T_1+T_2)r + V(1-h^*)\frac{l}{2} = -m(1-h^*)^3\frac{l^2}{12}\ddot{\theta}$$

6. Use the above equation in combination with the following two equations:

$$T_1 - T_2 = mg(1-h^*)\cos\theta + m(1-h^*)(1+h^*)\frac{h}{2}\dot{\theta}^2,$$

$$\ddot{\theta} = \frac{3}{2}\frac{g}{l}\sin\theta,$$

to show that

$$h^*_{max} \approx \frac{1}{3}$$

 **Flipped Classroom Exercise 3.6**

Show that the term $\int_{\mathcal{V}} \mathbf{r}_G \times ({}^N\boldsymbol{\alpha}^B \times \mathbf{r}_G)dm$ is nothing but $I_{33G}\ddot{\theta}$ for planar rigid-body motion. Here, $\mathbf{r}_G$ is the position vector between any point and the center of mass of the rigid body and $\theta$ is the rotation angle.

1. Since we are considering planar motion, choose $\mathbf{r}_G = x_G\hat{b}_1 + y_G\hat{b}_2$ and ${}^N\boldsymbol{\omega}^B = \dot{\theta}\hat{b}_3$.
2. Calculate $\mathbf{r}_G \times ({}^N\boldsymbol{\alpha}^B \times \mathbf{r}_G)$

   Now, it is straightforward to show that $\int_{\mathcal{V}} \mathbf{r}_G \times ({}^N\boldsymbol{\alpha}^B \times \mathbf{r}_G)dm = I_{33G}\ddot{\theta}$.

## 3.6 Non-planar Rigid-body Motion

The free motion of a rigid body in space consists of six degrees of freedom; three translational and the others rotational. Newton's translational equations as derived in the previous section for planar motion stay the same when considering three translational coordinates instead of two. The third translational equation can be obtained by summing forces in the third direction as well.

For the rotational motion, however, the planar motion assumption introduces a simplification to the equations of motion. Specifically, when deriving the rotational equations for planar motion, we showed that the term

$$\int_{\mathcal{V}} \mathbf{r}_G \times ({}^N\omega^B \times ({}^N\omega^B \times \mathbf{r}_G))dm, \tag{3.30}$$

in Equation (3.22) vanishes because, for planar rigid-body dynamics, $\mathbf{r}_G$ is parallel to $({}^N\omega^B \times ({}^N\omega^B \times \mathbf{r}_G))$. For non-planar motion, however, this is not the case. Thus, the integral of Equation (3.30) must be accounted for when deriving the rotational equations for non-planar motion.

Before adding this term to the right-hand side of the non-planar rotational equations, we simplify it using some mathematical identities. To this end, we start by applying the mathematical identity $\mathbf{A} \times (\mathbf{B} \times \mathbf{C}) = \mathbf{B}(\mathbf{A} \cdot \mathbf{C}) - \mathbf{C}(\mathbf{A} \cdot \mathbf{B})$ to Equation (3.30), and obtain

$$\mathbf{r}_G \times ({}^N\omega^B \times ({}^N\omega^B \times \mathbf{r}_G)) = {}^N\omega^B(\mathbf{r}_G \cdot ({}^N\omega^B \times \mathbf{r}_G)) - ({}^N\omega^B \times \mathbf{r}_G)(\mathbf{r}_G \cdot {}^N\omega^B).$$

Noting that $\mathbf{r}_G \cdot ({}^N\omega^B \times \mathbf{r}_G) = 0$, the previous equation simplifies to

$$\mathbf{r}_G \times ({}^N\omega^B \times ({}^N\omega^B \times \mathbf{r}_G)) = -(({}^N\omega^B \times \mathbf{r}_G)(\mathbf{r}_G \cdot {}^N\omega^B).$$

Next, using the identity $(\mathbf{A} \times \mathbf{B})(\mathbf{B} \cdot \mathbf{A}) = -\mathbf{A} \times (\mathbf{A}(\mathbf{B} \cdot \mathbf{B}) - \mathbf{B}(\mathbf{B} \cdot \mathbf{A}))$, we can write

$$\mathbf{r}_G \times ({}^N\omega^B \times ({}^N\omega^B \times \mathbf{r}_G)) = {}^N\omega^B \times ({}^N\omega^B(\mathbf{r}_G \cdot \mathbf{r}_G) - \mathbf{r}_G(\mathbf{r}_G \cdot {}^N\omega^B)).$$

Using the previous equation, the expression in Equation (3.30) becomes

$$\int_{\mathcal{V}} \mathbf{r}_G \times ({}^N\omega^B \times ({}^N\omega^B \times \mathbf{r}_G))dm = {}^N\omega^B \times \int_{\mathcal{V}} ({}^N\omega^B(\mathbf{r}_G \cdot \mathbf{r}_G) - \mathbf{r}(\mathbf{r}_G \cdot {}^N\omega^B))dm.$$

Letting $\mathbf{r}_G = x_G\hat{b}_1 + y_G\hat{b}_2 + z_G\hat{b}_3$ and ${}^N\omega^B = \omega_1\hat{b}_1 + \omega_2\hat{b}_2 + \omega_3\hat{b}_3$, then expanding the term $({}^N\omega^B(\mathbf{r}_G \cdot \mathbf{r}_G) - \mathbf{r}(\mathbf{r}_G \cdot {}^N\omega^B))$, it can be shown that

$${}^N\omega^B \times \int_{\mathcal{V}} ({}^N\omega^B(\mathbf{r}_G \cdot \mathbf{r}_G) - \mathbf{r}_G(\mathbf{r}_G \cdot {}^N\omega^B)) = {}^N\omega^B \times (I_G\,{}^N\omega^B), \tag{3.31}$$

where $I_G$ is the moment of inertia matrix about the center of mass. Using Equation (3.31), the rotational equations derived in the previous section for planar motion can be extended to non-planar motion as follows:

1. Point $Q$ is an inertial point in space. In this case, the moment $\mathbf{M}_Q$ exerted by the external forces on point $Q$ is described by:

$$\mathbf{M}_Q = m\,\mathbf{QG} \times {}^N\mathbf{a}^{G/Q} + I_G\,{}^N\alpha^B + {}^N\omega^B \times (I_G\,{}^N\omega^B). \tag{3.32}$$

2. Point $O$ is an accelerating point. In this case, the moment $\mathbf{M}_O$ exerted by the external forces on point $O$ is described by:

$$\mathbf{M}_O = m\,\mathbf{OG} \times {}^N\mathbf{a}^{O/Q} + I_O\,{}^N\alpha^B + {}^N\omega^B \times (I_O\,{}^N\omega^B). \tag{3.33}$$

3. Point $G$ is the center of mass of the rigid body. In this case, the moment $\mathbf{M}_G$ exerted by the external forces on point $G$ is described by

$$\mathbf{M}_G = I_G\,{}^N\boldsymbol{\alpha}^B + {}^N\boldsymbol{\omega}^B \times (I_G\,{}^N\boldsymbol{\omega}^B). \tag{3.34}$$

4. Point $S$ a fixed point on the rigid body. In this case, the moment $\mathbf{M}_S$ exerted by the external forces on point $S$ is described by

$$\mathbf{M}_S = I_S\,{}^N\boldsymbol{\alpha}^B + {}^N\boldsymbol{\omega}^B \times (I_S\,{}^N\boldsymbol{\omega}^B). \tag{3.35}$$

### 3.6.1 Euler Rotational Equations

When letting $\mathbf{M}_G = M_1\hat{b}_1 + M_2\hat{b}_3 + M_3\hat{b}_3$, ${}^N\boldsymbol{\omega}^B = \omega_1\hat{b}_1 + \omega_2\hat{b}_3 + \omega_3\hat{b}_3$, and ${}^N\boldsymbol{\alpha}^B = \alpha_1\hat{b}_1 + \alpha_2\hat{b}_3 + \alpha_3\hat{b}_3$, Equation (3.34) can be expressed in matrix form as follows:

$$\begin{bmatrix} M_{1G} \\ M_{2G} \\ M_{3G} \end{bmatrix} = \begin{pmatrix} I_{11G} & I_{12G} & I_{13G} \\ I_{21G} & I_{22G} & I_{23G} \\ I_{31G} & I_{32G} & I_{33G} \end{pmatrix} \begin{bmatrix} \alpha_1 \\ \alpha_2 \\ \alpha_3 \end{bmatrix} + \begin{pmatrix} 0 & -\omega_3 & \omega_2 \\ \omega_3 & 0 & -\omega_1 \\ -\omega_2 & \omega_1 & 0 \end{pmatrix} \begin{pmatrix} I_{11G} & I_{12G} & I_{13G} \\ I_{21G} & I_{22G} & I_{23G} \\ I_{31G} & I_{32G} & I_{33G} \end{pmatrix} \begin{bmatrix} \omega_1 \\ \omega_2 \\ \omega_3 \end{bmatrix}.$$

By aligning the rotating $B$-frame with the principal axes of the body, all the product of inertia terms vanish and the previous equation reduces to

$$\begin{aligned} M_{1G} &= I_{11G}\alpha_1 + (I_{33G} - I_{22G})\omega_3\omega_2, \\ M_{2G} &= I_{22G}\alpha_2 + (I_{11G} - I_{33G})\omega_1\omega_3, \\ M_{3G} &= I_{33G}\alpha_3 + (I_{22G} - I_{11G})\omega_1\omega_2. \end{aligned} \tag{3.36}$$

The previous set of equation are widely known as *Euler's rotational equations*. Euler's equations are critical in understanding the rotational motion of non-planar rigid bodies. One important notion that can be quickly deduced from Euler's equations is that a rigid body does not require an angular acceleration about a given axis to produce a moment about that axis or vice versa. In fact, as can be clearly seen in the previous equation, a moment can be produced about one axis of a rotating rigid body if the rigid body has angular velocity components about the other two axes provided that these two axes do not possess equal principal moments of inertia. This concept explains why a motorcycle can be steered to rotate in one direction by simply tilting it sideways in that direction.

To explain this better, examine Figure 3.20 in conjunction with the second Euler equation. In a motorcycle, the wheel rotates about $\hat{b}_1$ with some angular velocity $\omega_1$. When the rider tilts the motorcycle sideways, he produces another angular velocity about the $\hat{b}_3$ axis, as shown in the figure. According to the second Euler equation, this will produce a moment

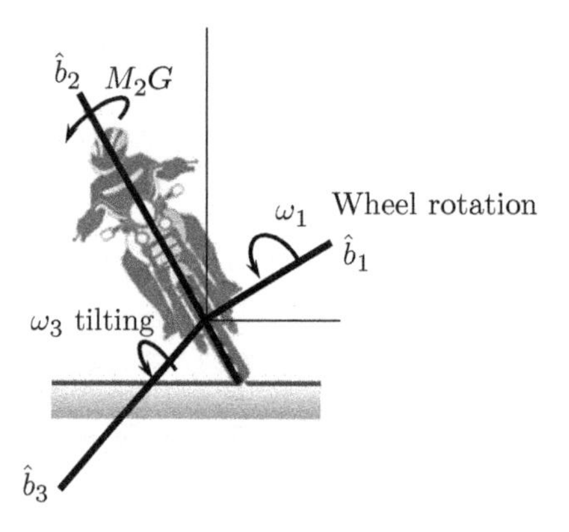


**Figure 3.20** Steering a motorcycle follows Euler's rotational equations.

$M_{2G} = (I_{11G} - I_{33G})\omega_1\omega_3$ about the $\hat{b}_2$ axis, which will force the motorcycle to rotate about the second axis.

### Summary of Equations for Non-planar Rigid-body Motion

Free non-planar rigid-body motion has six degrees of freedom that are described by six independent equations. Three of these equations result directly from Newton's second law of dynamics, which, for rigid bodies, can be written as

$$\mathbf{F} = m^N\mathbf{a}^{G/Q},$$

where $\mathbf{F}$ is the net external force vector acting on the body, $m$ is the mass of the rigid body, and ${}^N\mathbf{a}^{G/Q}$ is the acceleration of the center of mass $G$ of the rigid body with respect to an inertial point $Q$.

The remaining equations can be obtained by summing moments about a point on or outside the body. The following points are very common:

1. Point $Q$ is an inertial point in space. In this case, the moment $\mathbf{M}_Q$ exerted by the external forces on point $Q$ is described by

$$\mathbf{M}_Q = m\,\mathbf{QG} \times {}^N\mathbf{a}^{G/Q} + I_G\,{}^N\boldsymbol{\alpha}^B + {}^N\boldsymbol{\omega}^B \times (I_G\,{}^N\boldsymbol{\omega}^B).$$

2. Point $O$ is an accelerating point on the body. In this case, the moment $\mathbf{M}_O$ exerted by the external forces on point $O$ is described by

$$\mathbf{M}_O = m\,\mathbf{OG} \times {}^N\mathbf{a}^{O/Q} + I_O\,{}^N\boldsymbol{\alpha}^B + {}^N\boldsymbol{\omega}^B \times (I_O\,{}^N\boldsymbol{\omega}^B).$$

3. Point $G$ is the center of mass of the rigid body. In this case, the moment $\mathbf{M}_G$ exerted by the external forces on point $G$ is described by

$$\mathbf{M}_G = I_G\,{}^N\boldsymbol{\alpha}^B + {}^N\boldsymbol{\omega}^B \times (I_G\,{}^N\boldsymbol{\omega}^B).$$

4. Point $S$ is a fixed point on the rigid body. In this case, the moment $\mathbf{M}_S$ exerted by the external forces on point $S$ is described by

$$\mathbf{M}_S = I_S\,{}^N\boldsymbol{\alpha}^B + {}^N\boldsymbol{\omega}^B \times (I_S\,{}^N\boldsymbol{\omega}^B).$$

When aligning the rotating body frame with the principal axes of rotation, all the product of inertia terms vanish and the moment equation about the center of mass, $G$ reduces to Euler's form:

$$\begin{aligned}
M_{1G} &= I_{11G}\alpha_1 + (I_{33G} - I_{22G})\omega_3\omega_2,\\
M_{2G} &= I_{22G}\alpha_2 + (I_{11G} - I_{33G})\omega_1\omega_3,\\
M_{3G} &= I_{33G}\alpha_3 + (I_{22G} - I_{11G})\omega_1\omega_2,
\end{aligned} \tag{3.37}$$

where $\mathbf{M}_G = M_{1G}\hat{b}_1 + M_{2G}\hat{b}_3 + M_{3G}\hat{b}_3$, ${}^N\boldsymbol{\omega}^B = \omega_1\hat{b}_1 + \omega_2\hat{b}_3 + \omega_3\hat{b}_3$, and ${}^N\boldsymbol{\alpha}^B = \alpha_1\hat{b}_1 + \alpha_2\hat{b}_3 + \alpha_3\hat{b}_3$.

**Using Euler's Rotational Equations**

When using Euler's rotational equations, it is best to express all the vectors and calculate the inertia matrix in the body rotating frame. This is because all of these quantities are much simpler to describe in this frame.

**Example 3.10 A Rotating T-shaped Structure**

A bar of mass $m$ and length $l$ is welded to another bar of mass $2m$ and length $2l$ to form the T-shaped structure shown in Figure 3.21. The whole system is pinned to a rotating post at point $O$. The post rotates at a constant rate $\Omega$. Find the equations of motion describing the system's dynamics.

We define the frames necessary to describe the kinematics of the system. The $N$-frame is inertial in space, the $A$-frame is formed such that ${}^N\boldsymbol{\omega}^A = \Omega\hat{a}_2$, and the $B$-frame is formed such that ${}^A\boldsymbol{\omega}^B = \dot{\theta}\hat{b}_3$.

Next, we draw the free-body diagram shown in Figure 3.22 to describe the forces acting on the body. There are three reaction forces at point $O$: $O_1\hat{b}_1$, $O_2\hat{b}_2$, $O_3\hat{b}_3$, and two reaction moments $M_1\hat{b}_1$ and $M_2\hat{b}_2$. In addition, there is the weight $-3mg\,\hat{n}_2$ of the body placed at the center of mass, $G$ where $G$ is located at $\mathbf{OG} = -\frac{4}{3}l\,\hat{b}_2$.

Since all the unknown reactions appear at point $O$, it is most convenient to sum moments about this point to eliminate as many unknowns as possible. Before summing moments, it is important to find the inertia matrix about a set of axes whose origin is located at point $O$, as described in the $B$-frame. This is given by

$$I_O = ml^2\begin{pmatrix} \frac{20}{3} & 0 & 0\\ 0 & \frac{1}{12} & 0\\ 0 & 0 & \frac{27}{4}\end{pmatrix}.$$

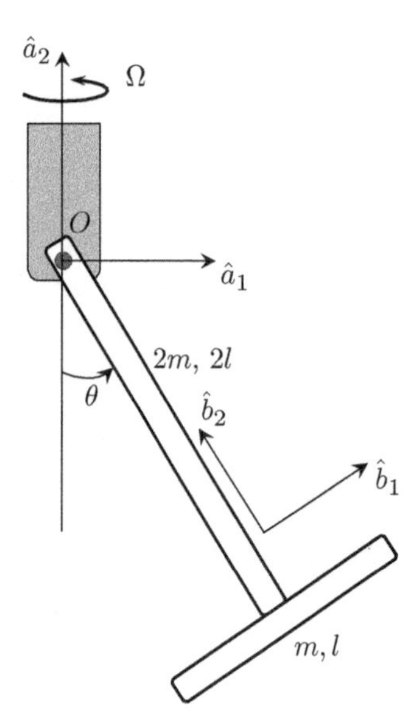


**Figure 3.21** Non-planar motion of a T-shaped rod.

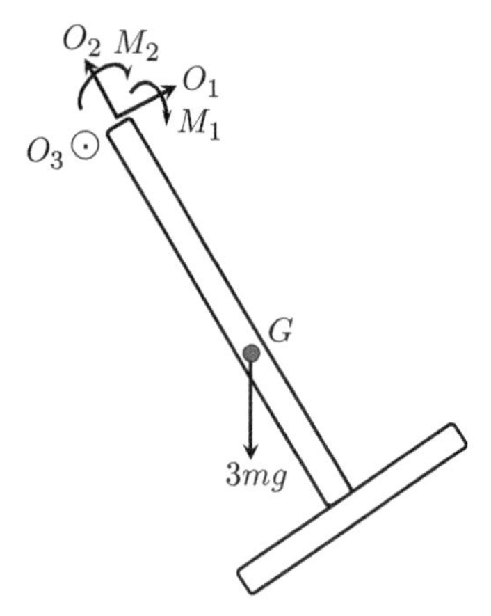


**Figure 3.22** Free body diagram of the T-shaped rod.

Since the inertia matrix is diagonal, we can use Euler's equations in their simplified form to obtain the equations of motion. To solve this problem, we only need to use the third equation, which states that

$$M_{3O} = I_{33O}\alpha_3 + (I_{22O} - I_{11O})\omega_1\omega_2,$$

where $\omega_1$ and $\omega_2$ are the components of the vector ${}^N\boldsymbol{\omega}^B$ when described in the $B$-frame: ${}^N\boldsymbol{\omega}^B = \Omega\sin\theta\hat{b}_1 + \Omega\cos\theta\hat{b}_2 + \dot{\theta}\hat{b}_3$. Thus, $\omega_1 = \Omega\sin\theta$ and $\omega_2 = \Omega\cos\theta$.

Similarly, ${}^N\boldsymbol{\alpha}^B = \Omega\dot{\theta}\cos\theta\hat{b}_1 - \Omega\dot{\theta}\sin\theta\hat{b}_2 + \ddot{\theta}\hat{b}_3$. Hence $\alpha_3 = \ddot{\theta}$. Using the obtained values of $I_{11O}$, $I_{22O}$, $I_{33O}$, $\omega_1$, $\omega_2$, and $\alpha_1$, and summing moments about point $O$, it follows from Euler's third equation that

$$-4mgl\sin\theta = \frac{27}{4}ml^2\ddot{\theta} - ml^2\left(\frac{20}{3} - \frac{1}{12}\right)\Omega^2\sin\theta\cos\theta,$$

which upon simplification yields

$$\ddot{\theta} + \left(\frac{16}{27}\frac{g}{l} - \frac{79}{81}\Omega^2\cos\theta\right)\sin\theta = 0.$$

As an additional exercise, let us try to resolve this problem by expressing Euler's equations in the $A$-frame. Describing the inertia matrix in the $A$-frame requires rotating the inertia matrix described in the $B$-frame using the transformation $I_O^* = L\,I_O\,L^T$, where $L$ is the rotation matrix from the $B$- to the $A$-frame. This can be obtained by rotating the $B$-frame by $-\theta\hat{n}_3$ using

$$L = \begin{pmatrix} \cos\theta & -\sin\theta & 0 \\ \sin\theta & \cos\theta & 0 \\ 0 & 0 & 1 \end{pmatrix}.$$

Using this transformation matrix, the inertia matrix about point $O$ as described in the $A$-frame becomes

$$I_O^* = ml^2\begin{pmatrix} \frac{20}{3}\cos^2\theta + \frac{1}{12}\sin^2\theta & \frac{79}{12}\sin\theta\cos\theta & 0 \\ \frac{79}{12}\sin\theta\cos\theta & \frac{1}{12}\cos^2\theta + \frac{20}{3}\sin^2\theta & 0 \\ 0 & 0 & \frac{27}{4} \end{pmatrix}.$$

Note that the inertia matrix described in the $A$-frame is not diagonal. As such, we cannot use Euler's equations in their simplified form. Next, we find the angular velocity and acceleration vectors and describe them in the $A$-frame. To this end, we use

$${}^N\boldsymbol{\omega}^B = \Omega\hat{a}_2 + \dot{\theta}\hat{a}_3, \qquad {}^N\boldsymbol{\alpha}^B = \ddot{\theta}\hat{a}_3.$$

It follows that $\omega_1 = 0$, $\omega_2 = \Omega$, $\omega_3 = \dot{\theta}$, $\alpha_1 = \alpha_2 = 0$, and $\alpha_3 = \ddot{\theta}$.

The moment about point $O$ is due to the weight and is given by $\mathbf{M}_O = -4mgl\sin\theta\,\hat{a}_3$. Thus, $M_{1O} = 0$, $M_{2O} = 0$, $M_{3O} = -4mgl\sin\theta$.

Next, we sum moments about point $O$ using

$$\begin{bmatrix} M_{1O} \\ M_{2O} \\ M_{3O} \end{bmatrix} = \begin{pmatrix} I_{11O} & I_{12O} & I_{13O} \\ I_{21O} & I_{22O} & I_{23O} \\ I_{31O} & I_{32O} & I_{33O} \end{pmatrix}\begin{bmatrix} \alpha_1 \\ \alpha_2 \\ \alpha_3 \end{bmatrix} + \begin{pmatrix} 0 & -\omega_3 & \omega_2 \\ \omega_3 & 0 & -\omega_1 \\ -\omega_2 & \omega_1 & 0 \end{pmatrix}\begin{pmatrix} I_{11O} & I_{12O} & I_{13O} \\ I_{21O} & I_{22O} & I_{23O} \\ I_{31O} & I_{32O} & I_{33O} \end{pmatrix}\begin{bmatrix} \omega_1 \\ \omega_2 \\ \omega_3 \end{bmatrix}.$$

This yields

$$\ddot{\theta} + \left(\frac{16}{27}\frac{g}{l} - \frac{79}{81}\Omega^2 \cos\theta\right)\sin\theta = 0,$$

which is the same answer we obtained when using the body-rotating $B$-frame. However, the process of obtaining the same answer is considerably more involved.

**Example 3.11 Constant Precession**

Consider a top made of a uniform disk of radius $R$, pinned to the origin by a massless stick (which is perpendicular to the disk) of length $l$, as shown in Figure 3.23. Paint a dot on the top at its highest point, and label this as point $P$. You wish to set up uniform circular precession, with the stick making a constant angle $\theta$ with the vertical, and with $P$ always being the highest point on the top. What relation between $R$ and $l$ must be satisfied for this motion to be possible?

For the top to perform constant precession with angular velocity $\Omega$ about the vertical axis, the net angular velocity has to be of magnitude $\Omega$ and has to always point upwards. By defining the $N$-frame as an inertial frame whose origin is at point $O$, the $A$-frame as a rotating frame, such that ${}^N\boldsymbol{\omega}^A = \Omega\hat{a}_2$, and the $B$-frame by performing a $-\theta$ rotation about the $\hat{a}_3$ axis, we can describe the angular velocity vector necessary to maintain such motion as ${}^N\boldsymbol{\omega}^B = \Omega\,\hat{a}_2$. Note that the term associated with $\dot{\theta}$ vanishes because the constant precession requires a constant $\theta$.

To solve the problem, we sum moments about point $O$ using the equation

$$\mathbf{M}_O = I_O\,{}^N\boldsymbol{\alpha}^B + {}^N\boldsymbol{\omega}^B \times (I_O\,{}^N\boldsymbol{\omega}^B),$$

where $I_O$ is the moment of inertia of the disk about point $O$ and can be written as

$$I_O = m\begin{pmatrix} \frac{R^2}{4} + l^2 & 0 & 0 \\ 0 & \frac{R^2}{2} & 0 \\ 0 & 0 & \frac{R^2}{4} + l^2 \end{pmatrix}.$$

Since the inertia matrix is diagonal, the moment equation can be decoupled into Euler's form. The angular velocity vector of the disk described in the $B$-frame is ${}^N\boldsymbol{\omega}^B = -\Omega\sin\theta\hat{b}_1 + \Omega\cos\theta\hat{b}_2$. The angular acceleration vector is zero because $\Omega$ and $\theta$ are constant. Using the third Euler's equation along $\hat{b}_3$, we obtain, by referring to Figure 3.24:

$$-mgl\sin\theta = m\Omega^2\left(l^2 - \frac{R^2}{4}\right)\cos\theta\sin\theta.$$

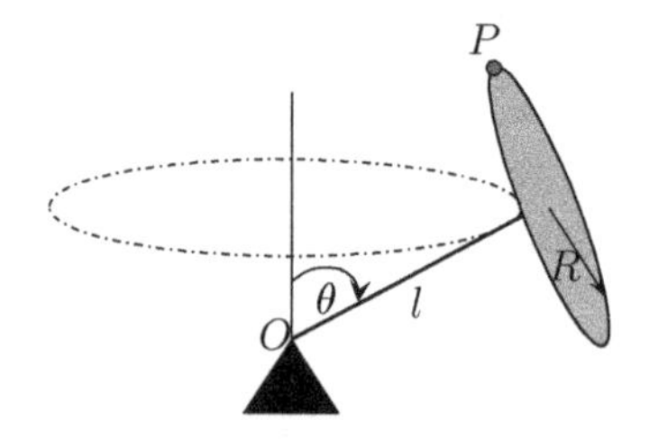


**Figure 3.23** A top undergoing constant precession motion.

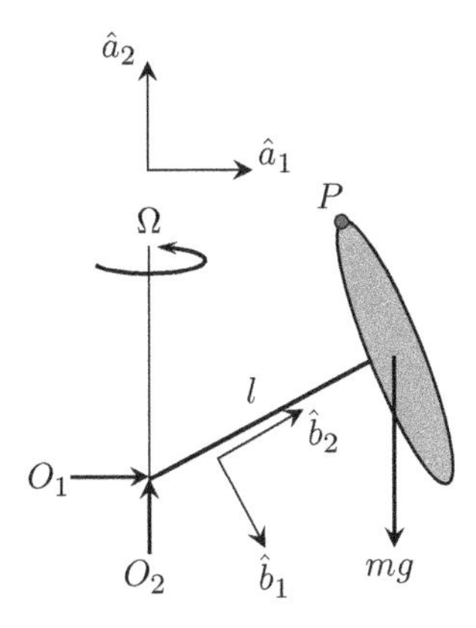


**Figure 3.24** Free-body diagram.

It follows from the previous equation that

$$\Omega = \sqrt{\frac{4gl}{(R^2 - 4l^2)\cos\theta}}.$$

Thus, $R > 2l$ for such motion to be possible.

### Example 3.12 Losing Contact with a Surface

In Figure 3.25, rod $QB$ rotates with angular velocity $\Omega = t^2$ where $t$ represents time. Pinned at $B$ is another thin rod of mass $m$ and length $l$. This can swing in a vertical plane which is rotating with rod $QB$. Initially, the thin rod remains in contact with the smooth surface, but at a critical time $t_{cr}$ it rises. The system starts from rest and all frictional forces can be neglected. Find $t_{cr}$.

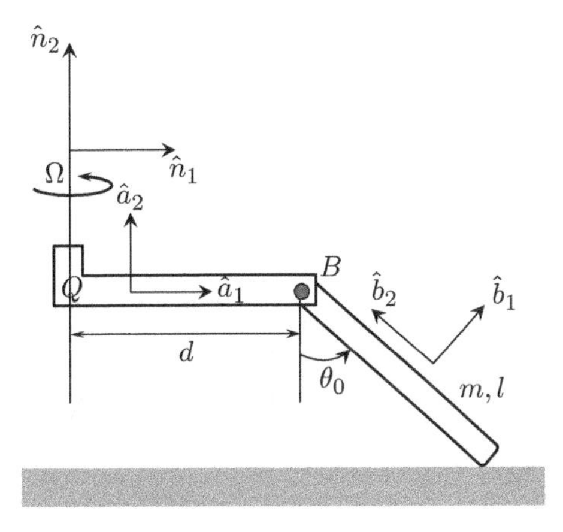


**Figure 3.25** Rotating rod which loses contact with the horizontal surface.

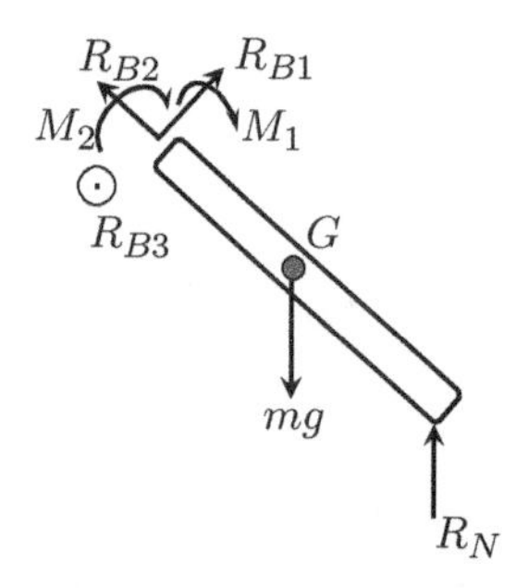


**Figure 3.26** Free-body diagram.

To solve this problem, we first define three different frames to describe the kinematics, as shown in Figure 3.25. The $N$-frame is inertial in space, the $A$-frame is formed such that ${}^N\omega 84^A = \Omega \hat{a}_2$, and the $B$-frame is formed by rotating the $A$-frame through an angle $\theta$ about $\hat{a}_3$.

To use Newton's second law or Euler's rotational equations, we must first identify the forces and moments acting on the body. To this end, we draw a free body diagram, as shown in Figure 3.26. There are three reaction forces at point $B$, namely $R_{B1}\hat{b}_1$, $R_{B2}\hat{b}_2$, $R_{B3}\hat{b}_3$, and two reaction moments, namely $M_1\hat{b}_1$ and $M_2\hat{b}_2$. In addition, there is the weight $-mg\hat{a}_2$ located at the center of mass and the reaction $R_N\hat{a}_2$ exerted by the ground on the rod.

We also define the angular velocity and angular acceleration vectors of the rod in the $B$-frame; such that

$$
\begin{aligned}
{}^N\boldsymbol{\omega}^B &= \Omega \sin\theta_0 \hat{b}_1 + \Omega \cos\theta_0 \hat{b}_2, \\
{}^N\boldsymbol{\alpha}^B &= \dot{\Omega} \sin\theta_0 \hat{b}_1 + \dot{\Omega} \cos\theta_0 \hat{b}_2.
\end{aligned}
$$

In this problem it is sufficient to use Euler's equations to obtain the equation of motion of the system. To this end, we need to sum moments about a certain point. Here, we choose point $B$ because most of the unknowns are located at point $B$. Since point $B$ is a moving point, the moment equation is given by

$$
\mathbf{M}_B = m\,\mathbf{BG} \times {}^N\mathbf{a}^{B/Q} + I_B\,{}^N\boldsymbol{\alpha}^B + {}^N\boldsymbol{\omega}^B \times (I_B\,{}^N\boldsymbol{\omega}^B), \tag{3.38}
$$

where, as shown in Figure 3.25, $Q$ is an inertial point, $\mathbf{BG} = -\frac{l}{2}\hat{b}_2$, and ${}^N\mathbf{a}^{B/Q} = -d\Omega^2 \cos\theta_0 \hat{b}_1 + d\Omega^2 \sin\theta_0 \hat{b}_2$.

We also obtain the moment of inertia matrix about point $B$ described in the $B$-frame as:

$$
I_B = ml^2 \begin{pmatrix} \frac{1}{3} & 0 & 0 \\ 0 & 0 & 0 \\ 0 & 0 & \frac{1}{3} \end{pmatrix}.
$$

Summing moments in the $\hat{b}_3$ direction, yields

$$
-mg\frac{l}{2}\sin\theta_0 + R_N l \sin\theta_0 = -m\frac{ld}{2}\Omega^2 \cos\theta_0 - \frac{1}{3}ml^2\Omega^2 \sin\theta_0 \cos\theta_0.
$$

At the moment the rod lifts off the ground, $R_N$ vanishes. This yields:

$$\Omega^2 = \frac{g \sin\theta_0}{d\cos\theta_0 + \frac{2}{3}l\cos\theta_0\sin\theta_0},$$

or

$$t_{cr} = \left(\frac{g \sin\theta_0}{d\cos\theta_0 + \frac{2}{3}l\cos\theta_0\sin\theta_0}\right)^{1/4}.$$

**Flipped Classroom Exercise 3.7**

A slender bar of mass $m$ is suspended by a massless cable from a pivot at $A$. It executes a steady precession about the vertical axis at angular speed $\Omega$ while maintaining the orientation shown. Determine the angular velocity $\Omega$ and the angle $\lambda$ of the steady precession.

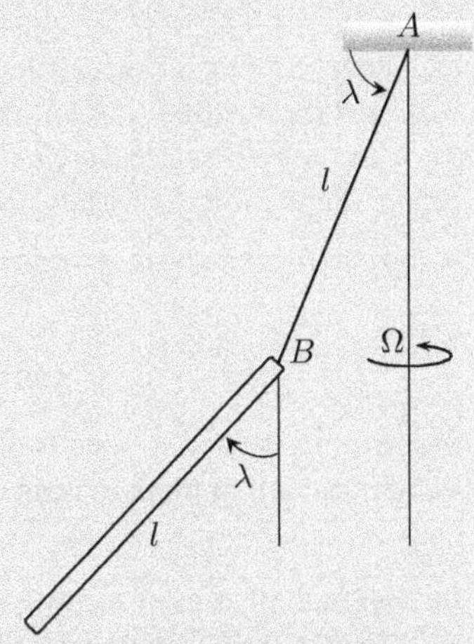


1. Define your inertial and rotating frames such that ${}^N\omega^A = \Omega\hat{a}_2$, and ${}^A\omega^B = -\dot{\lambda}\hat{b}_2$ where $N$ is the inertial frame.
2. Find the angular velocity and angular acceleration vectors in the $B$-frame.
3. Find the inertia matrix of the rod as described in the rotating frame.
4. Draw a free-body diagram. Note that there are only two forces acting on the rod.
5. Implement Newton's translational equations along the different directions.
6. Use the third Euler rotational equation to show that, for constant precision, $\lambda \approx 47^\circ$ and $\Omega \approx 1.06\sqrt{\frac{g}{l}}$.

**Flipped Classroom Exercise 3.8**

Consider a cylinder of mass $m$, radius $r$, and length $4r$. Find the analytical solution for the angular velocity vector $\omega(t)$ when the cylinder is subject to the initial angular velocity vector $\omega(t=0) = \frac{3}{5}\omega_0\hat{b}_1 + \frac{4}{5}\omega_0\hat{b}_3$.

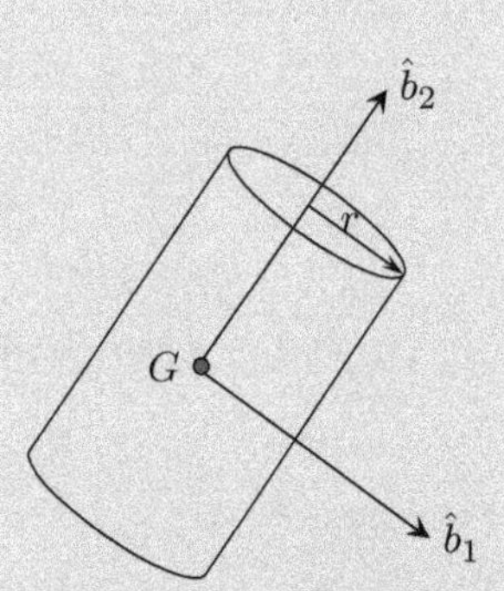


You only need to use Euler's rotational equations to solve this problem.

1. Find the inertia matrix about the center of mass of the cylinder. Note that it is a diagonal matrix.
2. Apply Euler's rotational equations along the three directions about the center of mass. Note that there are no external moments acting on the body. This will lead to three first-order differential equations describing the angular velocity.
3. Solve the resulting differential equations using the given initial condition.

## Exercises

3.1 Four masses ($m$, $m$, $2m$, $4m$) are located, respectively, at ($2a$, $2a$, 0), ($a$, $2a$, $a$), ($a$, $a$, $3a$), ($a$, $a$, $a$). Find the center of mass of these masses, then find the inertia matrix about the center of mass.

3.2 The system shown in the Figure 3.27 consists of two welded rods, each of mass $m$ and length $l$. The system is pinned at point $O$ to a column that rotates with a constant angular velocity $\Omega$. Find the equation of motion of the system.

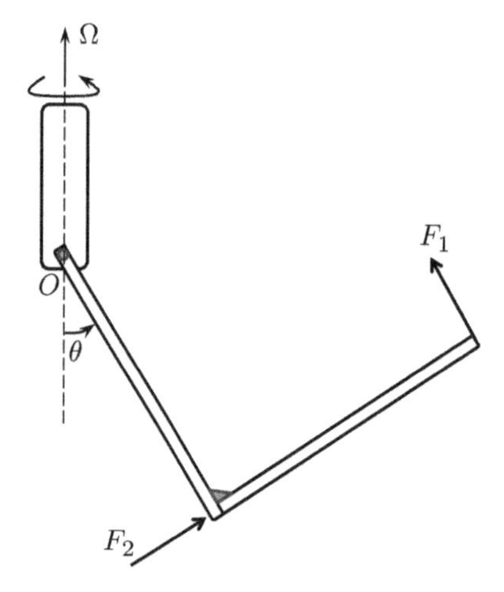


**Figure 3.27** Exercise 3.2.

3.3 Consider a top made of a uniform thin plate of length $2a$, width $a$, and connected to the origin by a massless stick (which is perpendicular to the disk) of length $l$, as shown in Figure 3.28. Mark a dot on the top at its highest point, and label this as point $P$. You wish to set up uniform circular precession, with the stick making a constant angle $\theta$ with the vertical, and with $P$ always being the highest point on the top. What relation between $a$ and $l$ must be satisfied for this motion to be possible? What is the frequency of precession? Is it possible to create such a motion when $\theta = \pi/2$?

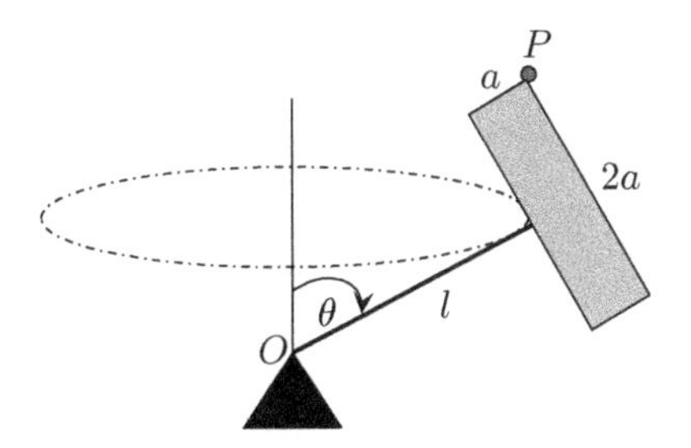


**Figure 3.28** Exercise 3.3.

3.4 A thin disk of mass $m$ and radius $r$ rolls without slipping inside a rough circular surface of radius $R$, as shown in Figure 3.29.

(a) Using the moment of inertia definition in polar coordinates, find the moment of inertia of the disk about an axis passing through the center of gravity of the disk and always pointing towards you.

(b) Derive the differential equation governing the angle $\theta$.

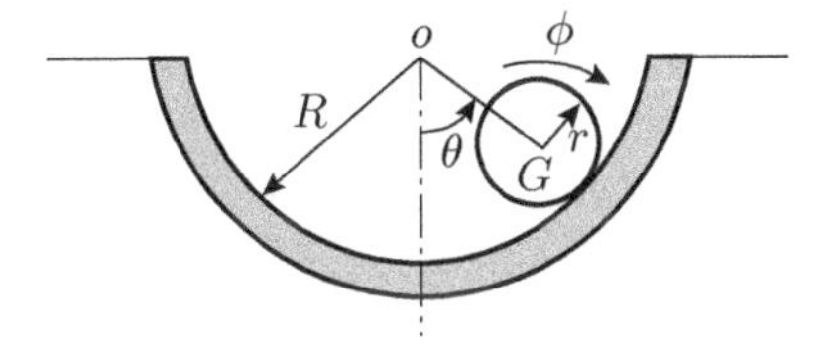


**Figure 3.29** Exercise 3.4.

3.5 Consider a double pendulum undergoing planar motion, as shown in Figure 3.30. Obtain the equations of motion.

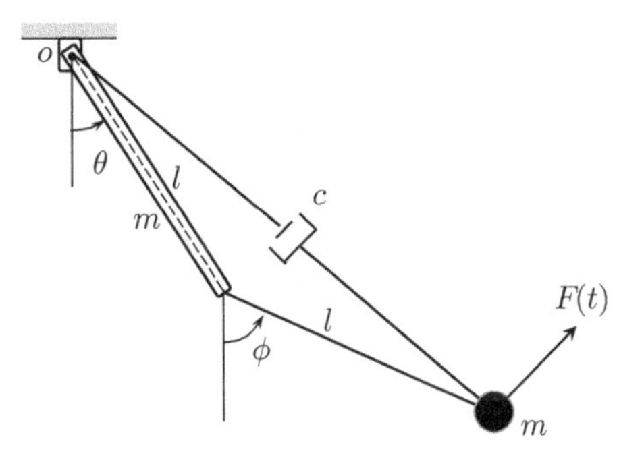


**Figure 3.30** Exercise 3.5.

3.6 The thin plate shown in Figure 3.31 has a uniform area density $\rho$. Find its center of mass and the moment of inertia matrix about the set of axes shown.

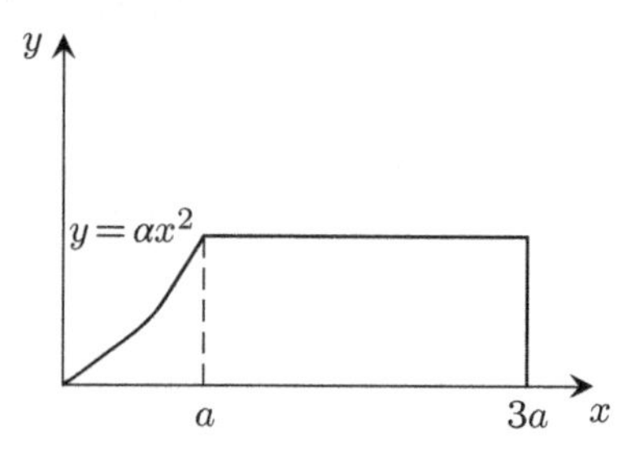


**Figure 3.31** Exercise 3.6.

3.7 The triangular plate shown in Figure 3.32 is pinned at point $O$ to a rotating rod. The rod rotates with a constant angular velocity $\Omega$, and the triangle rises by performing a rotation $\theta$ about its shorter side. Find the equation of motion of the system knowing that the plate is thin and has a uniform area density $\rho$.

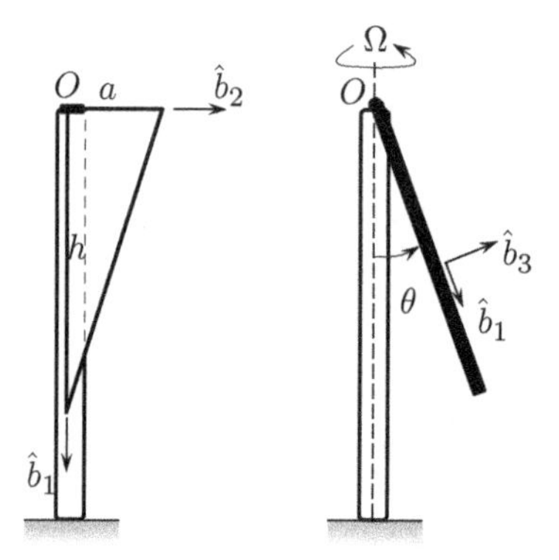


**Figure 3.32** Exercise 3.7.

3.8 For the rod of mass, $m$ and length $3l$ shown in Figure 3.33, find the shear force, axial load, and internal moment at point $B$ as function of $\theta$. The rod is released from rest when $\theta = \frac{\pi}{3}$.

3.9 For the system shown in Figure 3.34, the triangular part of the bent thin plate is parallel to the $y$–$z$ plane. The rectangular part is on the $x$–$y$ plane. Find the moment of inertia matrix about an axis passing through point $O$.

3.10 A uniform rod of mass $m$ and length $\sqrt{2}R$ slides inside a smooth circular surface of radius $R$ as shown in Figure 3.35. Derive the equations of motion of the rod in terms of the angle $\theta$.

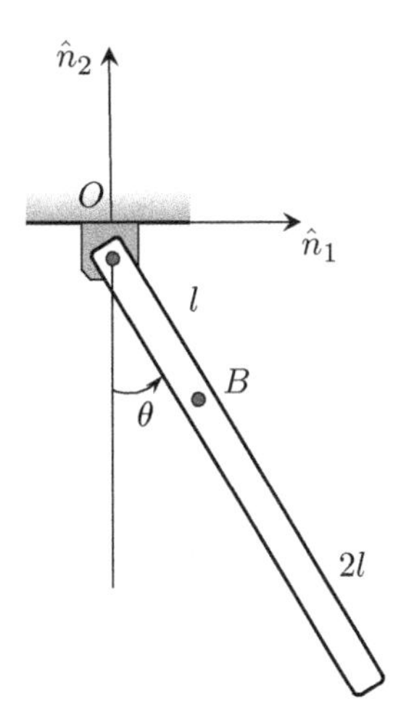


**Figure 3.33** Exercise 3.8.

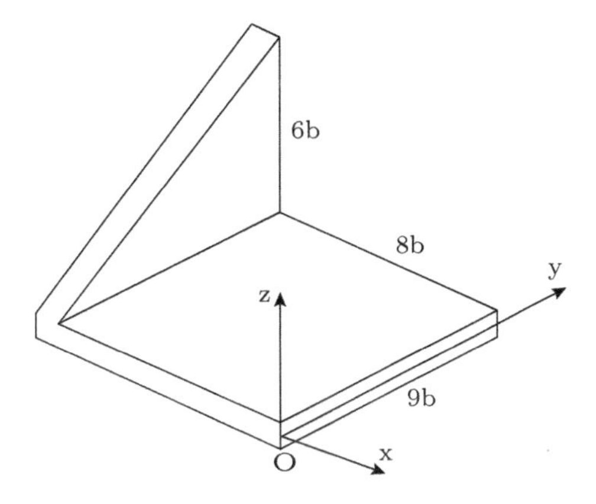


**Figure 3.34** Exercise 3.9.

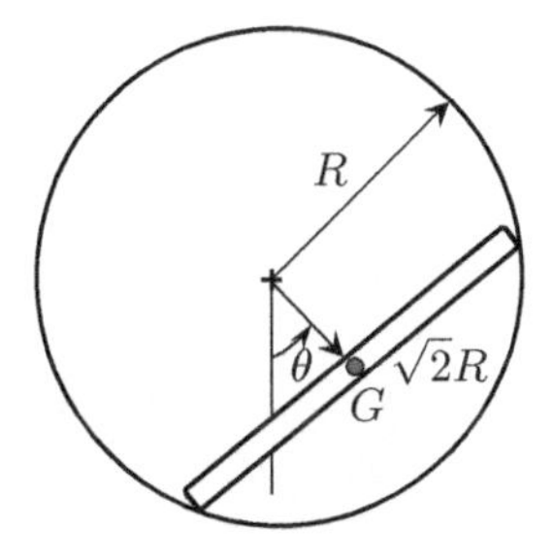


**Figure 3.35** Exercise 3.10.

3.11 A uniform rod of mass $m$ and length $l$ rests on two pulleys that rotate in the direction shown in Figure 3.36. Denoting by $\mu_k$ the coefficient of kinetic friction between the rod and the pulleys, determine the equation of motion when the rod is given an initial displacement to the right.

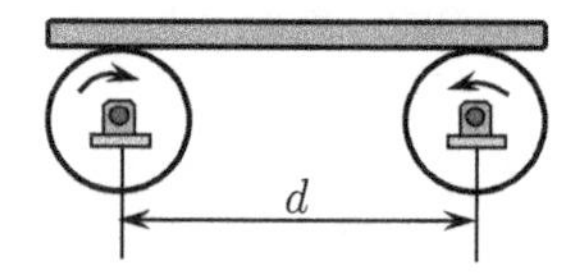


**Figure 3.36** Exercise 3.11.

3.12 A rod of mass $m$ and length $l$ is welded at its base to the massless ring shown in Figure 3.37. The ring rolls without slipping on a circular surface of radius $R$. Determine the equation of motion.

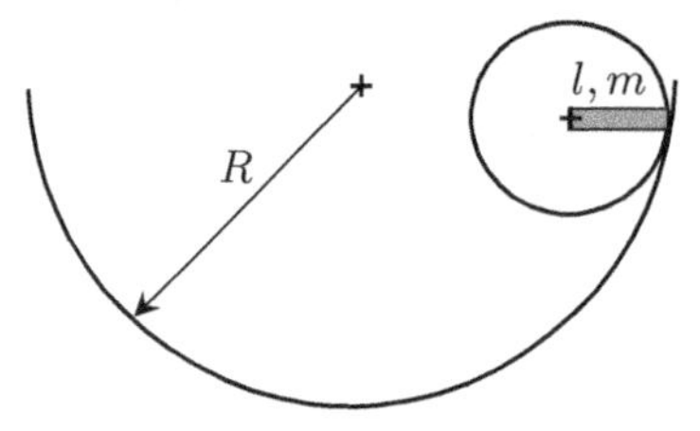


**Figure 3.37** Exercise 3.12.

3.13 A rod and a tube of the same mass and length are given a small nudge to the right as shown in Figure 3.38. Which one hits the ground first?

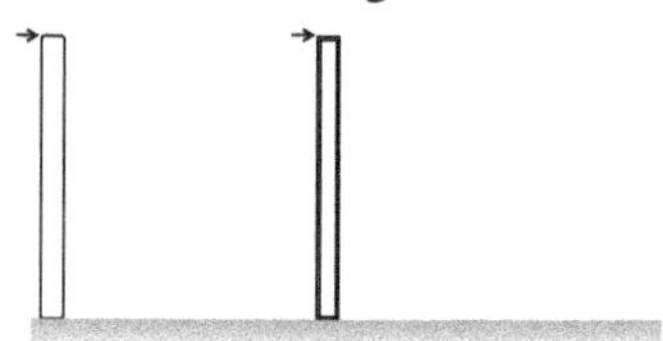

**Figure 3.38** Exercise 3.13.

3.14 A thin rod of mass $m$ and length $l$ is pinned to the back of a truck, as shown in Figure 3.39. The truck moves in the horizontal direction with an acceleration $a = g/4\ \mathrm{m/s^2}$. If the rod was released from rest when $\theta = 0$, find its angular velocity just before it hits the back of the truck.

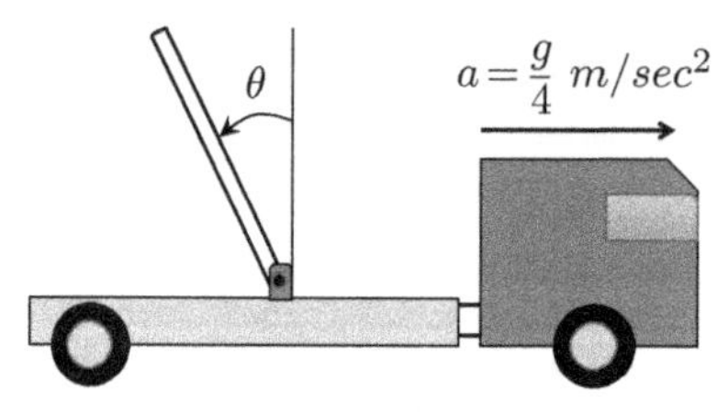


**Figure 3.39** Exercise 3.14.

3.15 A massless rod $AP$ of length $l$ is attached at point $A$ to the center of a homogeneous circular disk of radius $r$, and at point $P$ to a smooth universal joint, so that it can rotate freely in all directions. The disk rolls on a rough horizontal surface, as shown Figure 3.40. Determine the force exerted by the disk on the floor if $AP$ rotates with constant angular velocity $\Omega$ about the vertical.

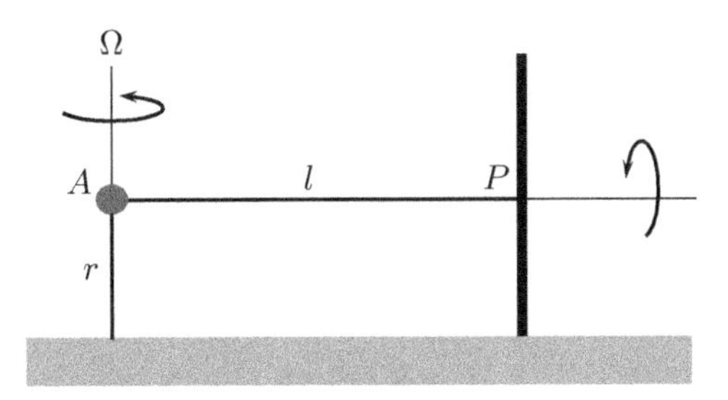


**Figure 3.40** Exercise 3.15.

3.16 Figure 3.41 shows a coin, which rolls with negligible rolling resistance on a horizontal table along a circle of radius $R$. The coin can be considered as a thin homogeneous disk of radius $r$ and mass $m$. How long does it take for the coin to roll once around the circle?

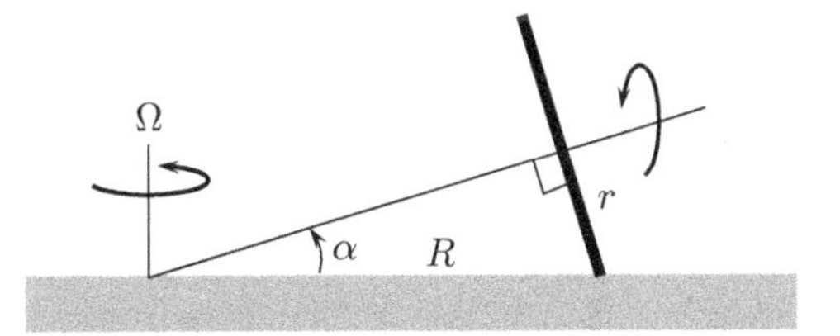


**Figure 3.41** Exercise 3.16.

3.17 A vertical pole $OA$ of length $r$ is attached to a horizontal ceiling, as shown Figure 3.42. Point $A$ is connected to a massless rod $AP$ of length $R$ by means of a smooth universal joint $A$. A circular homogeneous disk of mass $m$ and radius $r$ is mounted with its center at point $P$, such that $AP$ is always perpendicular to the disk. The disk rolls without slipping along a circular path of radius $R$. The angular velocity of the disk about point $P$ is $\omega$. Find the force that the disk exerts on the ceiling and the smallest $\omega$ needed to sustain contact with the ceiling.

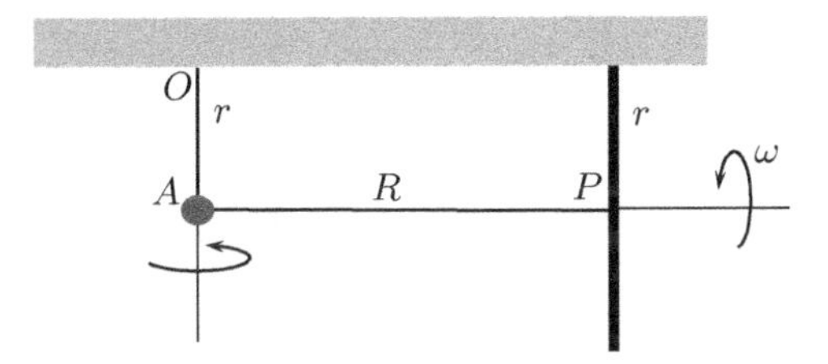


**Figure 3.42** Exercise 3.17.

3.18 A car with a gas turbine engine passes over the brow of a hill at speed $v$. The hilltop has a radius of curvature $R$ (Figure 3.43). The axis of the gas turbine is oriented along the length of the car. Its rotor has moment of inertia $J_e$ with respect to the turbine axis, and angular velocity $\omega$ relative to the car, so that the angular velocity vector $\omega$ points in the forward direction. The driver notices that the car has a tendency to turn sideways. Explain the reason for this tendency and identify which way the car tends to turn. Find the moment that the driver needs to apply to counter this tendency assuming that the car has mass $m$ and moment of inertia about its center of mass $I_{33G} = I_{car}$.

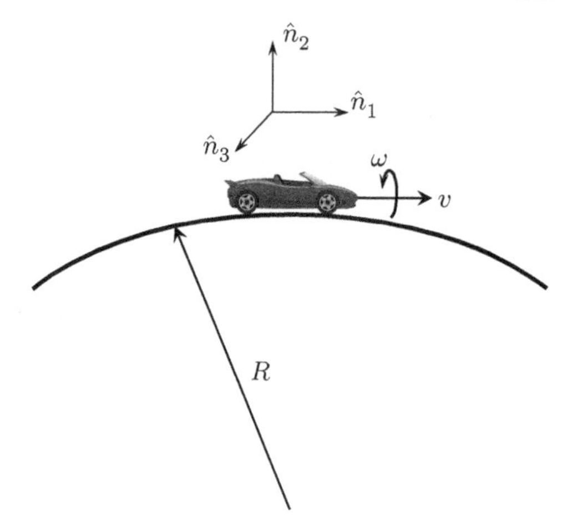


**Figure 3.43** Exercise 3.18.

3.19 The uniform disk of radius $R$ and mass $M$ can rotate freely about point $O$, as shown in Figure 3.44. A thick cable of length $l$ and mass $m$ is wound around the cylinder in a single layer. Find the equation of motion of the cable as function of the free length $x$. Assume the cable does not oscillate and that the winding/unwinding of the cable occurs without slipping.

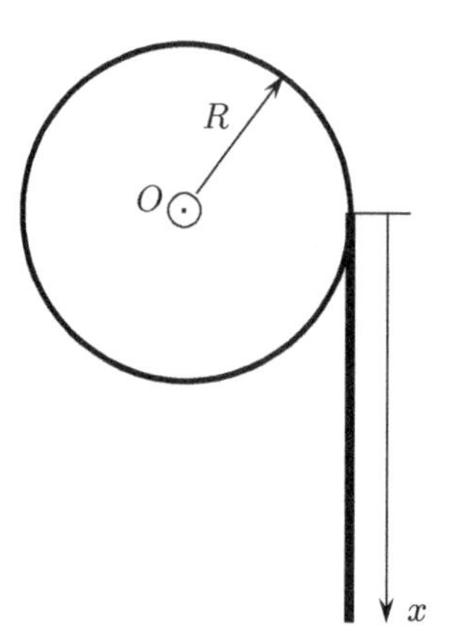


**Figure 3.44** Exercise 3.20.

3.20 A cube of uniform density $\rho$, mass $m$, and edge length $l$ is divided into eight identical cubes. Seven of the resulting cubes are then glued together leaving out one of the corner cubes. Find the moment of inertia matrix of the new body about a set of axes parallel to the sides of the cube and whose origin is located at the center of the original system.

## Reference

1. E. Mach (1919) *The Science of Mechanics*, pp. 173–187. Open Court Publishing.

# 4

# System Constraints and Virtual Displacement

Most dynamical systems do not move freely in space but are subject to constraints that alter, restrict, and shape their motion. In this chapter, we classify these constraints and understand their influence on the dynamics.

## 4.1 Constraints

Constraints acting on a dynamical system reduce the minimum number of independent coordinates necessary to describe its motion; in other words, its degrees of freedom. To better understand the influence of constraints on the motion of a dynamical system, consider the planar motion of a rigid pendulum. The pendulum consists of one particle of mass $m$, connected through a rigid cable of length $l$ to a fixed support. Generally, the *free* motion of a particle is fully described by three coordinates $(x, y, z)$. However, we learned in earlier chapters that the dynamics of a planar rigid pendulum can be fully characterized by a single angle. Thus, in reality, one needs only one independent coordinate to describe its planar motion. This is a result of the action of two constraints that restrict the free motion of the particle and, therefore, reduce the number of degrees of freedom to only one. For the planar pendulum, the two constraints are as follows:

- The particle is not free to move in all directions but is rather constrained to move in the $x$–$y$ plane; this means that $z = 0$ is a geometric constraint acting on the particle.
- For a rigid cable, the position of the particle is constrained to the equation $x^2 + y^2 = l^2$, which acts as a constraint preventing the particle from moving arbitrarily in the $x$–$y$ plane.

*Dynamics of Particles and Rigid Bodies: A Self-Learning Approach*, First Edition. Mohammed F. Daqaq.

Companion website: www.wiley.com/go/daqaq/dynamics

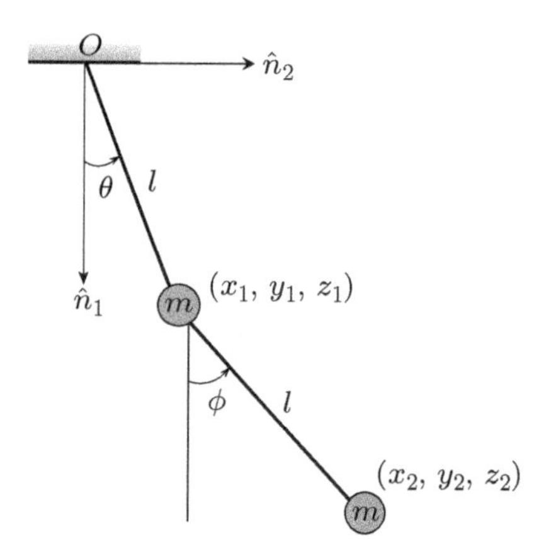


**Figure 4.1** Dynamics of a double pendulum.

These two constraints eliminate two degrees of freedom. In general, as explained in Section 2.4, a system of $\mathcal{N}$ particles has $3\mathcal{N} - C$ degrees of freedom, where $C$ is the number of constraints.

**Example 4.1 Double Pendulum Planar Dynamics**

Consider the dynamics of the double pendulum shown in Figure 4.1. Find the number of coordinates, constraints, and degrees of freedom.

The double pendulum system has two particles, and therefore six coordinates. The coordinates are $(x_1, y_1, z_1)$ for the first particle and $(x_2, y_2, z_2)$ for the second particle. Since the motion is planar, the first and second particles are subject, respectively, to the constraints $z_1 = 0$ and $z_2 = 0$. Also, since the lengths of the cables are fixed, the constraints $x_1^2 + y_1^2 = l^2$ and $(x_2 - x_1)^2 + (y_2 - y_1)^2 = l^2$ must be satisfied. Hence, there are four constraints. This yields $3 \times 2 - 4 = 2$ degrees of freedom.

### 4.1.1 *Classification of Constraints*

Constraints can be classified into two general categories: *holonomic* and *non-holonomic*. A constraint is said to be holonomic when it can be described in the following form:

$$f(x_1, x_2, x_3, \dots, x_N, t) = 0, \tag{4.1}$$

where $x_i$ are the coordinates used to describe the dynamics. Holonomic constraints do not depend on the derivative of the coordinates, $\dot{x}_i$. As such, they are commonly referred to as *geometric* constraints because they describe regions in space restricted to the path of the system. Even when a holonomic constraint depends on the derivative of the coordinates, $\dot{x}_i$, the derivatives should be removable by direct integration. Any constraint that cannot be described by Equation (4.1) is said to be a non-holonomic or a *kinematic* constraint. Kinematic constraints place limitations on the velocity of the system.

The concept of holonomic constraints leads to the definition of *generalized coordinates*. These are the set of coordinates chosen to describe the configuration of the system. The term "generalized" is used to distinguish these coordinates from the Cartesian coordinates $(x, y, z)$ that are typically used to describe the configuration of the system. The derivatives of the generalized coordinates are called *generalized velocities*. The reader should bear in mind that generalized coordinates need not be the minimum number of coordinates needed to describe the dynamics of the system.

To clarify the concept of a generalized coordinate, consider again the planar dynamics of the simple rigid pendulum when described in terms of the typical $x$ and $y$ Cartesian coordinates. In such a case, the equations of motion can be written as

$$-T\frac{x}{l} = m\ddot{x}, \qquad T\frac{y}{l} - mg = m\ddot{y}, \tag{4.2}$$

where $T$ is the tension in the cable and $l$ is its length. Next, using the holonomic constraints $y - l\cos\theta = 0$ and $x - l\sin\theta = 0$, we reduce the dynamics of the system to one coordinate, $\theta$. This yields:

$$-T\sin\theta = ml(\cos\theta\ddot{\theta} - \sin\theta\dot{\theta}^2), \qquad T\cos\theta = ml(\sin\theta\ddot{\theta} + \cos\theta\dot{\theta}^2) + mg.$$

Rearranging and simplifying, we obtain

$$\ddot{\theta} + \frac{g}{l}\sin\theta = 0.$$

The angle $\theta$ in this example is a generalized coordinate because it can be used to describe the dynamics, yet it is not a traditional Cartesian coordinate. As such, using the geometric constraints, we were able to reduce the description of the dynamics from two dependent Cartesian coordinates $(x, y)$ into one independent generalized coordinate, $\theta$.

Regardless of their type, a number $\mathcal{K}$ of constraints expressed in terms of $\mathcal{M}$ generalized coordinates can be described in the following general differential form, also known as the *Pfaffian* form:

$$\sum_{j}^{\mathcal{M}} c_{ij}\mathrm{d}q_j + c_i\mathrm{d}t = 0, \qquad i = 1, 2, 3, \ldots, \mathcal{K}, \tag{4.3}$$

where $j$ refers to the $j$th generalized coordinates, $q_j$, and $i$ refers to the $i$th constraint. If the previous equation is integrable to the form

$$f(q_1, q_2, \ldots, q_{\mathcal{M}}, t) = 0, \qquad i = 1, 2, 3, \ldots, \mathcal{K}, \tag{4.4}$$

then the system is holonomic; otherwise the system is non-holonomic.

Both holonomic and non-holonomic constraints can be further classified into two types: *scleronomic* if the constraint does not depend on time explicitly, and *rheonomic* if it does. These phrases originate from the Greek language, where *sclero* means rigid and *rheo* means flowing.

It is also possible to define constraints as *catastatic* when $c_i = 0$ and as *acatastatic* when $c_i \neq 0$. Note that a catastatic constraint is not equivalent to a *scleronomic* constraint because $c_{ij}$ can be time dependent even when $c_i = 0$.

**Summary of Constraint Classifications**

Any system of constraints can be described mathematically in the following form:

$$\sum_{j}^{\mathcal{M}} c_{ij}\mathrm{d}q_j + c_i\mathrm{d}t = 0, \qquad i = 1, 2, 3, \ldots \mathcal{K}.$$

The constraint is said to be

- holonomic: when it is integrable to the form $f(q_1, q_2, \cdots, q_{\mathcal{M}}, t) = 0$.
- non-holonomic: when it is not integrable to the form $f(q_1, q_2, \cdots, q_{\mathcal{M}}, t) = 0$.
- scleronomic: when the constraint does not depend on time.
- rheonomic: when the constraint depends on time.
- catastatic: when $c_i = 0$.
- acatastatic: when $c_i \neq 0$.

The knowledge of whether the Pfaffian form is integrable to the general holonomic form can be established by noting that, for integrable constraints, there must exist a scalar function, $\phi$, such that

$$\sum_{j=1}^{\mathcal{M}} \frac{\partial \phi}{\partial q_j} dq_j + \frac{\partial \phi}{\partial t} = 0. \tag{4.5}$$

It follows from Equation (4.3) that

$$c_{ij} = \frac{\partial \phi}{\partial q_j}, \qquad c_i = \frac{\partial \phi}{\partial t}, \qquad i = 1, 2, \ldots, \mathcal{K}. \tag{4.6}$$

Differentiating the scalar function $\phi$ with respect to $q_j$ then $q_k$, one can write

$$\frac{\partial^2 \phi}{\partial q_k \partial q_j} = \frac{\partial c_{ij}}{\partial q_k} = \frac{\partial c_{ik}}{\partial q_j}. \tag{4.7}$$

It follows that

$$\frac{\partial c_{ij}}{\partial q_k} = \frac{\partial c_{ik}}{\partial q_j}, \qquad k = 1, 2, \ldots, \mathcal{M}. \tag{4.8}$$

Similarly, differentiating the scalar function $\phi$ with respect to $q_j$ then $t$, one can write

$$\frac{\partial^2 \phi}{\partial q_j \partial t} = \frac{\partial c_{ij}}{\partial t} = \frac{\partial c_i}{\partial q_j}. \tag{4.9}$$

It follows that

$$\frac{\partial c_{ij}}{\partial t} = \frac{\partial c_i}{\partial q_j}. \tag{4.10}$$

Thus, for a constraint to be holonomic, Equations (4.8) and (4.10) must be satisfied.

 **Integrability of Constraints**

A constraint given by the general form

$$\sum_{j}^{\mathcal{M}} c_{ij}\mathrm{d}q_j + c_i\mathrm{d}t = 0, \qquad i = 1, 2, 3, \dots, \mathcal{K},$$

where $\mathcal{M}$ is the number of generalized coordinates and $\mathcal{K}$ is the number of constraints, is said to be holonomic if it is integrable to the form

$$f(q_1, q_2, \dots, q_{\mathcal{M}}, t) = 0, \qquad i = 1, 2, 3, \dots, \mathcal{K}.$$

The integrability of the constraint is guaranteed if the following conditions are satisfied:

$$\frac{\partial c_{ij}}{\partial q_k} = \frac{\partial c_{ik}}{\partial q_j}, \qquad k = 1, 2, 3, \dots, \mathcal{M}$$

$$\frac{\partial c_{ij}}{\partial t} = \frac{\partial c_i}{\partial q_j}.$$

**Example 4.2 Classification of Constraints**

Classify the following constraints:

(a) $(q_1 - q_4)\mathrm{d}q_1 + (q_4 - q_1)\mathrm{d}q_4 + (q_2 - q_3)\mathrm{d}q_3 + (q_3 - q_2)\mathrm{d}q_2 = 0$
Here, $\mathcal{M} = 4$, $\mathcal{K} = 1$. Checking the conditions for integrability using $i = 1$:

$$\frac{\partial c_{1j}}{\partial q_k} = \frac{\partial c_{1k}}{\partial q_j}, \qquad j, k = 1, 2, 3, 4,$$

yields

$$\frac{\partial c_{11}}{\partial q_2} = \frac{\partial c_{12}}{\partial q_1} = 0, \qquad \frac{\partial c_{11}}{\partial q_3} = \frac{\partial c_{13}}{\partial q_1} = 0,$$

$$\frac{\partial c_{11}}{\partial q_4} = \frac{\partial c_{14}}{\partial q_1} = -1, \qquad \frac{\partial c_{12}}{\partial q_3} = \frac{\partial c_{13}}{\partial q_2} = 1,$$

$$\frac{\partial c_{12}}{\partial q_4} = \frac{\partial c_{14}}{\partial q_2} = 0, \qquad \frac{\partial c_{13}}{\partial q_4} = \frac{\partial c_{14}}{\partial q_3} = 0.$$

Since the conditions are satisfied, the constraint is holonomic. Furthermore, since the constraint does not depend explicitly on time, it is scleronomic.

(b) $x_1x_2^2\dot{x}_1 + x_1^2x_2\dot{x}_2 + \sin t = 0$
To check for integrability, we first put the constraint in the differential form:

$$x_1x_2^2\mathrm{d}x_1 + x_1^2x_2\mathrm{d}x_2 + \sin t\mathrm{d}t = 0$$

Here, $\mathcal{M} = 2$, and $\mathcal{N} = 1$. Applying the conditions for integrability using $i = 1$ yields

$$\frac{\partial c_{11}}{\partial x_2} = \frac{\partial c_{12}}{\partial x_1} = 2x_1x_2,$$
$$\frac{\partial c_{11}}{\partial t} = \frac{\partial c_1}{\partial x_1} = 0,$$
$$\frac{\partial c_{12}}{\partial t} = \frac{\partial c_1}{\partial x_2} = 0.$$

Hence, the constraint is holonomic and rheonomic by the fact that it depends explicitly on time.

**Example 4.3 Skating Boot**

Consider the motion of a skating boot, modeled using two particles, each of mass $m$, connected through a massless rod of length $d$, as shown in Figure 4.2. The boot slides on ice such that the center of mass $G$ is constrained to move in the direction of $\mathbf{e}_t$, which is tangential to the path. The motion of the skate is restricted to the plane shown. Find all the constraints and classify them. How many degrees of freedom does the system have?

The free motion of each particle is described by three coordinates: $(x_1, y_1, z_1)$ for the first particle and $(x_2, y_2, z_2)$ for the second particle. The system is subject to the following constraints:

1. $z_1 = 0$.
2. $z_2 = 0$.
3. $(x_2 - x_1)^2 + (y_2 - y_1)^2 = d^2$.
4. ${}^N\mathbf{v}^{G/O}.\mathbf{e}_n = 0$, because the skating boot always moves perpendicular to $\mathbf{e}_n$.

Here

$${}^N\mathbf{v}^{G/O} = \frac{1}{2}((\dot{x}_1 + \dot{x}_2)\hat{n}_1 + (\dot{y}_1 + \dot{y}_2)\hat{n}_2)$$

$$\mathbf{e}_n = \sin\theta\hat{n}_1 - \cos\theta\hat{n}_2 = \frac{y_2 - y_1}{d}\hat{n}_1 - \frac{x_2 - x_1}{d}\hat{n}_2.$$

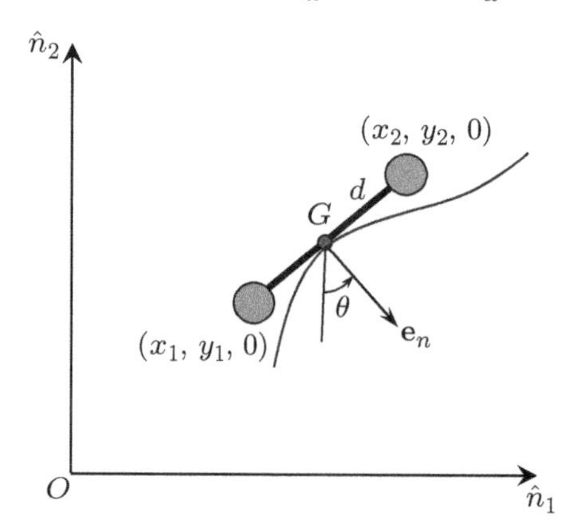


**Figure 4.2** Constraints associated with skating motion.

Therefore the system has $6 - 4 = 2$ degrees of freedom.

We define the generalized coordinates such that $q_1 = x_1, q_2 = y_1, q_3 = z_1, q_4 = x_2, q_5 = y_2,$ $q_6 = z_2$ and use them to describe the constraints in the differential form:

$$\begin{aligned}
&\mathrm{d}q_3 = 0, \qquad \mathrm{d}q_6 = 0,\\
&(q_1 - q_4)\mathrm{d}q_1 + (q_2 - q_5)\mathrm{d}q_2 + (q_4 - q_1)\mathrm{d}q_4 + (q_5 - q_2)\mathrm{d}q_5 = 0,\\
&(q_5 - q_2)\mathrm{d}q_1 + (q_1 - q_4)\mathrm{d}q_2 + (q_5 - q_2)\mathrm{d}q_4 + (q_1 - q_4)\mathrm{d}q_5 = 0.
\end{aligned}$$

One can show that the first three constraints are holonomic scleronomic. However, the fourth constraint is not integrable because $\partial c_{41}/\partial q_2 = -1$ while $\partial c_{42}/\partial q_1 = 1$. Therefore the fourth constraint is non-holonomic scleronomic.

**Flipped Classroom Exercise 4.1**

Consider the two-dimensional dynamics of the container crane shown in the figure [1]. Containers are grabbed using a spreader bar hoisted by means of four cables, two of which are shown. The cables are spaced a distance $d$ at the trolley and a distance $w$ at the spreader bar. Find two constraints relating the angle $\theta$ to the angle $\phi$ and the length $l$ to the length $L$. Classify these constraints.

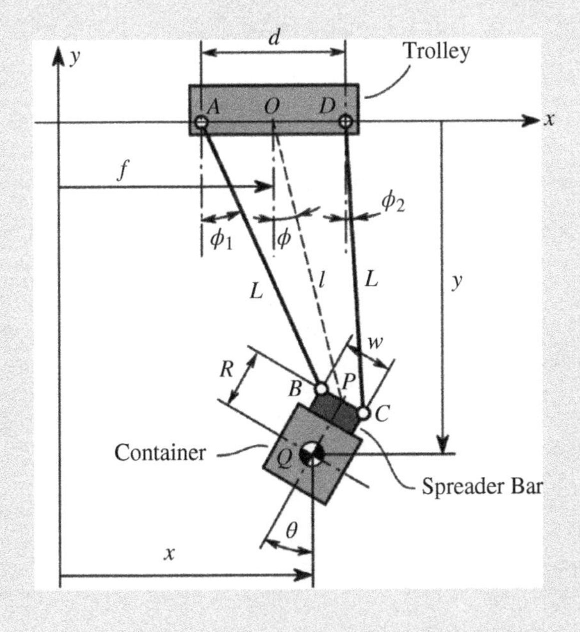

To find the constraints equations, do the following:

1. Write the closing position vector from point $O$ to point $O$ by passing through point $A$, $B$, and $P$. Use this position vector to show that:

$$l\sin\phi - \frac{1}{2}w\cos\theta + \frac{1}{2}d = L\sin\phi_1,$$

$$l\cos\phi - \frac{1}{2}w\sin\theta = L\cos\phi_1.$$

2. Write the closing position vector from point $O$ to point $O$ by passing through point $D$, $C$, and $P$. Use this position vector to show that:

$$l\sin\phi + \frac{1}{2}w\cos\theta - \frac{1}{2}d = L\sin\phi_2,$$

$$l\cos\phi + \frac{1}{2}w\sin\theta = L\cos\phi_2.$$

3. Manipulate the previous equations to show that

$$\theta = -\phi + \arcsin\left(\frac{d}{w}\sin\phi\right),$$

$$l^2 = L^2 - \frac{1}{4}(d^2 + w^2 - 2dw\cos\theta).$$

4. Put each of the previous constraints in the differential form and classify them.

## 4.2 Actual and Virtual Displacements

Consider the motion of a particle on a smooth surface, as shown in Figure 4.3. The equation of the surface is given by $f(x, y, z) = 0$. Our goal is to find the work done by the reaction force $\mathbf{R}_n$ on the particle. To this end, we note that the position of the particle in space $(x(t), y(t), z(t))$ must satisfy the constraint equation. Therefore we can differentiate the constraint equation and write

$$\frac{\partial f}{\partial x}\frac{\mathrm{d}x}{\mathrm{d}t} + \frac{\partial f}{\partial y}\frac{\mathrm{d}y}{\mathrm{d}t} + \frac{\partial f}{\partial z}\frac{\mathrm{d}z}{\mathrm{d}t} = 0, \tag{4.15}$$

or

$$\frac{\partial f}{\partial x}\mathrm{d}x + \frac{\partial f}{\partial y}\mathrm{d}y + \frac{\partial f}{\partial z}\mathrm{d}z = 0, \tag{4.16}$$

which can also be expressed as

$$\nabla f.\mathrm{d}\mathbf{r} = 0, \tag{4.17}$$

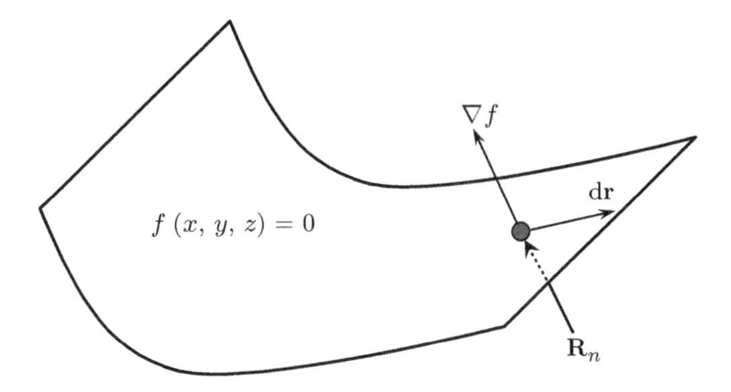


**Figure 4.3** Motion of a particle on a constraint surface.

where $\nabla f = \frac{\partial f}{\partial x}\hat{n}_1 + \frac{\partial f}{\partial y}\hat{n}_2 + \frac{\partial f}{\partial z}\hat{n}_3$ is the gradient of $f$ and always points in the direction normal to the surface, and $\mathbf{dr} = \mathrm{d}x\hat{n}_1 + \mathrm{d}y\hat{n}_2 + \mathrm{d}z\hat{n}_3$ is a differential displacement in the direction of the vector $\mathbf{r}$.

Since the constraint force $\mathbf{R}_n$ is also normal to the constraint surface; in other words, it is parallel to $\nabla f$, we conclude that

$$\mathbf{R}_n.\mathbf{dr} = 0. \tag{4.18}$$

The previous equation implies that the work done by the constraint force along an actual differential displacement is zero.

Next, we consider a time-varying constraint surface whose equation is described by

$$f(x, y, z, t) = 0. \tag{4.19}$$

Again, we differentiate the constraint with respect to time and find that

$$\frac{\partial f}{\partial x}\frac{\mathrm{d}x}{\mathrm{d}t} + \frac{\partial f}{\partial y}\frac{\mathrm{d}y}{\mathrm{d}t} + \frac{\partial f}{\partial z}\frac{\mathrm{d}z}{\mathrm{d}t} + \frac{\partial f}{\partial t} = 0, \tag{4.20}$$

which can be expressed as

$$\nabla f.\mathbf{dr} + \partial f = 0. \tag{4.21}$$

Since $\nabla f \parallel \mathbf{R}_n$, the first term in the previous equation represents the work done by the constraint forces along a differential displacement, which in this case does not vanish. Thus, we conclude that, when the constraint surface varies with time, the work done by the constraint force along a differential displacement is not zero.

The previous discussion can be clarified further by considering the curve shown in Figure 4.4. For simplicity, the surface is replaced by the curve $f(x, y, t) = 0$, which moves in the horizontal direction with a velocity $\mathbf{u}_s(t)$. A particle slides on the surface as the surface is simultaneously moving to the right. A snapshot of the complete motion after time $\Delta t$ is also shown in the figure. When inspecting the direction of the reaction force $\mathbf{R}_n$ and the absolute velocity $\mathbf{u}$ of the particle, it becomes evident that $\mathbf{R}_n$ is not

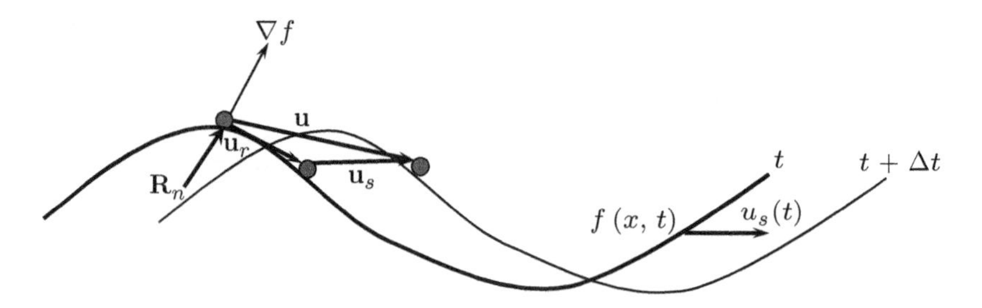


**Figure 4.4** Planar motion of a particle on a moving surface.

perpendicular to $\mathbf{u}$. Furthermore, since $\mathbf{u}$ is parallel to the actual differential displacement $\mathrm{d}\mathbf{r}$, one correctly concludes that the work done by the constraint force along $\mathrm{d}\mathbf{r}$ is indeed not equal zero.

On the other hand, we note that the constraint force is always perpendicular to the relative velocity $\mathbf{u}_r$. As such, to force the work done by the constraint forces to vanish, we introduce the concept of the virtual displacement $\delta\mathbf{r}$. The virtual displacement is one that is parallel to the relative velocity $\mathbf{u}_r$ and, as a result, is always normal to the constraint force. As such, the work done by the constraint force along the virtual displacement, also known as the *virtual work* $\delta\mathcal{W}$, is always zero. Bear in mind that, for scleronomic constraints, actual and virtual displacements are equivalent.

The virtual displacement can also be thought of as the displacement achieved when the constraint is held constant in time. Thus, for the position vector $\mathbf{r} = \mathbf{r}(q_1, q_2, q_3, \cdots, q_{\mathcal{M}}, t)$, the virtual displacement can be expressed as

$$\delta\mathbf{r} = \sum_{j=1}^{\mathcal{M}} \frac{\partial \mathbf{r}}{\partial q_j}\delta q_j, \tag{4.22}$$

where $t$ is held constant. On the other hand, the actual differential displacement can be written as

$$\mathrm{d}\mathbf{r} = \sum_{j=1}^{\mathcal{M}} \frac{\partial \mathbf{r}}{\partial q_j}\delta q_j + \frac{\partial \mathbf{r}}{\partial t}. \tag{4.23}$$

 **Virtual Displacement**

The virtual displacement is defined as the displacement along which the work done by constraint forces is zero. The virtual displacement is not necessary along the actual dynamical path of the particle; in other words, the actual displacement. Nonetheless, the virtual and actual displacements align when the constraints are scleronomic in nature.

## 4.3 Virtual Work

The virtual work is defined as the work done by all forces acting on the system along the virtual displacement. For a system of $\mathcal{N}$ particles, we can use Newton's second law to write

$$\sum_{i=1}^{\mathcal{N}}(\mathbf{F}_i + \mathbf{f}_i) = \sum_{i=1}^{\mathcal{N}} m_i \ddot{\mathbf{r}}_i, \tag{4.24}$$

where $\mathcal{N}$ is the number of particles, $m_i$ denotes the mass of the $i$th particle, $\mathbf{F}_i$ and $\mathbf{f}_i$ are the external and constraint forces acting on the system, respectively, and $\mathbf{r}_i = \mathbf{r}_i(q_1, q_2, \dots, q_{\mathcal{M}}, t)$ represents the position vector of the $i$th particle, represented in terms of the generalized coordinates $q_j$. Treating the inertial forces as part of the external forces, we let

$$\sum_{i=1}^{\mathcal{N}}(\mathbf{F}_i + \mathbf{f}_i - m_i \ddot{\mathbf{r}}_i) = 0. \tag{4.25}$$

Taking the dot product of the previous equation with the virtual displacement $\delta\mathbf{r}_i$, we obtain

$$\sum_{i=1}^{\mathcal{N}}(\mathbf{F}_i + \mathbf{f}_i - m_i \ddot{\mathbf{r}}_i)\delta\mathbf{r}_i = 0. \tag{4.26}$$

Noting that the work done by constraint forces along the virtual displacement is zero, the previous equation simplifies to

$$\delta\mathcal{W} = \sum_{i=1}^{\mathcal{N}}(\mathbf{F}_i - m_i \ddot{\mathbf{r}}_i)\delta\mathbf{r}_i = 0 \tag{4.27}$$

Equation (4.27) is known as the *principle of virtual work* and states that the work done by the external and inertial forces along the virtual displacement is equal to zero.

In the absence of inertial forces, the previous equation reduces to

$$\sum_{i=1}^{\mathcal{N}} \mathbf{F}_i \delta\mathbf{r}_i = 0. \tag{4.28}$$

One important consequence that arises from the principle of virtual work is the analysis of the static equilibrium of bodies in a very simple manner.

 **Principle of Virtual Work**

The principle of virtual work, also known as the principle of least action, states that the work done by the external and inertial forces along the virtual displacement is equal to zero.

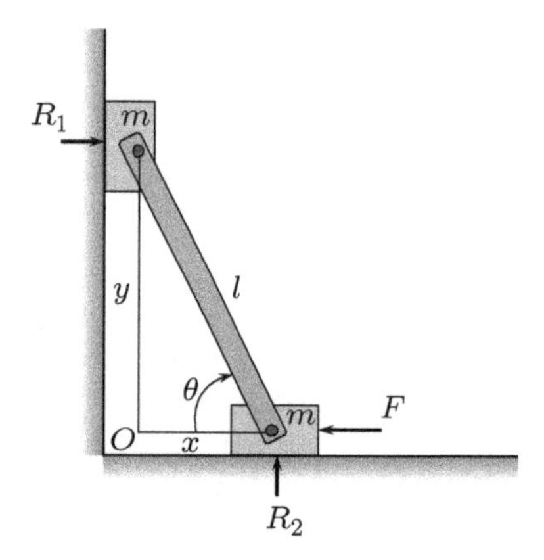


**Figure 4.5** Analyzing the static equilibrium of a system using the principle of virtual work.

### Example 4.4 Static Equilibrium

For the system shown in Figure 4.5, find the force $F$ necessary to achieve static equilibrium. Assume that all forces are smooth and that the connecting rod is massless.

First, we solve this problem using Newton's second law. There are four different forces acting on the system; these are the reaction forces $R_1\hat{n}_1$ and $R_2\,\hat{n}_2$, the weight $-2mg\,\hat{n}_2$, and the external force $-F\,\hat{n}_1$. Summing forces in the $\hat{n}_1$ and $\hat{n}_2$ directions, and moments about point $O$, we obtain, respectively:

$$R_1 - F = 0, \qquad R_2 - 2mg = 0, \qquad -R_1 l \sin\theta + R_2 l \cos\theta - mgl\cos\theta = 0.$$

Upon solving these equations, we obtain $F = mg\cot\theta$.

Next, we solve the problem using the principle of virtual work. We note that the constraint forces does not do any work along the virtual displacement and that the sum of the virtual work done by weight and the external force along the virtual displacements is zero; that is

$$mg\delta y - F\delta x = 0. \tag{4.29}$$

Noting that $y = l\sin\theta$, $x = -l\cos\theta$ or $\delta y = l\cos\theta\delta\theta$, and $\delta x = l\sin\theta\delta\theta$. This yields

$$F = mg\cot\theta. \tag{4.30}$$

### Flipped Classroom Exercise 4.2

Consider the system shown in the figure. Find the angle $\theta$ at equilibrium.

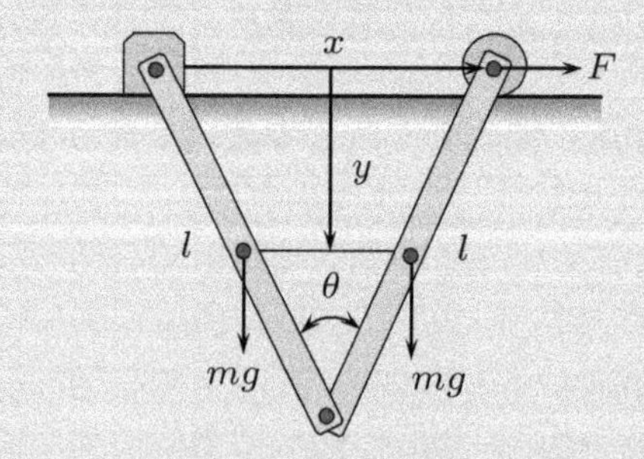

To solve this problem, use the principle of virtual work. To this end, carry out the following steps:

1. Identify the external forces acting on the system.
2. Express the virtual work done by these forces along the virtual displacement.
3. Write the constraint equations and express them in terms of the virtual angle $\delta\theta$.
4. Show that the equilibrium angle is

$$\theta = 2\arctan\frac{2F}{mg}.$$

## Exercises

4.1 A particle is constrained to move in a plane such that its orientation is always proportional to the time elapsed since the beginning of its motion. Describe and classify the constraints acting on the particle.

4.2 A particle is constrained to move such that the magnitude of its velocity is proportional to the total time elapsed since the beginning of its motion. Describe and classify the constraints acting on the particle.

4.3 A particle moves under the influence of gravitational acceleration such that the square of its velocity is constant at all times. Describe and classify the constraints acting on the particle.

4.4 Classify the following constraints:

(a) $(y^2 - 2x)\dot{x} + (2xy - 1)\dot{y} = 0$
(b) $x(1 - \sin y)\dot{y} - (\cos x - \cos y - y)\dot{x} = 0$
(c) $(3x^2y - 1)\dot{x} + (x^3 + 6y - y^2)\dot{y} = \sin t$.

4.5 A fox tries to hunt a rabbit, as shown in Figure 4.6. Describe the initial position of the fox as $(x_0, y_0)$ and assume that the rabbit runs in the $x$-direction with a constant velocity $v_R$. On the other hand, the fox runs in such a way that its instantaneous velocity is always in the direction of the vector between the rabbit and itself. Show that the path followed by the fox satisfies the constraint

$$\dot{y}_F(V_Rt - x_F) + y_F\dot{x}_F = 0,$$

where $(\dot{x}_F, \dot{y}_F)$ are the components of the fox velocity in the $x$- and $y$-directions, respectively. What type of constraint is this?

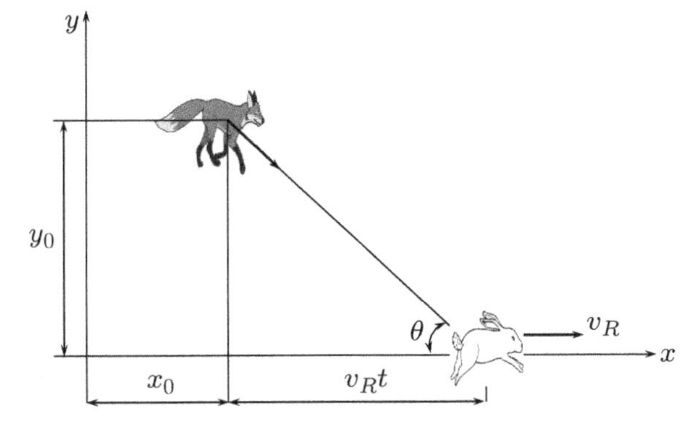


**Figure 4.6** Exercise 4.5.

4.6 A force $F$ is applied on the sliding collar shown in Figure 4.7 when link $OQ$ is in the horizontal position. Find the magnitude of $F$ necessary to keep link $QP$ from rotating under the influence of the moment $M_O$. Assume all sliding surfaces are smooth.

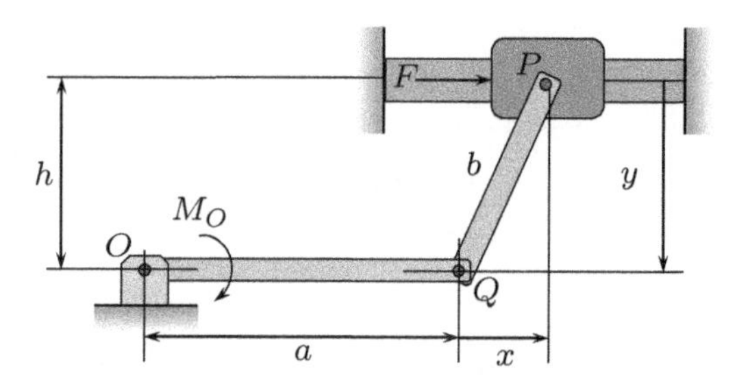


**Figure 4.7** Exercise 4.6.

## Reference

1. Daqaq M.F. and Masoud Z.N. (2006) "Nonlinear input-shaping controller for quay-side container cranes", *Nonlinear Dynamics,* 45, 149.

# 5

# Dynamics of Particles: Analytical Approach

In this chapter, the analytical approach to dynamics is presented. In particular, Lagrange's equations for conservative and non-conservative systems and Hamilton's principle are derived and applied to the modeling of the motion of particles.

## 5.1 The Brachistochrone Problem

The Newtonian vectorial approach we have used so far to analyze the dynamics of particles and rigid bodies is based on summing forces and moments (vectorial quantities) to obtain the equations of motion. In contrast, the *analytical approach* is based on using work and energy (scalar quantities) to analyze motion. As will be shown through this and subsequent chapters, this approach is very useful when analyzing complex dynamical systems, especially when we are only interested in the equation of motion and not the forces causing it.

The development of the analytical approach to dynamics is based on the calculus of variations, a branch of mathematics dealing with finding the extrema of functionals (integrals of a function). It can be argued that calculus of variations started with the *brachistochrone* or *shortest-time* problem,[1] which considers the motion of a bead of mass $m$ sliding on a frictionless wire under the influence of gravity. The objective is to find the shape of the path that minimizes the time required for the bead to go between two arbitrary points $A$ and $B$.

The brachistochrone problem was first solved by Johann Bernoulli, who then posted it as a challenge to European mathematicians in the 1696 edition of the *Acta Eruditorum* [1]. As shown in Figure 5.1, it aims at finding the path $y(x)$ that permits a bead of mass, $m$, to slide on a frictionless wire between points $A$ and $B$ in the shortest possible time. In this section,

[1] In the Greek language *brachis* stands for shortest and *chronos* stands for time.

*Dynamics of Particles and Rigid Bodies: A Self-Learning Approach*, First Edition. Mohammed F. Daqaq.

Companion website: www.wiley.com/go/daqaq/dynamics

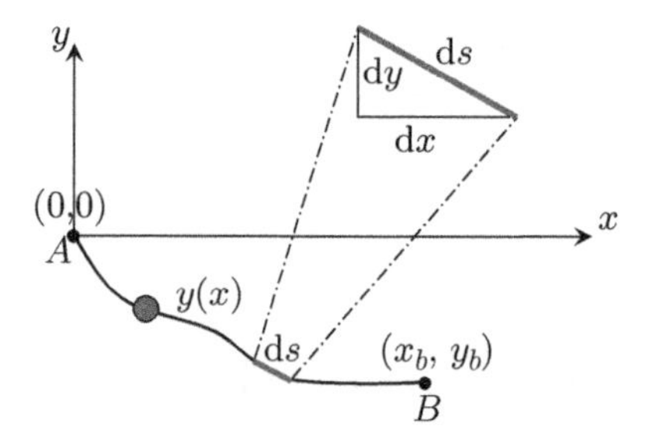


**Figure 5.1** The brachistochrone problem.

we tackle the brachistochrone problem and show how it relates to other general problems in dynamics.

To this end, we first relate the function $y(x)$ to the velocity of the particle $v$ using Newton's second law, which yields $v = \sqrt{2gy}$. As such, the differential time, $\mathrm{d}t$ that the particle takes to move a differential distance $\mathrm{d}s$ along the unknown path can be written as

$$\mathrm{d}t = \frac{\mathrm{d}s}{v}, \tag{5.1}$$

Note that $\mathrm{d}s = \sqrt{(\mathrm{d}x)^2 + (\mathrm{d}y)^2} = \sqrt{1 + \left(\frac{\mathrm{d}y}{\mathrm{d}x}\right)^2}\,\mathrm{d}x$. It follows from Equation (5.1) that

$$\mathrm{d}t = \sqrt{\frac{1 + \left(\frac{\mathrm{d}y}{\mathrm{d}x}\right)^2}{2gy}}\mathrm{d}x. \tag{5.2}$$

The total time, $T$, required to go from point $A$, which is assumed to be at $(0, 0)$, to point $B$, which is located at an arbitrary point $(x_b, y_b)$, can hence be written as

$$T = \int_0^{x_b} \sqrt{\frac{1 + \left(\frac{\mathrm{d}y}{\mathrm{d}x}\right)^2}{2gy}}\mathrm{d}x. \tag{5.3}$$

The goal of the brachistochrone problem is therefore to find the function $y(x)$ that minimizes the functional $T$. Nevertheless, before delving into the solution of this specific problem, we treat the general case of finding the function $y(x)$ that corresponds to an extremum of the general functional

$$J = \int_{x_A}^{x_B} L(x, y, y')\mathrm{d}x. \tag{5.4}$$

where the prime is a derivative with respect to $x$. To this end, suppose that we have a function $Y(x)$ that is slightly varied from the function $y(x)$ we are seeking. Both functions have the same start and end point, $y(x_A) = Y(x_A)$ and $y(x_B) = Y(x_B)$, as shown in Figure 5.2. For any other point between $x_A$ and $x_B$, the two functions are related via

$$Y(x) = y(x) + h(x), \tag{5.5}$$

where $h(x)$ is a small variation around $y(x)$ such that $h(x_A) = h(x_B) = 0$. The functional $J$ exhibits an extremum when its variation with respect to $h(x)$ vanishes; that is

$$\begin{aligned}\Delta J = J(y(x) + h(x)) - J(y(x)) &= \int_{x_A}^{x_B} L(x, y + h, y' + h')\mathrm{d}x - \int_{x_A}^{x_B} L(x, y, y')\mathrm{d}x \\ &= \int_{x_A}^{x_B} [L(x, y + h, y' + h') - L(x, y, y')]\mathrm{d}x = 0,\end{aligned} \tag{5.6}$$

where the prime is a derivative with respect to $x$. Expanding the previous equation in a Taylor series and keeping only linear terms, yields

$$\Delta J = \int_{x_A}^{x_B} \left[\frac{\partial L(x, y, y')}{\partial y}h - \frac{\partial L(x, y, y')}{\partial y'}h'\right] \mathrm{d}x = 0. \tag{5.7}$$

The second term on the right-hand side of the previous equation can be integrated by parts, resulting in the following simplified version:

$$\int_{x_A}^{x_B} \left(\frac{\partial L}{\partial y} - \frac{\mathrm{d}}{\mathrm{d}x}\left(\frac{\partial L}{\partial y'}\right)\right) h(x)\mathrm{d}x = 0. \tag{5.8}$$

Since $h(x)$ is an arbitrary non-zero variation, the only plausible way to force the previous integral to zero is by letting

$$\boxed{\left(\frac{\partial L}{\partial y} - \frac{\mathrm{d}}{\mathrm{d}x}\left(\frac{\partial L}{\partial y'}\right)\right) = 0} \tag{5.9}$$

Equation (5.9), known as the *Euler–Lagrange equation*, can be used to solve for $y(x)$, which renders the integral of Equation (5.4) an extremum.

Next, we reconsider the brachistochrone problem for which, as given by Equation (5.3), the function $L$ takes the form

$$L = \sqrt{\frac{1 + y'^2}{2gy}}. \tag{5.10}$$

The next step would be to take the previous equation and plug it into the Euler–Lagrange equation (5.9), and solve for $y(x)$. To this end, we first note that $L$ in this problem does not depend explicitly on $x$. This allows us to take a shortcut, where we let

$$\left(\frac{\partial L}{\partial y} - \frac{\mathrm{d}}{\mathrm{d}x}\left(\frac{\partial L}{\partial y'}\right)\right) y' = 0, \qquad \frac{\partial L}{\partial y}y' - y'\frac{\mathrm{d}}{\mathrm{d}x}\left(\frac{\partial L}{\partial y'}\right) = 0. \tag{5.11}$$

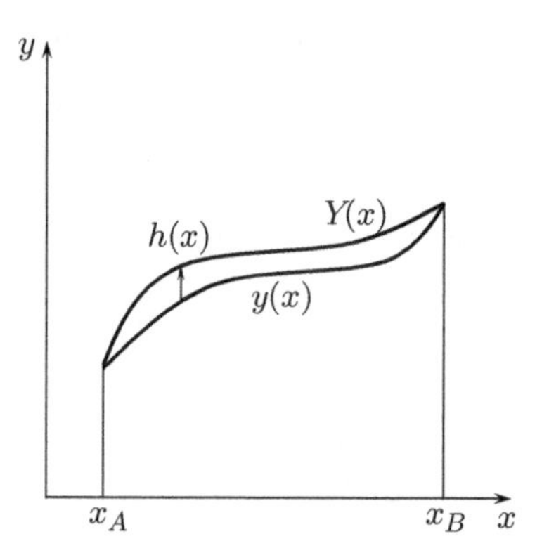


**Figure 5.2** Actual and virtual paths.

Using the fact that

$$\frac{\mathrm{d}L}{\mathrm{d}x} = \frac{\partial L}{\partial y}y' + y''\frac{\partial L}{\partial y'}, \tag{5.12}$$

it follows that

$$\frac{\mathrm{d}L}{\mathrm{d}x} - y''\frac{\partial L}{\partial y'} - y'\frac{\mathrm{d}}{\mathrm{d}x}\left(\frac{\partial L}{\partial y'}\right) = 0, \qquad \frac{\mathrm{d}}{\mathrm{d}x}\left(L - y'\frac{\partial L}{\partial y'}\right) = 0. \tag{5.13}$$

Thus,

$$L - y'\frac{\partial L}{\partial y'} = C. \tag{5.14}$$

Substituting Equation (5.10) in Equation (5.14) yields

$$\sqrt{\frac{1+y'^2}{y}} - \frac{y'^2}{\sqrt{y(1+y'^2)}} = C. \tag{5.15}$$

Squaring the previous equation and simplifying leads to

$$y(1+y'^2) = \frac{1}{C^2} = C_0. \tag{5.16}$$

Solving for $x$, yields

$$x = \int \sqrt{\frac{y}{C_0 - y}}\mathrm{d}y. \tag{5.17}$$

To carry out the previous integration, we substitute $y = C_0 \sin^2\theta$ into Equation (5.17), which yields $x = C_0(\theta - \frac{1}{2}\sin 2\theta) + C_1$. The constant $C_1$ can be obtained by letting $\theta = 0$ when $x = 0$, which yields $C_1 = 0$.

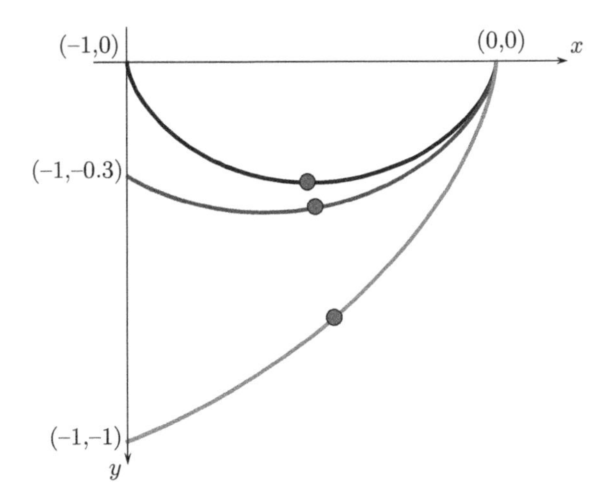


**Figure 5.3** Family of cycloidal curves resulting from the solution of the brachistochrone problem.

The constant $C_0$ is obtained by enforcing the functions $x$ and $y$ to satisfy the end conditions $(x_b, y_b)$. This yields:

$$x_b = C_0 \left( \theta_b - \frac{1}{2} \sin 2\theta_b \right),$$
$$y_b = C_0 \sin^2 \theta_b. \tag{5.18}$$

For known values of $x_b$ and $y_b$, the previous set of equations can be solved for $C_0$ and $\theta_b$. This can then be used to obtain the following parametric curve, which minimizes the travel time of the bead:

$$x = C_0 \left( \theta - \frac{1}{2} \sin 2\theta \right),$$
$$y = C_0 \sin^2 \theta \qquad \theta \in [0, \theta_b]. \tag{5.19}$$

This set of parametric equations results in a curve known as a *cycloid*. Figure 5.3 shows a family of cycloids for a single starting condition $(0, 0)$ and different end conditions $(x_b, y_b)$.

**Example 5.1 Shortest Path between Two Points**

Show that the shortest path between two points is a line.

Our goal is to prove that the function $y(x)$ that results in the shortest distance between the two points $(0, 0)$ and $(x_b, y_b)$ is a linear function. To this end, we take a differential element $\mathrm{d}s = \sqrt{\mathrm{d}x^2 + \mathrm{d}y^2} = \sqrt{1 + y'^2}\mathrm{d}x$ on the curve and use it to define the distance between the two points as

$$d = \int_0^{x_b} \sqrt{1 + y'^2} \mathrm{d}x.$$

This implies that $L = \sqrt{1+y'^2}$, which, upon substituting into the Euler–Lagrange equation, yields

$$\frac{\mathrm{d}}{\mathrm{d}x}\left(\frac{\partial L}{\partial y'}\right) = 0, \qquad \frac{y'}{\sqrt{1+y'^2}} = C_0,$$

or

$$y(x) = \sqrt{\frac{C^2}{1-C^2}}x + B.$$

Letting $A = \sqrt{\frac{C^2}{1-C^2}}$ and using $y(0) = 0, y(x_b) = y_b$ yields

$$y = \frac{y_b}{x_b}x,$$

which is clearly a linear function of $x$.

 **Flipped Classroom Exercise 5.1**

Determine the function $y(x)$ which, when rotated about the $x$-axis as shown in the figure, yields the surface of revolution with the smallest surface area.

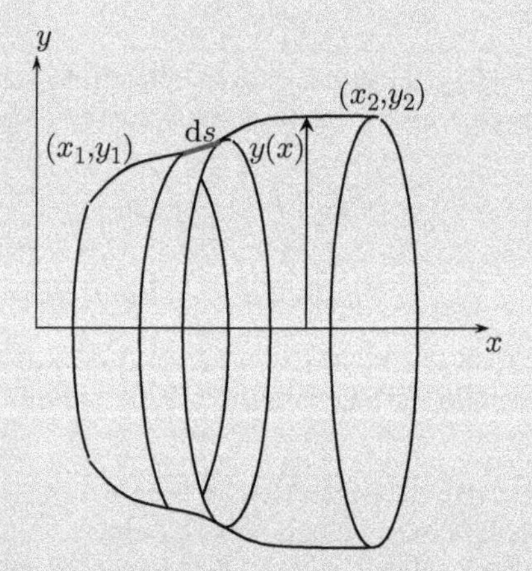


To find the desired function $y(x)$ carry out the following steps:

1. Start by defining a differential surface area $\mathrm{d}A = 2\pi y(x)\mathrm{d}s$ where $\mathrm{d}s$ is the length of a differential segment along the function $y(x)$.
2. Show that the total area of the surface of revolution can be written as

$$A = 2\pi \int_{x_1}^{x_2} y\sqrt{1+y'^2}\mathrm{d}x.$$

3. Using the Euler–Lagrange equation, show that the function $y(x)$ that minimizes the area of revolution is governed by the following differential equation:

$$yy'' - y'^2 - 1 = 0.$$

4. To solve the previous equation, let $g(x) = y'$ and show that

$$y(x) = a\cosh\frac{x+b}{a},$$

where $a$ and $b$ are determined from the boundary conditions $y(x_1) = y_1, y(x_2) = y_2$. The following figure depicts a family of curves that minimizes the area of revolution for different end conditions.

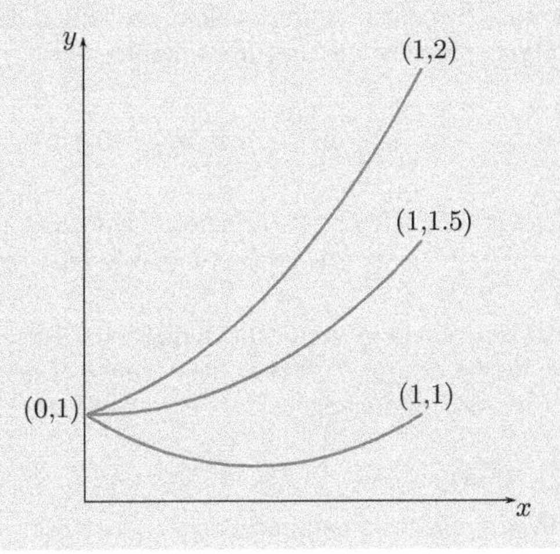


## 5.2 Lagrange's Equation for a Conservative System

The relationship between the Euler–Lagrange equation, which is used to find the extrema of functionals, and dynamics lies in the principle of least action. The concept behind this principle started with the work of Maupertuis, who said that "nature is thrifty in all its actions" [2]. Following this concept, Maupertuis attempted to obtain the equation of motion of a simple pendulum by minimizing its kinetic energy. However, he quickly realized that the minimization of the kinetic energy of the pendulum would not yield the correct dynamical configuration of the system.

Despite Maupertuis' failure to reproduce the equation governing the motion of the pendulum by minimizing its kinetic energy, it turned out the nature does indeed minimize a function to determine the dynamic configuration of any system. In order to determine which quantity nature minimizes to define the actual configuration of dynamical systems, we start by applying Newton's second law of dynamics to $\mathcal{N}$ particles as follows:

$$\sum_{i=1}^{\mathcal{N}}(\mathbf{F}_i + \mathbf{f}_i) = \sum_{i=1}^{\mathcal{N}} m_i\ddot{\mathbf{r}}_i, \tag{5.20}$$

where $m_i$ denotes the mass of the $i$th particle, $\mathbf{F}_i$ and $\mathbf{f}_i$ are the external and internal constraint forces acting on the system, respectively, and $\mathbf{r}_i = \mathbf{r}_i(q_1, q_2, \dots, q_{\mathcal{M}})$ represents the position vector of the $i$th particle, represented in terms of $\mathcal{M}$ generalized coordinates $q$.

D'Alembert noted that inertial forces can be treated similarly to external forces, and that therefore one could write

$$\sum_{i=1}^{\mathcal{N}} (\mathbf{F}_i + \mathbf{f}_i - m_i \ddot{\mathbf{r}}_i) = 0. \tag{5.21}$$

Multiplying Equation (5.21) by the virtual displacement $\delta \mathbf{r}_i$, we obtain

$$\sum_{i=1}^{\mathcal{N}} (\mathbf{F}_i + \mathbf{f}_i - m_i \ddot{\mathbf{r}}_i) \delta \mathbf{r}_i = 0. \tag{5.22}$$

Noting that the work done by constraint forces along the virtual displacement is zero, regardless of the constraint type, the previous equation simplifies to

$$\sum_{i=1}^{\mathcal{N}} (\mathbf{F}_i - m_i \ddot{\mathbf{r}}_i) \delta \mathbf{r}_i = 0. \tag{5.23}$$

Equation (5.23) is known as d'Alembert's principle . However, it was first described in its variational form by Lagrange . As such, it is sometimes referred to as the *d'Alembert–Lagrange principle* [3].

Considering a conservative and irrotational external force vector field $\mathbf{F}_i$, we can describe the virtual work done by the external forces as the negative of the variation of the associated potential energy $\mathcal{U}$ of the system; that is

$$\mathbf{F}_i . \delta \mathbf{r}_i = -\delta \mathcal{U}. \tag{5.24}$$

Furthermore, the term $m_i \ddot{\mathbf{r}}_i$ can be simplified using

$$\begin{aligned} \frac{\mathrm{d}}{\mathrm{d}t}(\dot{\mathbf{r}}_i \delta \mathbf{r}_i) &= \ddot{\mathbf{r}}_i \delta \mathbf{r}_i + \dot{\mathbf{r}}_i \delta \dot{\mathbf{r}}_i, \\ &= \ddot{\mathbf{r}}_i \delta \mathbf{r}_i + \delta \left( \frac{1}{2} \dot{\mathbf{r}}_i \dot{\mathbf{r}}_i \right). \end{aligned} \tag{5.25}$$

Hence, we can write

$$\ddot{\mathbf{r}}_i \delta \mathbf{r}_i = \frac{\mathrm{d}}{\mathrm{d}t}(\dot{\mathbf{r}}_i \delta \mathbf{r}_i) - \delta \left( \frac{1}{2} \dot{\mathbf{r}}_i \dot{\mathbf{r}}_i \right). \tag{5.26}$$

Substituting Equation (5.26) back into Equation (5.24), yields

$$-\delta \mathcal{U} + \sum_{i=1}^{\mathcal{N}} \left( -m_i \frac{\mathrm{d}}{\mathrm{d}t}(\dot{\mathbf{r}}_i \delta \mathbf{r}_i) \delta \mathbf{r}_i + \delta \left( \frac{1}{2} m_i \dot{\mathbf{r}}_i \dot{\mathbf{r}}_i \right) \right) = 0. \tag{5.27}$$

Note that $\frac{1}{2} m_i \dot{\mathbf{r}}_i \dot{\mathbf{r}}_i$ is the kinetic energy $\mathcal{T}$ of the $i$th particle. It follows that

$$\delta \mathcal{T} - \delta \mathcal{U} = \sum_{i=1}^{\mathcal{N}} \left( m_i \frac{\mathrm{d}}{\mathrm{d}t}(\dot{\mathbf{r}}_i \delta \mathbf{r}_i) \right). \tag{5.28}$$

Integrating Equation (5.28) over an arbitrary time period $(t_1, t_2)$ and forcing the virtual displacement to vanish at the boundaries – i.e. $\delta\mathbf{r}_i(t_1) = \delta\mathbf{r}_i(t_2) = 0$ – yields the following equation:

$$\int_{t_1}^{t_2} (\delta\mathcal{T} - \delta\mathcal{U})\mathrm{d}t = 0 \tag{5.29}$$

Equation (5.29) is referred to as the *Hamilton variational principle* for conservative systems.

The kinetic energy of the system is a function of the generalized coordinates and velocities; that is, $T = \mathcal{T}(q_1, q_2, \dots, q_\mathcal{M}, \dot{q}_1, \dot{q}_2, \dots, \dot{q}_\mathcal{M})$. The potential energy, on the other hand, is only a function of the generalized coordinates; that is, $\mathcal{U} = \mathcal{U}(q_1, q_2, \dots, q_\mathcal{M})$. Taking the variation of $\mathcal{T}$ and $U$, we can write

$$\begin{aligned}\delta\mathcal{T}(q_1, q_2, \dots, q_\mathcal{M}, \dot{q}_1, \dot{q}_2, \dots, \dot{q}_\mathcal{M}) &= \frac{\partial\mathcal{T}}{\partial q_1}\delta q_1 + \frac{\partial\mathcal{T}}{\partial q_2}\delta q_2 + \dots + \frac{\partial\mathcal{T}}{\partial q_\mathcal{M}}\delta q_\mathcal{M} \\ &\quad + \frac{\partial\mathcal{T}}{\partial \dot{q}_1}\delta \dot{q}_1 + \frac{\partial\mathcal{T}}{\partial \dot{q}_2}\delta \dot{q}_2 + \dots + \frac{\partial\mathcal{T}}{\partial \dot{q}_\mathcal{M}}\delta \dot{q}_\mathcal{M} \\ &= \sum_{j=1}^{\mathcal{M}} \left( \frac{\partial\mathcal{T}}{\partial q_j}\delta q_j + \frac{\partial\mathcal{T}}{\partial \dot{q}_j}\delta \dot{q}_j \right),\end{aligned} \tag{5.30}$$

$$\delta\mathcal{U}(q_1, q_2, \dots, q_\mathcal{M}) = \frac{\partial\mathcal{U}}{\partial q_1}\delta q_1 + \frac{\partial\mathcal{U}}{\partial q_2}\delta q_2 + \dots + \frac{\partial\mathcal{U}}{\partial q_\mathcal{M}}\delta q_\mathcal{M} = \sum_{j=1}^{\mathcal{M}} \frac{\partial\mathcal{U}}{\partial q_j}\delta q_j, \tag{5.31}$$

It follows from Equation (5.29) that

$$\int_{t_1}^{t_2} \sum_{j=1}^{\mathcal{M}} \left( \frac{\partial\mathcal{T}}{\partial q_j}\delta q_j + \frac{\partial\mathcal{T}}{\partial \dot{q}_j}\delta \dot{q}_j - \frac{\partial\mathcal{U}}{\partial q_j}\delta q_j \right) \mathrm{d}t = 0. \tag{5.32}$$

The middle term in the previous equation can be integrated by parts to obtain

$$\int_{t_1}^{t_2} \frac{\partial\mathcal{T}}{\partial \dot{q}_j}\delta \dot{q}_j \mathrm{d}t = \left.\frac{\partial\mathcal{T}}{\partial q_j}\delta q_j\right|_{t_1}^{t_2} - \int_{t_1}^{t_2} \frac{\mathrm{d}}{\mathrm{d}t}\left( \frac{\partial\mathcal{T}}{\partial \dot{q}_j} \right) \delta q_j \mathrm{d}t. \tag{5.33}$$

Forcing $\delta q_j(t_1) = \delta q_j(t_2) = 0$, yields

$$\int_{t_1}^{t_2} \frac{\partial\mathcal{T}}{\partial \dot{q}_j}\delta \dot{q}_j \mathrm{d}t = -\int_{t_1}^{t_2} \frac{\mathrm{d}}{\mathrm{d}t}\left( \frac{\partial\mathcal{T}}{\partial \dot{q}_j} \right) \delta q_j \mathrm{d}t. \tag{5.34}$$

It follows then from Equation (5.32) that

$$\int_{t_1}^{t_2} \sum_{j=1}^{\mathcal{M}} \left( \frac{\partial\mathcal{T}}{\partial q_j} - \frac{\mathrm{d}}{\mathrm{d}t}\left( \frac{\partial\mathcal{T}}{\partial \dot{q}_j} \right) - \frac{\partial\mathcal{U}}{\partial q_j} \right) \delta q_j \mathrm{d}t = 0. \tag{5.35}$$

Since $\delta q_j$ is an arbitrary non-zero function and $t_1$ and $t_2$ are also arbitrary, the only way to guarantee that the previous integral vanishes is by letting

$$\left( \frac{\partial\mathcal{T}}{\partial q_j} - \frac{\mathrm{d}}{\mathrm{d}t}\left( \frac{\partial\mathcal{T}}{\partial \dot{q}_j} \right) - \frac{\partial\mathcal{U}}{\partial q_j} \right) = 0, \qquad j = 1, 2, 3, \dots, \mathcal{M}. \tag{5.36}$$

Letting $\mathcal{L} = \mathcal{T} - \mathcal{V}$ in the previous equation leads to

$$\frac{\mathrm{d}}{\mathrm{d}t}\left(\frac{\partial \mathcal{L}}{\partial \dot{q}_j}\right) - \frac{\partial \mathcal{L}}{\partial q_j} = 0, \qquad j = 1, 2, 3, \ldots, \mathcal{M}. \tag{5.37}$$

When comparing Equation (5.9) to Equation (5.37), it becomes evident that they are the same, with $L(y, y', x)$ replacing $\mathcal{L}(q_j, \dot{q}_j, t)$. As such, it can be concluded that the path $\mathbf{r}(q_1, q_2, \ldots, q_{\mathcal{M}})$ that describes the dynamic configuration of a particle in space is the one that minimizes the integral of the function $\mathcal{L}$ on any arbitrary time interval, $[t_1, t_2]$. The function $\mathcal{L}$ is known as the *Lagrangian*, and Equation (5.37) is the widely-celebrated Lagrange equation for conservative systems described in terms of the generalized coordinates, $q$.

**The Lagrangian**

The path that describes the configuration of a number of particles in space is the one which minimizes the integral of the Lagrangian $\mathcal{L} = \mathcal{T} - \mathcal{U}$ on any arbitrary time interval, $[t_1, t_2]$.

**Procedure for Applying Lagrange's Equation to Conservative Systems**

When applying Lagrange's equation to obtain the equations of motion of a conservative system of particles, it might be useful to follow this procedure:

1. Define the type and number of generalized coordinates, $q_j$, you want to use to describe the configuration of the particles in space.
2. Express the velocity, ${}^{N}\mathbf{v}^{P_i/O}$, of each particle, $P_i$ with respect to an inertial point $O$ in terms of the generalized coordinates and their derivatives.
3. Find the kinetic energy of the particles using

$$\mathcal{T} = \sum_{i=1}^{\mathcal{N}} \frac{1}{2} m_i \, {}^{N}\mathbf{v}^{P_i/O} \cdot {}^{N}\mathbf{v}^{P_i/O}. \tag{5.38}$$

4. Choose a reference potential energy level (datum) and find the sum of the potential energy of all particles:

$$\mathcal{U} = \sum_{i=1}^{\mathcal{N}} \mathcal{U}_i. \tag{5.39}$$

5. Determine the Lagrangian $\mathcal{L} = \mathcal{T} - \mathcal{U}$, then apply Lagrange's equation for each generalized coordinate $q_j$.

$$\frac{\mathrm{d}}{\mathrm{d}t}\left(\frac{\partial \mathcal{L}}{\partial \dot{q}_j}\right) - \frac{\partial \mathcal{L}}{\partial q_j} = 0, \qquad j = 1, 2, 3, \ldots, \mathcal{M} \tag{5.40}$$

**Example 5.2 Motion of a Double Pendulum**

Consider the double pendulum system shown in Figure 5.4, which consists of two particles each of mass $m$ and length $l$. Find the equations of motion of the system considering planar motion only.

To find the equations of motion, we follow the procedure outlined above.

1. We use $\theta$ and $\phi$ as the generalized coordinates to describe the configuration of the particles.
2. To find the velocity of the particles, we define angular velocity vectors ${}^N\boldsymbol{\omega}^B = \dot{\theta}\hat{b}_3$ and ${}^N\boldsymbol{\omega}^C = \dot{\phi}\hat{c}_3$ where the $B$- and $C$-frames are body-rotating frames. The position vector and velocity of the first particle can therefore be written as

$$\mathbf{OP}_1 = l\hat{b}_1, \qquad {}^N\mathbf{v}^{P_1/O} = l\dot{\theta}\hat{b}_2.$$

The position vector and velocity of the second particle can be written as

$$\mathbf{OP}_2 = l\hat{b}_1 + l\hat{c}_1, \qquad {}^N\mathbf{v}^{P_2/O} = \dot{\theta}\hat{b}_3 \times l\hat{b}_1 + \dot{\phi}\hat{c}_3 \times l\hat{c}_1 = l\dot{\theta}\hat{b}_2 + l\dot{\phi}\hat{c}_2.$$

3. The kinetic energy can then be written as

$$\mathcal{T} = \frac{1}{2}m{}^N\mathbf{v}^{P_1/O}.{}^N\mathbf{v}^{P_1/O} + \frac{1}{2}m{}^N\mathbf{v}^{P_2/O}.{}^N\mathbf{v}^{P_2/O},$$

$$\mathcal{T} = \frac{1}{2}ml^2\dot{\theta}^2 + \frac{1}{2}m(l\dot{\theta}\hat{b}_2 + l\dot{\phi}\hat{c}_2).(l\dot{\theta}\hat{b}_2 + l\dot{\phi}\hat{c}_2).$$

Note that part of the ${}^N\mathbf{v}^{P_2/O}$ is described in the $B$-frame, while the rest is described in the $C$-frame. To simplify the process of finding the dot product, we describe $\hat{c}_2$ in terms of the $B$ coordinates using $\hat{c}_2 = -\sin(\phi - \theta)\hat{b}_1 + \cos(\phi - \theta)\hat{b}_2$. This yields

$$\mathcal{T} = \frac{1}{2}m\,(2l^2\dot{\theta}^2 + l^2\dot{\phi}^2 + 2l^2\dot{\theta}\dot{\phi}\cos(\phi - \theta)).$$

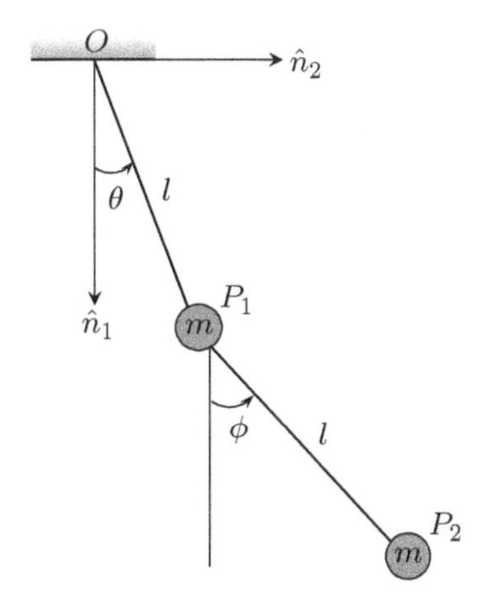


**Figure 5.4** Motion of a double pendulum.

4. We choose our datum to be a horizontal line passing through point $O$ and write the potential energy as

$$\mathcal{U} = -mgl(2\cos\theta + \cos\phi).$$

5. Using the Lagrangian $\mathcal{L} = \mathcal{T} - \mathcal{U}$ and applying Lagrange's equation along the coordinates $\theta$ and $\phi$ we obtain

$$2\ddot{\theta} + \ddot{\phi}\cos(\phi - \theta) - \dot{\phi}^2 \sin(\phi - \theta) + 2\frac{g}{l}\sin\theta = 0,$$

$$\ddot{\phi} + \ddot{\theta}\cos(\phi - \theta) - \dot{\theta}^2 \sin(\phi - \theta) + \frac{g}{l}\sin\phi = 0.$$

**Example 5.3 Inclined Pendulum**

Mass $M$ slides on an inclined stationary smooth surface, which makes an angle $\alpha$ with the horizontal. Attached to the sliding mass is a pendulum of mass $m$ and fixed length $l$. Find the equations of motion for the two masses shown in Figure 5.5.

Following the same procedure:

1. The motion of the system can be fully described by two generalized coordinates $x$ and $\theta$.
2. The position and velocity of mass $M$ with respect to the inertial frame can be written as

$$\mathbf{OM} = x\cos\alpha\hat{n}_1 - x\sin\alpha\hat{n}_2,$$

$$^{N}\mathbf{v}^{M/O} = \dot{x}\cos\alpha\hat{n}_1 - \dot{x}\sin\alpha\hat{n}_2.$$

The position and velocity of mass $m$ with respect to the inertial frame can be written as

$$\mathbf{Om} = (x\cos\alpha + l\sin\theta)\hat{n}_1 - (x\sin\alpha + l\cos\theta)\hat{n}_2,$$

$$^{N}\mathbf{v}^{m/O} = (\dot{x}\cos\alpha + l\dot{\theta}\cos\theta)\hat{n}_1 - (\dot{x}\sin\alpha - l\dot{\theta}\sin\theta)\hat{n}_2.$$

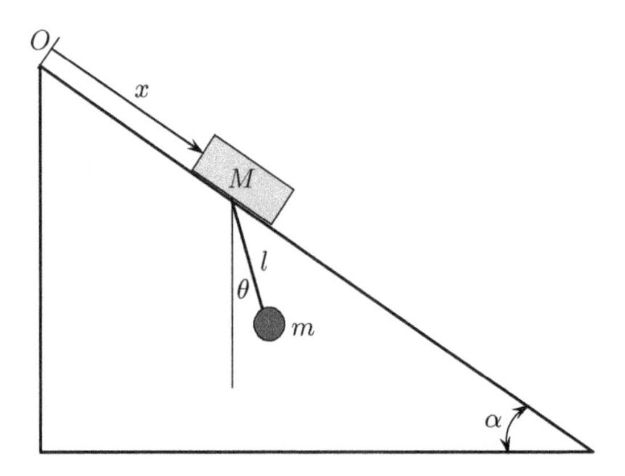


**Figure 5.5** Motion of an inclined pendulum.

3. The kinetic energy of the system can be written as

$$\mathcal{T} = \frac{1}{2}M^N\mathbf{v}^{M/O}.{}^N\mathbf{v}^{M/O} + \frac{1}{2}m^N\mathbf{v}^{m/O}.{}^N\mathbf{v}^{m/O}$$
$$= \frac{1}{2}M\dot{x}^2 + \frac{1}{2}m(\dot{x}^2 + l^2\dot{\theta}^2 + 2l\dot{x}\dot{\theta}\cos(\alpha+\theta)).$$

4. Choosing the top of the inclined surface as our datum, the potential energy of the system can be written as

$$\mathcal{U} = -Mgx\sin\alpha - mg(x\sin\alpha + l\cos\theta).$$

5. Using the Lagrangian $\mathcal{L} = \mathcal{T} - \mathcal{U}$ and applying Lagrange's equation along the $x$ and $\theta$ generalized coordinates, we obtain

$$\ddot{x} + \frac{m}{m+M}l(\ddot{\theta}\cos(\theta+\alpha) - \dot{\theta}^2\sin(\theta+\alpha)) - g\sin\alpha = 0,$$
$$\ddot{\theta} + \frac{\ddot{x}}{l}\cos(\alpha+\theta) - \frac{\dot{x}\dot{\theta}}{l}\sin(\alpha+\phi) + \frac{g}{l}\sin\theta = 0.$$

**Example 5.4 Rotating Hoop and a Pulley**

A mass $M$ is attached to a massless hoop, as shown in Figure 5.6. The mass $M$ is tied to a cable which winds around the hoop, then rises vertically up and over a massless pulley. A mass $m$ is attached to the other end of the cable. Find the equation of motion for the system in terms of the angle $\theta$.

1. The angle $\theta$ is the only generalized coordinate needed to describe the motion of the system.

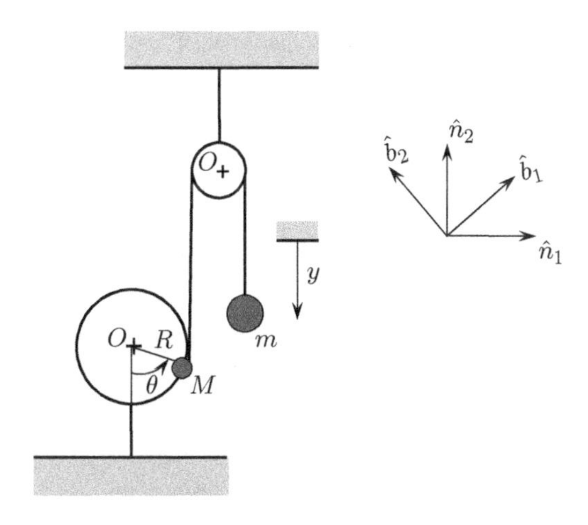


**Figure 5.6** Motion of a rotating hoop.

2. The velocity of $M$ and $m$ are, respectively,

$$^N\mathbf{v}^{M/O} = R\dot{\theta}\hat{b}_1,$$
$$^N\mathbf{v}^{m/O} = -\dot{y}\hat{n}_2 = -R\dot{\theta}\hat{n}_2.$$

3. The kinetic and potential energy are, respectively,

$$\mathcal{T} = \frac{1}{2}(M+m)R^2\dot{\theta}^2,$$
$$\mathcal{U} = MgR(1-\cos\theta) - mgR\theta.$$

4. Defining the Lagrangian $\mathcal{L}$ and applying Lagrange's equation using $q = \theta$, we obtain

$$(m+M)R\ddot{\theta} + g(M\sin\theta - m) = 0.$$

 **Flipped Classroom Exercise 5.2**

For the spherical pendulum with an elastic cable, find the equations of motion using Lagrange's approach.

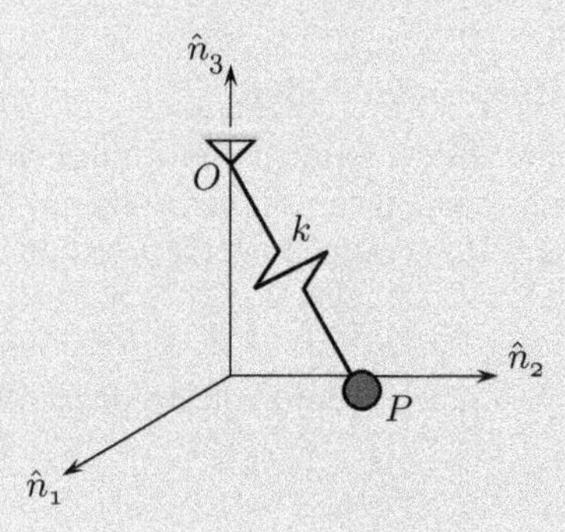


1. Define the rotating frames such that $^N\boldsymbol{\omega}^C = \dot{\phi}\hat{n}_3$ and $^C\boldsymbol{\omega}^B = -\dot{\theta}\hat{c}_2$.
2. Using $\phi$, $\theta$ and $r$ as your generalized coordinates, show that

$$^N\mathbf{v}^{P/O} = (l_0 + r)\dot{\theta}\hat{b}_1 + (l_0 + r)\dot{\phi}\sin\theta\hat{b}_2 - \dot{r}\hat{b}_3.$$

3. Write the potential and kinetic energy of the system and use the resulting Lagrangian to implement Lagrange's equations along the three generalized coordinates $q_1 = \theta$, $q_2 = \phi$, and $q_3 = r$. Show that

$$(l_0 + r)\ddot{\theta} + 2\dot{r}\dot{\theta} - (l_0 + r)\dot{\phi}\sin\theta\cos\theta + g\sin\theta = 0,$$
$$m(l_0 + r)^2\dot{\phi}\sin^2\theta = C_0,$$
$$\ddot{r} - (l_0 + r)\dot{\phi}^2\sin^2\theta + (l_0 + r)\dot{\theta}^2 + \frac{k}{m}r - g\cos\theta = 0,$$

where $C_0$ is a constant.

 **Flipped Classroom Exercise 5.3**

Starting with Newton's second law, show that, for a conservative system with scleronomic constraints, the sum of the kinetic and potential energy is always constant; in other words, $\mathcal{T} + \mathcal{U} = C$.

To solve this problem, carry the following procedure:

1. Consider $\mathcal{N}$ particles each of mass $m_i$, located at $\mathbf{r}_i$, and subject to a net external force $\mathbf{F}_i$ and a net internal force $\mathbf{f}_i$.
2. Apply Newton's second law to this system of particles.
3. Treat the inertial force as part of the applied forces by moving it to the left-hand side of the equation.
4. Multiply the resulting equation by $\mathrm{d}\mathbf{r}_i$, where $\mathrm{d}\mathbf{r}_i$ is a differential displacement in the direction of the actual displacement.
5. Eliminate the constraint forces, $\mathbf{f}_i$ by using the fact that $\mathbf{f}_i.\mathrm{d}\mathbf{r}_i = 0$ for a system subjected to scleronomic constraints.
6. Express the external force vector $\mathbf{F}_i$ in terms of the potential energy $\mathcal{U}$, knowing that for a conservative system, $\mathbf{F}_i.\mathrm{d}\mathbf{r}_i = -\mathrm{d}\mathcal{U}_i$.
7. Express the inertial forces in terms of the kinetic energy. This requires some algebraic manipulation.
8. Show that $\mathcal{T} + \mathcal{U} = C$.

## 5.3 Lagrange's Equation for Non-conservative Systems

For non-conservative forces that dissipate or channel energy from/to the system, $\mathbf{F}_i.\delta\mathbf{r}_i \neq -\delta\mathcal{U}_i$. In other words, the force is not equal to the negative of the gradient of the potential energy. To account for such forces in Lagrange's equation, we divide the forces into conservative $\mathbf{F}_i^*$, non-conservative $\mathbf{F}_i$, and constraint forces $\mathbf{f}_i$. We apply Newton's second law on a system of $\mathcal{N}$ particles and write

$$\sum_{i=1}^{\mathcal{N}}(\mathbf{F}_i^* + \mathbf{F}_i + \mathbf{f}_i) = \sum_{i=1}^{\mathcal{N}} m_i\ddot{\mathbf{r}}_i, \tag{5.41}$$

where $m_i$ denotes the mass of the $i$th particle, and $\mathbf{r}_i = \mathbf{r}_i(q_1, q_2, \dots, q_{\mathcal{M}})$ is the position vector of the $i$th particle described in terms of the generalized coordinates $q$.

Multiplying Equation (5.41) by the virtual displacement $\delta\mathbf{r}_i$, we obtain

$$\sum_{i=1}^{\mathcal{N}}(\mathbf{F}_i^* + \mathbf{F}_i + \mathbf{f}_i - m_i\ddot{\mathbf{r}}_i)\delta\mathbf{r}_i = 0. \tag{5.42}$$

Noting that the work done by constraint forces along the virtual displacement is zero, Equation (5.42) simplifies to

$$\sum_{i=1}^{\mathcal{N}}(\mathbf{F}_i^* + \mathbf{F}_i - m_i\ddot{\mathbf{r}}_i)\delta\mathbf{r}_i = 0. \tag{5.43}$$

For conservative forces, the force vector is the negative of the gradient of the potential energy. Therefore we can write $\mathbf{F}_i^*.\delta\mathbf{r}_i = -\delta\mathcal{U}_i$. For the non-conservative forces, we let $\mathbf{F}_i.\delta\mathbf{r}_i = \delta\bar{\mathcal{W}}_i$ where $\delta\bar{\mathcal{W}}_i$ is the virtual work done by the non-conservative forces on the $i$th particle such that $\delta\bar{\mathcal{W}} = \sum_{i=1}^{\mathcal{N}} \delta\bar{\mathcal{W}}_i$. Furthermore, as illustrated in Section 5.2, the term $\ddot{\mathbf{r}}_i.\delta\mathbf{r}_i$ can be rewritten as

$$\ddot{\mathbf{r}}_i\delta\mathbf{r}_i = \frac{\mathrm{d}}{\mathrm{d}t}(\dot{\mathbf{r}}_i\delta\mathbf{r}_i) - \delta\mathcal{T}.$$

Substituting this, together with the potential energy expression, and the virtual work terms back into Equation (5.43) yields

$$-\delta\mathcal{U} + \delta\bar{\mathcal{W}} + \delta\mathcal{T} - m_i\sum_{i=1}^{\mathcal{N}}\frac{\mathrm{d}}{\mathrm{d}t}(\dot{\mathbf{r}}_i\delta\mathbf{r}_i)\delta\mathbf{r}_i = 0, \tag{5.44}$$

which upon integrating over an arbitrary period of time $(t_1, t_2)$ and forcing the virtual displacement on the boundaries to vanish – $\delta\mathbf{r}_i(t_1) = \delta\mathbf{r}_i(t_2) = 0$ – yields

$$\int_{t_1}^{t_2}(\delta\mathcal{T} - \delta\mathcal{U} + \delta\bar{\mathcal{W}})\mathrm{d}t = 0. \tag{5.45}$$

Using $\mathcal{L} = \mathcal{T} - \mathcal{U}$, we obtain

$$\boxed{\int_{t_1}^{t_2}(\delta\mathcal{L} + \delta\bar{\mathcal{W}})\mathrm{d}t = 0} \tag{5.46}$$

Equation (5.46) is known as *Hamilton's extended principle* and can be directly utilized to obtain the equations of motion for any dynamical system. Nonetheless, because of complexities involving the integration over time, Equation (5.46) is rarely used in its current form to obtain the equations of motion for particles and rigid bodies. However, it is commonly used to derive the equations of motion for elastic systems.

To derive Lagrange's equations from Hamilton's extended principle, we let

$$\delta\mathbf{r}_i = \sum_{j=1}^{\mathcal{M}}\frac{\partial\mathbf{r}_i}{\partial q_j}\delta q_j, \tag{5.47}$$

where $\mathcal{M}$ is the number of generalized coordinates used to describe the motion. Using the previous equation, the virtual work done by non-conservative forces becomes

$$\delta\bar{\mathcal{W}} = \sum_{j=1}^{\mathcal{M}}\sum_{i=1}^{\mathcal{N}}\mathbf{F}_i.\frac{\partial\mathbf{r}_i}{\partial q_j}\delta q_j. \tag{5.48}$$

Denoting the term $\sum_{i=1}^{\mathcal{N}}\mathbf{F}_i.\frac{\partial\mathbf{r}_i}{\partial q_j}$ as $Q_j$, we obtain

$$\delta\bar{\mathcal{W}} = \sum_{j=1}^{\mathcal{M}}Q_j\delta q_j \tag{5.49}$$

where the $Q_j$ are known as the *generalized forces*. These quantities do not necessarily have units of force but can also have units of moment, depending on the generalized coordinate associated with them. For instance, if the generalized coordinate is the deflection of a point in space, then the generalized force has units of force. On the other hand, if the generalized coordinate is a rotation angle, then the associated generalized force has units of moment.

Substituting Equation (5.49) into Equation (5.46), we obtain

$$\int_{t_1}^{t_2}\left(\delta\mathcal{L}+\sum_{j=1}^{\mathcal{M}}Q_j\mathrm{d}q_j\right)\mathrm{d}t=0. \tag{5.50}$$

Since, $\mathcal{L}=\mathcal{L}(q_1,q_2,\ldots,q_{\mathcal{M}},\dot{q}_1,\dot{q}_2,\ldots,\dot{q}_{\mathcal{M}},t)$, we can write

$$\delta\mathcal{L}=\sum_{j=1}^{\mathcal{M}}\left(\frac{\partial\mathcal{L}}{\partial q_j}\delta q_j+\frac{\partial\mathcal{L}}{\partial\dot{q}_j}\delta\dot{q}_j\right). \tag{5.51}$$

Upon substituting Equation (5.51) into Equation (5.50), and integrating the term associated with $\delta\dot{q}_j$ once by parts, we obtain

$$\int_{t_1}^{t_2}\sum_{j=1}^{\mathcal{M}}\left(\frac{\mathrm{d}}{\mathrm{d}t}\left(\frac{\partial\mathcal{L}}{\partial\dot{q}_j}\right)-\frac{\partial\mathcal{L}}{\partial q_j}-Q_j\right)\delta q_j\mathrm{d}t=0. \tag{5.52}$$

Since the $\delta q_j$ are arbitrary non-zero virtual displacements, the only plausible way to force the integral in the previous equation to zero is by setting the coefficients of the $\delta q_j$ to zero. This yields

$$\frac{\mathrm{d}}{\mathrm{d}t}\left(\frac{\partial\mathcal{L}}{\partial\dot{q}_j}\right)-\frac{\partial\mathcal{L}}{\partial q_j}=Q_j,\qquad Q_j=\sum_{i=1}^{\mathcal{N}}\mathbf{F}_i\cdot\frac{\partial\mathbf{r}_i}{\partial q_j},\qquad j=1,2,3,\ldots,\mathcal{M} \tag{5.53}$$

Equation (5.53) is commonly referred to as Lagrange's equation for non-conservative systems and can be used directly to obtain the equations of motion of particles and rigid bodies involving non-conservative forces.

In many instances involving Equation (5.53), it can be convenient to express the generalized forces $Q_j$ in the following form:

$$Q_j=\sum_{i=1}^{\mathcal{N}}\mathbf{F}_i\cdot\frac{\partial\mathbf{r}_i}{\partial q_j}=\sum_{i=1}^{\mathcal{N}}\mathbf{F}_i\cdot\frac{\partial\dot{\mathbf{r}}_i}{\partial\dot{q}_j}. \tag{5.54}$$

Equation (5.54) is sometimes referred to as the *cancellation of dots identity* and can be proven as set out below.

*Proof:* Differentiate $\mathbf{r}_i=\mathbf{r}_i(q_1,q_2,\ldots,q_{\mathcal{M}},t)$ once with respect to time and obtain

$$\dot{\mathbf{r}}_i=\sum_{j=1}^{\mathcal{M}}\frac{\partial\mathbf{r}_i}{\partial q_j}\dot{q}_j+\frac{\partial\mathbf{r}_i}{\partial t}. \tag{5.55}$$

This implies that

$$\dot{\mathbf{r}}_i=\dot{\mathbf{r}}_i(q_1,q_2,\ldots,q_{\mathcal{M}},\dot{q}_1,\dot{q}_2,\ldots,\dot{q}_{\mathcal{M}},t). \tag{5.56}$$

Next, differentiate Equation (5.56) once with respect to time to obtain

$$\ddot{\mathbf{r}}_i = \sum_{j=1}^{\mathcal{M}} \left( \frac{\partial \dot{\mathbf{r}}_i}{\partial q_j} \dot{q}_j + \frac{\partial \dot{\mathbf{r}}_i}{\partial \dot{q}_j} \ddot{q}_j \right) + \frac{\partial \dot{\mathbf{r}}_i}{\partial t}. \tag{5.57}$$

Now, differentiate Equation (5.55) once with respect to time and obtain

$$\ddot{\mathbf{r}}_i = \sum_{j=1}^{\mathcal{M}} \left( \frac{\partial \dot{\mathbf{r}}_i}{\partial q_j} \dot{q}_j + \frac{\partial \mathbf{r}_i}{\partial q_j} \ddot{q}_j \right) + \frac{\partial \dot{\mathbf{r}}_i}{\partial t}. \tag{5.58}$$

Comparing Equation (5.57) to Equation (5.58), one can conclude that

$$\frac{\partial \mathbf{r}_i}{\partial q_j} \equiv \frac{\partial \dot{\mathbf{r}}_i}{\partial \dot{q}_j}. \tag{5.59}$$

■

## 5.3.1 Viscous Damping

One of the most common mechanisms used to model energy dissipation in dynamics is the linear viscous damping mechanism, as discussed earlier in Section 2.2. A linear viscous damping force acting on the $i$th particle is proportional to the velocity of the particle, and can be modeled as

$$\mathbf{F}_i = -c_i \dot{\mathbf{r}}_i, \tag{5.60}$$

where $c_i$ is the damping coefficient. Therefore, the generalized forces associated with the linear dissipative damping model are

$$Q_j = -\sum_{i=1}^{\mathcal{N}} c_i \dot{\mathbf{r}}_i \frac{\partial \mathbf{r}_i}{\partial q_j} = -\sum_{i=1}^{\mathcal{N}} c_i \dot{\mathbf{r}}_i \frac{\partial \dot{\mathbf{r}}_i}{\partial \dot{q}_j}, \qquad j = 1, 2, 3, \ldots, \mathcal{M}. \tag{5.61}$$

In order to directly include the generalized force associated with linear dissipative damping in Lagrange's equation, we define the following dissipative power function:

$$\mathcal{D} = \sum_{i=1}^{\mathcal{N}} \frac{1}{2} c_i \dot{\mathbf{r}}_i . \dot{\mathbf{r}}_i. \tag{5.62}$$

Differentiating the previous equation with respect to $\dot{q}_j$, we get

$$\frac{\partial \mathcal{D}}{\partial \dot{q}_j} = \sum_{i=1}^{\mathcal{N}} c_i \dot{\mathbf{r}}_i \frac{\partial \dot{\mathbf{r}}_i}{\partial \dot{q}_j}, \qquad j = 1, 2, 3, \ldots, \mathcal{M}. \tag{5.63}$$

This implies that $Q_j = \frac{\partial \mathcal{D}}{\partial \dot{q}_j}$, which permits rewriting Lagrange's equation in the following form to account for linear viscous damping directly:

$$\frac{\mathrm{d}}{\mathrm{d}t} \left( \frac{\partial \mathcal{L}}{\partial \dot{q}_j} \right) + \frac{\partial \mathcal{D}}{\partial \dot{q}_j} - \frac{\partial \mathcal{L}}{\partial q_j} = Q_j, \quad Q_j = \sum_{i=1}^{\mathcal{N}} \mathbf{F}_i . \frac{\partial \mathbf{r}_i}{\partial q_j}, \quad j = 1, 2, 3, \ldots, \mathcal{M} \tag{5.64}$$

Note that $\mathcal{D}$ has units of power rather than energy.

**Applying Lagrange's Equation to Systems Involving Non-conservative Forces**

The following is a general procedure which you can follow to obtain the equations of motion of a non-conservative system of particles using Lagrange's equation.

1. Choose the type and number of generalized coordinates $q_j$ you want to use to describe the configuration of the particles in space.
2. Express the velocity ${}^N\mathbf{v}^{P_i/O}$ of each particle $P_i$ with respect to an inertial point $O$ in terms of the generalized coordinates and their derivatives.
3. Find the kinetic energy of the system using

$$\mathcal{T} = \sum_{i=1}^{\mathcal{N}} \frac{1}{2} m_i \; {}^N\mathbf{v}^{P_i/O} . {}^N\mathbf{v}^{P_i/O}. \tag{5.65}$$

4. Choose a reference potential energy level (datum) and find the sum of the potential energy of all particles with respect to the chosen datum.

$$\mathcal{U} = \sum_{i=1}^{\mathcal{N}} \mathcal{U}_i. \tag{5.66}$$

5. Define the dissipative power of any forces following the linear viscous damping model using

$$\mathcal{D} = \sum_{i=1}^{\mathcal{N}} \frac{1}{2} c_i \dot{\mathbf{r}}_i . \dot{\mathbf{r}}_i. \tag{5.67}$$

6. Obtain the generalized forces associated with each generalized coordinate using

$$Q_j = \sum_{i=1}^{\mathcal{N}} \mathbf{F}_i . \frac{\partial \mathbf{r}_i}{\partial q_j} \quad \text{or} \quad Q_j = \sum_{i=1}^{\mathcal{N}} \mathbf{F}_i . \frac{\partial \dot{\mathbf{r}}_i}{\partial \dot{q}_j}. \tag{5.68}$$

7. Apply Lagrange's equation along each of the generalized coordinates, $q_j$ using

$$\frac{\mathrm{d}}{\mathrm{d}t}\left(\frac{\partial \mathcal{L}}{\partial \dot{q}_j}\right) + \frac{\partial \mathcal{D}}{\partial \dot{q}_j} - \frac{\partial \mathcal{L}}{\partial q_j} = Q_j, \qquad j = 1, 2, 3, \ldots, \mathcal{M}. \tag{5.69}$$

**Example 5.5 Mass–Spring–Damper System with Two Degrees of Freedom**

Use Lagrange's equation to obtain the equations of motion for the system shown Figure 5.7.

This problem can be easily solved using Newton's second law. However, the problem statement demands using Lagrange's approach for the purpose of clarifying the process of dealing with the generalized forces. To this end, we follow the procedure described above:

1. We choose $x_1$ and $x_2$ as the generalized coordinates to describe the motion of the two masses.

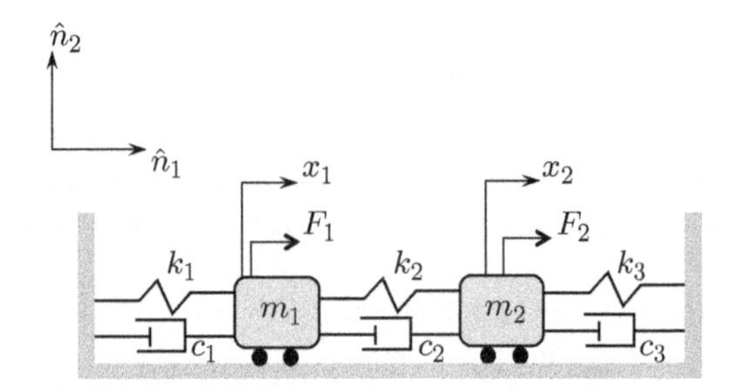


**Figure 5.7** Motion of a spring–mass–damper system with two degrees of freedom.

2. The velocity of $m_1$ with respect to an inertial point is $\dot{x}_1\hat{n}_1$ and that of $m_2$ is $\dot{x}_2\hat{n}_1$.
3. The kinetic energy is therefore:

$$\mathcal{T} = \frac{1}{2}m_1\dot{x}_1^2 + \frac{1}{2}m_2\dot{x}_2^2.$$

4. The potential energy is only due to the energy stored in the linear springs; there is no change of elevation with respect to any chosen datum:

$$\mathcal{U} = \frac{1}{2}k_1x_1^2 + \frac{1}{2}k_2(x_2 - x_1)^2 + \frac{1}{2}k_3x_2^2.$$

5. The dissipative power of the linear viscous damping forces can be written as

$$\mathcal{D} = \frac{1}{2}c_1\dot{x}_1^2 + \frac{1}{2}c_2(\dot{x}_2 - \dot{x}_1)^2 + \frac{1}{2}c_3\dot{x}_2^2.$$

6. The generalized forces can be obtained by noting that we have two generalized coordinates ($\mathcal{M} = 2$). Let $j = 1$ be associated with $x_1$ and $j = 2$ be associated with $x_2$. Also, we have two masses, which implies that $\mathcal{N} = 2$. Therefore, we can write:

For $j = 1:$

$$Q_1 = F_1\hat{n}_1.\frac{\partial x_1}{\partial x_1}\hat{n}_1 + F_2\hat{n}_1.\frac{\partial x_2}{\partial x_1}\hat{n}_1 = F_1,$$

For $j = 2:$

$$Q_2 = F_1\hat{n}_1.\frac{\partial x_1}{\partial x_2}\hat{n}_1 + F_2\hat{n}_1.\frac{\partial x_2}{\partial x_2}\hat{n}_1 = F_2.$$

7. Defining the Lagrangian $\mathcal{L} = \mathcal{T} - \mathcal{U}$ and implementing Lagrange's equation along the generalized coordinates $q_1 = x_1$ and $q_2 = x_2$, we obtain

$$m_1\ddot{x}_1 + c_1\dot{x}_1 - c_2(\dot{x}_2 - \dot{x}_1) + k_1x_1 - k_2(x_2 - x_1) = F_1,$$
$$m_2\ddot{x}_2 + c_3\dot{x}_2 + c_2(\dot{x}_2 - \dot{x}_1) + k_3x_2 + k_2(x_2 - x_1) = F_2.$$

#### Example 5.6 Forced Double Pendulum with Spring

For the system shown in Figure 5.8, the spring has an unstretched length $l_0$ when $\theta = \phi = 0$. The force $F_1$ is always normal to $OP_1$ while $F_2$ is always horizontal. Determine the equations of motion of the system.

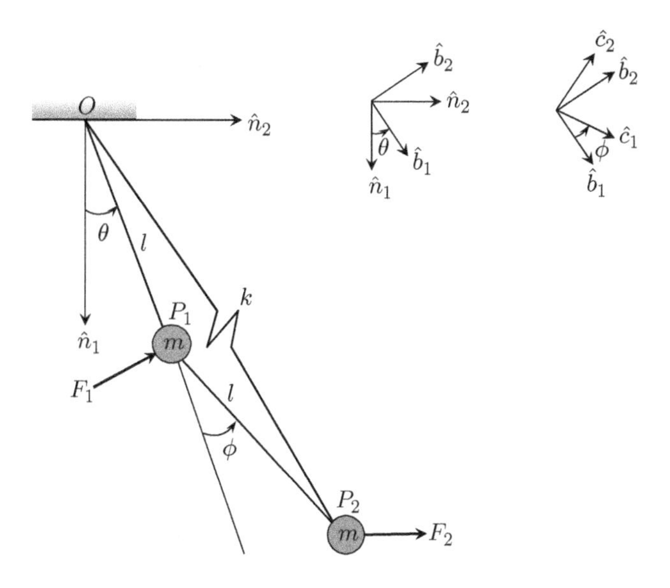


**Figure 5.8** Motion of a double pendulum with a connecting spring.

To describe the system dynamics, we choose the two generalized coordinates $\theta$ and $\phi$. We define the rotating $B$-frame such that ${}^N\boldsymbol{\omega}^B = \dot{\theta}\hat{b}_3$ and the rotating $C$-frame such that ${}^B\boldsymbol{\omega}^C = \dot{\phi}\hat{c}_3$. Using the definition of the rotating frames, the velocities of points $P_1$ and $P_2$ with respect to point $O$ are, respectively,

$$
\begin{aligned}
{}^N\mathbf{v}^{P_1/O} &= l\dot{\theta}\hat{b}_2, \\
{}^N\mathbf{v}^{P_2/O} &= -l(\dot{\theta}+\dot{\phi})\sin\phi\hat{b}_1 + (l\dot{\theta} + l(\dot{\theta}+\dot{\phi})\cos\phi)\hat{b}_2.
\end{aligned}
$$

The kinetic energy can therefore be written as

$$
\begin{aligned}
\mathcal{T} &= \frac{1}{2}m{}^N\mathbf{v}^{P_1/O}.{}^N\mathbf{v}^{P_1/O} + \frac{1}{2}m{}^N\mathbf{v}^{P_2/O}.{}^N\mathbf{v}^{P_2/O}, \\
\mathcal{T} &= \frac{1}{2}ml^2\dot{\theta}^2 + \frac{1}{2}m(l^2\dot{\theta}^2 + l^2(\dot{\theta}+\dot{\phi})^2 + 2l^2\dot{\theta}(\dot{\theta}+\dot{\phi})\cos\phi).
\end{aligned}
$$

The potential energy is due to the energy stored in the spring and the weight of the particles. To find the potential energy, $\mathcal{U}_s$, stored in the spring, we describe its deflection in terms of the generalized coordinates using $\delta = l(2-\sqrt{2(1+\cos\phi)})$. It follows that

$$
\mathcal{U}_s = \frac{1}{2}k\delta^2 = \frac{1}{2}kl^2(2-\sqrt{2(1+\cos\phi)})^2.
$$

By choosing a horizontal line passing through point $O$ as a reference point, the potential energy, $\mathcal{U}_w$, due to weight, can be written as

$$
\mathcal{U}_w = -mgl(2\cos\theta + \cos(\theta+\phi)).
$$

It follows that the total potential energy is

$$
\mathcal{U} = \frac{1}{2}kl^2(2-\sqrt{2(1+\cos\phi)})^2 - mgl(2\cos\theta + \cos(\theta+\phi)).
$$

To determine the generalized forces, we note that two generalized coordinates ($\mathcal{M} = 2$: $j = 1 \Rightarrow \theta$, $j = 2 \Rightarrow \phi$) were used to describe the dynamics of the system. Also, there are two particles; that is, $\mathcal{N} = 2$. The generalized force associated with $j = 1$ or $\theta$ can be written as:

$$Q_1 = \sum_{i=1}^{2} \mathbf{F}_i . \frac{\partial \mathbf{r}_i}{\partial \theta} = F_1 \hat{b}_2 . \frac{\partial \mathbf{r}_1}{\partial \theta} + F_2 \hat{n}_2 . \frac{\partial \mathbf{r}_2}{\partial \theta},$$

where $\mathbf{r}_i$ is the position vector of the point at which each force is applied. Therefore, $\mathbf{r}_1 = l\hat{b}_1$ and $\mathbf{r}_2 = l\hat{b}_1 + l\hat{c}_1$. Therefore, we can write

$$\frac{\partial \mathbf{r}_1}{\partial \theta} = \frac{\partial (l\hat{b}_1)}{\partial \theta} = l\hat{b}_2,$$

$$\frac{\partial \mathbf{r}_2}{\partial \theta} = \frac{\partial (l\hat{b}_1 + l(\cos\phi\hat{b}_1 + \sin\phi\hat{b}_2)}{\partial \theta} = l\hat{b}_2 + l\cos\phi\hat{b}_2 - l\sin\phi\hat{b}_1,$$

which yields

$$Q_1 = F_1 \hat{b}_2 . l\hat{b}_2 + F_2 \hat{n}_2 . (l\hat{b}_2 + l\cos\phi\hat{b}_2 - l\sin\phi\hat{b}_1),$$

$$Q_1 = F_1 l + F_2 l \cos\theta + F_2 l \cos(\theta + \phi).$$

When trying to reproduce $Q_1$, some readers may struggle to understand how the derivatives were calculated. Indeed, this process requires deep intuition into the differentiation of rotating vectors. To avoid getting into this issue, the reader may resort into describing all the vectors in the inertial frame. However, this is not advisable since the process of describing all the vectors in the inertial frame can be algebraically involved. Here, we recommended using the velocity version of the generalized force description – the“cancellation of dots” formula – which yields

$$Q_1 = \sum_{i=1}^{2} \mathbf{F}_i . \frac{\partial \dot{\mathbf{r}}_i}{\partial \dot{\theta}}$$

Using the previous equation, we can write

$$\frac{\partial \dot{\mathbf{r}}_1}{\partial \dot{\theta}} = \frac{\partial^N \mathbf{v}^{P_1/O}}{\partial \dot{\theta}} = l\hat{b}_2,$$

$$\frac{\partial \dot{\mathbf{r}}_2}{\partial \dot{\theta}} = \frac{\partial^N \mathbf{v}^{P_2/O}}{\partial \dot{\theta}} = l\hat{b}_2 + l\cos\phi\hat{b}_2 - l\sin\phi\hat{b}_1.$$

Note that in this case, the process of finding the derivatives is straightforward because $\dot{\theta}$ is in the direction of $\dot{\mathbf{r}}$. As such, it is advised to use the second approach unless you are very comfortable with the first approach.

Similarly, we obtain $Q_2$ using

$$Q_2 = \sum_{i=1}^{2} \mathbf{F}_i . \frac{\partial \dot{\mathbf{r}}_i}{\partial \dot{\phi}},$$

which yields

$$Q_2 = F_1\hat{b}_1.\frac{\partial}{\partial\dot{\phi}}(l\dot{\theta}\hat{b}_2) + F_2\hat{n}_2.\frac{\partial}{\partial\dot{\phi}}(-l(\dot{\theta}+\dot{\phi})\sin\phi\hat{b}_1 + (l\dot{\theta} + l(\dot{\theta}+\dot{\phi})\cos\phi)\hat{b}_2,$$

$$Q_2 = F_2 l\cos(\theta+\phi).$$

Now, one can implement Lagrange's equation along the two generalized coordinates to obtain the equations of motion.

**Example 5.7 Hamilton's Variational Principle**

The massless rod of length $l$ shown in Figure 5.9 is connected at one end to a spring of stiffness $k$ and, at the other end, to a mass $m$ . The ends of the rod are constrained to move in the smooth grooves shown. Use Hamilton's extended principle to obtain the equation of motion. The spring is unstretched when $\theta = 0$.

To implement Hamilton's extended principle, we need to obtain $\delta\mathcal{L}$ and $\delta\bar{\mathcal{W}}$. Since there are no external or dissipative forces acting on the system, then $\delta\bar{\mathcal{W}} = 0$. To obtain $\delta\mathcal{L}$, we first obtain $\mathcal{T}$ and $\mathcal{U}$.

The velocity of mass $m$ with respect to point $O$ is given by

$${}^{N}\mathbf{v}^{m/O} = -l\cos\theta\dot{\theta}\hat{n}_2.$$

The kinetic energy is therefore

$$\mathcal{T} = \frac{1}{2}ml^2\cos^2\theta\dot{\theta}^2.$$

The potential energy is

$$\begin{aligned}\mathcal{U} &= \mathcal{U}_s + \mathcal{U}_w \\ &= \frac{1}{2}kl^2(1-\cos\theta)^2 - mgl\sin\theta.\end{aligned}$$

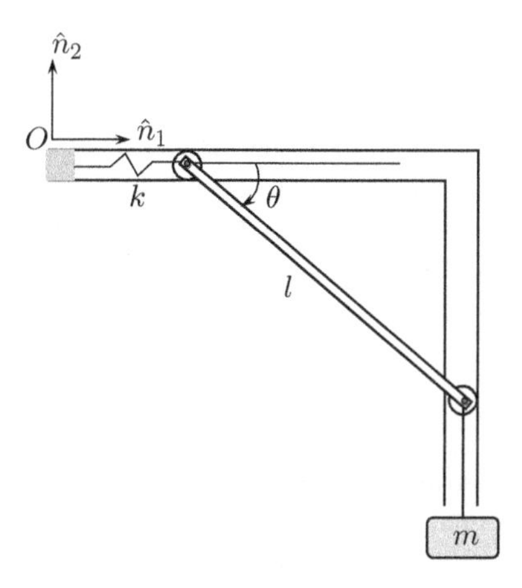


**Figure 5.9** Motion of a mass via Hamilton's variational principle.

It follows that

$$\mathcal{L} = \mathcal{T} - \mathcal{U} = \frac{1}{2}ml^2\left(\dot{\theta}^2\cos^2\theta - \frac{k}{m}(1-\cos\theta)^2 + 2\frac{g}{l}\sin\theta\right), \tag{5.70}$$

and

$$\begin{aligned}\delta\mathcal{L} &= \frac{\partial\mathcal{L}}{\partial\theta}\delta\theta + \frac{\partial\mathcal{L}}{\partial\dot{\theta}}\delta\dot{\theta},\\ &= -ml^2(\dot{\theta}^2\sin\theta\cos\theta + \frac{k}{m}(1-\cos\theta)\sin\theta - \frac{g}{l}\cos\theta)\delta\theta + ml^2\dot{\theta}\cos^2\theta\delta\dot{\theta}.\end{aligned}$$

Using Hamilton's principle, we obtain

$$\int_{t_1}^{t_2}\delta\mathcal{L}\,\mathrm{d}t = 0,$$

$$\int_{t_1}^{t_2}\left(\dot{\theta}^2\sin\theta\cos\theta + \frac{k}{m}(1-\cos\theta)\sin\theta - \frac{g}{l}\cos\theta)\delta\theta - \dot{\theta}\cos^2\theta\delta\dot{\theta}\right)\mathrm{d}t = 0.$$

Integrating the term involving $\delta\dot{\theta}$ once by parts, yields

$$-\int_{t_1}^{t_2}\dot{\theta}\cos^2\theta\delta\dot{\theta}\mathrm{d}t = -\dot{\theta}\cos^2\theta\delta\theta\Big|_{t_1}^{t_2} + \int_{t_1}^{t_2}\frac{\mathrm{d}}{\mathrm{d}t}(\dot{\theta}\cos^2\theta)\delta\theta\mathrm{d}t.$$

Since $\delta\theta$ vanishes at $t_1$ and $t_2$, the previous equation reduces to

$$-\int_{t_1}^{t_2}\dot{\theta}\cos^2\theta\delta\dot{\theta}\mathrm{d}t = \int_{t_1}^{t_2}\frac{\mathrm{d}}{\mathrm{d}t}(\dot{\theta}\cos^2\theta)\delta\theta\mathrm{d}t.$$

Substituting the previous equation back into Hamilton's principle yields

$$\int_{t_1}^{t_2}\left(\dot{\theta}^2\sin\theta\cos\theta + \frac{k}{m}(1-\cos\theta)\sin\theta - \frac{g}{l}\cos\theta + \frac{\mathrm{d}}{\mathrm{d}t}(\dot{\theta}\cos^2\theta)\right)\delta\theta\mathrm{d}t = 0.$$

Since $\delta\theta$ is an arbitrary non-zero angle, the only way to force the previous equation to zero is by letting

$$\frac{d}{dt}(\dot{\theta}\cos^2\theta) + \dot{\theta}^2\sin\theta\cos\theta + \frac{k}{m}(1-\cos\theta)\sin\theta - \frac{g}{L}\cos\theta = 0,$$

which represents the equation of motion of the mass. It is evident that Hamilton's principle can be applied directly to obtain the equations of motion. However, this process involved calculating the variation of different quantities and an integration by parts. It is therefore more convenient to use Lagrange's equations directly when studying the dynamics of particles and rigid bodies.

**Flipped Classroom Exercise 5.4**

Particle $P$ of mass $m$ slides without friction in a pipe that rotates with angle $\theta$ relative to a rotating post. The post rotates at a constant angular velocity $\Omega$. Define $r$ as the distance from point $O$ to point $P$ and your coordinates such that ${}^N\omega^A = \Omega\hat{a}_2$, ${}^A\omega^B = \dot{\theta}\hat{b}_3$ then find the equation of motion of the particle knowing that $F(t)$ is always in the direction of $\hat{a}_1$:

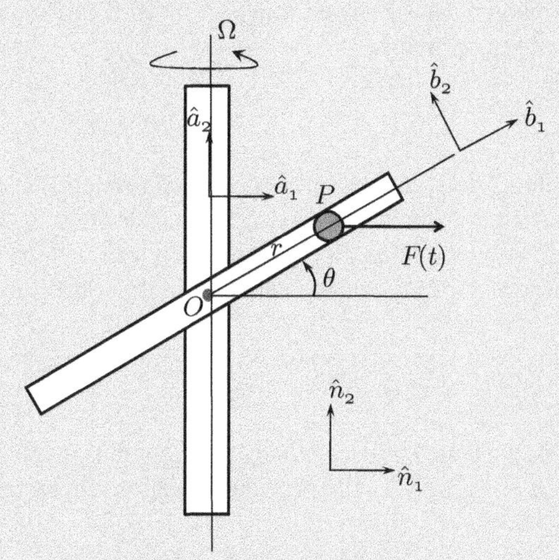


To find the equation of motion, carry out the following steps:

1. Define the generalized coordinates as $q_1 = r$ and $q_2 = \theta$.
2. Find the velocity of point $P$ with respect to $O$ in the inertial frame.
3. Find the kinetic energy of the particle.
4. Find the potential energy of the particle using a horizontal line passing through point $O$ as your reference.
5. Find the generalized forces $Q_1$ and $Q_2$, where $Q_1$ and $Q_2$ are, respectively, associated with the $r$ and $\theta$ generalized coordinates.
6. Apply Lagrange's equations along the coordinates $r$ and $\theta$ and show that the equations of motion are given by:

$$\ddot{\theta} + \left(\frac{2\dot{r}}{r} + \Omega^2 \cos\theta \sin\theta\right)\dot{\theta} + g\cos\theta = -\frac{F(t)}{m}\sin\theta,$$

$$\ddot{r} - r\dot{\theta}^2 - r\Omega^2\cos^2\theta + g\sin\theta = \frac{F(t)}{m}\cos\theta.$$

## 5.4 Lagrange's Equations with Constraints

You will have learned in undergraduate mathematics classes the process of optimizing functions subject to some constraints. You will also have learned that the process typically involves the use of Lagrange multipliers. Here, we review this process and build upon it to derive the Lagrange equations of motion for dynamical systems subject to constraints. These constraints can either be holonomic or non-holonomic.

To this end, consider a function $L = L(q_1, q_2, \dots, q_n)$ subject to a set of holonomic constraints of the form $f_i(q_1, q_2, \dots, q_n)$, $i = 1, 2, \dots, m$. In absence of the constraints, an extremum of $L$ can be found by setting the gradient of $L$ equal to zero, $\nabla L = 0$. This yields a set of $n$ equations, which can be solved together for the corresponding set of $q_i$ that render the function $L$ an extremum; that is

$$\frac{\partial L}{\partial q_i} = 0, \qquad i = 1, 2, \dots, n. \tag{5.71}$$

In the presence of constraints, the set $q_i$ must not only render the function $L$ an extremum, but must do so while simultaneously satisfying the constraints, $f_i$. This can only occur at the point/points where $L$ is tangential to all constraints. At such points, the gradient of $L$ must be parallel to the gradient of $f_i$. This can be expressed in the following mathematical form:

$$\nabla L = \sum_{i=1}^{m} \lambda_i \nabla f_i \tag{5.72}$$

where $\lambda_i$ are scaling constants known as the Lagrange multipliers. Letting $G(q_1, q_2, \dots, q_n, \lambda_1, \lambda_2, \dots, \lambda_m) = L - \sum_{i=1}^{m} \lambda_i f_i$, the previous equation can be rewritten as

$$\nabla_{q_1, q_2, \dots, q_n, \lambda_1, \lambda_2, \dots, \lambda_m} G = 0. \tag{5.73}$$

Equation (5.73) yields $n + m$ algebraic equations, which can be solved together for the unknowns $(q_1, q_2, \dots, q_n, \lambda_1, \lambda_2, \dots, \lambda_m)$.

**Example 5.8 Minimizing a Function Subject to a Single Constraint**

Find the point $(x_0, y_0, z_0)$ on the plane $2x + y - z - 5 = 0$ that is closest to the origin.

The goal of this problem is to find the coordinates of the point $(x_0, y_0, z_0)$ constrained to the plane $2x + y - z - 5 = 0$ that minimizes the distance between the origin and the plane.

The distance between any point in space and the origin is given by $d = \sqrt{x^2 + y^2 + z^2}$. Since the distance is positive, then the point that minimizes the distance is the same as the point that minimizes the square of the distance $d^2 = x^2 + y^2 + z^2$. Therefore, we can cast the problem as minimizing $x^2 + y^2 + z^2 = 0$ subject to $2x + y - z - 5 = 0$. Using Equation (5.73), we can express the optimization problem as:

$$\nabla_{x,y,z,\lambda} G = 0, \qquad G = x^2 + y^2 + z^2 + \lambda(2x + y - z - 5),$$

which yields the following set of algebraic equations:

$$\begin{aligned}
\frac{\partial G}{\partial x} &= 0, \qquad 2x + 2\lambda = 0,\\
\frac{\partial G}{\partial y} &= 0, \qquad 2y + \lambda = 0,\\
\frac{\partial G}{\partial z} &= 0 \qquad 2z - \lambda = 0,\\
\frac{\partial G}{\partial \lambda} &= 0 \qquad 2x + y - z - 5 = 0.
\end{aligned}$$

Solving this set of equations yields $(x_0, y_0, z_0) = (\frac{5}{3}, \frac{5}{6}, -\frac{5}{6})$.

**Example 5.9 Maximizing a Function Subject to Two Constraints**

The plane $x + y + z = 1$ cuts the cylinder $x^2 + y^2 - 1 = 0$ to form an ellipse. Find the points on the ellipse that lie furthest from the origin.

The distance from the origin to any point in space is given by $d^2 = x^2 + y^2 + z^2$. The problem is therefore to find the point $(x_0, y_0, z_0)$ that maximizes $d^2$ subject to the constraints: $x + y + z = 1$ and $x^2 + y^2 - 1 = 0$.

Using Equation (5.73), we can write

$$\nabla_{x,y,z,\lambda_1,\lambda_2} G = 0, \qquad G = x^2 + y^2 + z^2 + \lambda_1(x + y + z - 1) + \lambda_2(x^2 + y^2 - 1),$$

which yields the following set of algebraic equations:

$$\begin{aligned}
\frac{\partial G}{\partial x} &= 0, \qquad 2x + 2\lambda_1 + 2\lambda_2 x = 0,\\
\frac{\partial G}{\partial y} &= 0, \qquad 2y + \lambda_1 + 2\lambda_2 y = 0,\\
\frac{\partial G}{\partial z} &= 0 \qquad 2z + \lambda_1 = 0,\\
\frac{\partial G}{\partial \lambda_1} &= 0 \qquad x + y + z - 1 = 0,\\
\frac{\partial G}{\partial \lambda_2} &= 0 \qquad x^2 + y^2 - 1 = 0.
\end{aligned}$$

Solving the previous set of equations yields the following extrema $(0, 1, 0)$, $(1, 0, 0)$, $(\frac{\sqrt{2}}{2}, \frac{\sqrt{2}}{2}, 1 - \sqrt{2})$, $(-\frac{\sqrt{2}}{2}, -\frac{\sqrt{2}}{2}, 1 + \sqrt{2})$. Using the previous coordinates to calculate the distance $d$, we obtain the following values:

$$\begin{aligned}
&(0, 1, 0) \Rightarrow d = 1,\\
&(1, 0, 0) \Rightarrow d = 1,\\
&(\frac{\sqrt{2}}{2}, \frac{\sqrt{2}}{2}, 1 - \sqrt{2}) \Rightarrow d^2 = 4 - 2\sqrt{2},\\
&(-\frac{\sqrt{2}}{2}, -\frac{\sqrt{2}}{2}, 1 + \sqrt{2}) \Rightarrow d^2 = 4 + 2\sqrt{2}.
\end{aligned}$$

Hence, the first two points are minima and the last point is the global maximum which lies furthest from the origin.

Now that we have a good understanding of the process of optimization of functions subject to constraints, we turn our attention to studying the dynamics of particles under the influence of constraints. To this end, we consider a system of particles subject to $\mathcal{K}$ constraints of the form:

$$\sum_{j=1}^{\mathcal{M}} c_{lj}\dot{q}_j + c_l = 0, \qquad l = 1, 2, \ldots, \mathcal{K}, \tag{5.74}$$

where $\mathcal{M}$ is the number of generalized coordinates used to describe the dynamics.

Recall that the process of finding Lagrange's equations is a process of minimizing a functional involving the integral of the Lagrangian and the non-conservative work done on/by the system over an arbitrary period of time; that is

$$\int_{t_1}^{t_2} \delta(\mathcal{L} + \bar{\mathcal{W}})\mathrm{d}t = 0. \tag{5.75}$$

Note that $\delta$ in the previous equation is equivalent to the gradient in the case of regular functions. Therefore, now that we have $\mathcal{K}$ constraints acting on the system, we must guarantee that the gradient of the constraints is parallel to the gradient of the functional. This can be done by augmenting Equation (5.75) with the constraints by using $\mathcal{K}$ Lagrange multiplier, as follows:

$$\int_{t_1}^{t_2} \delta\left(\mathcal{L} + \bar{\mathcal{W}} + \left(\sum_{l=1}^{\mathcal{K}} \lambda_l \sum_{j=1}^{\mathcal{M}} c_{lj}\dot{q}_j + c_l\right)\right)\mathrm{d}t = 0. \tag{5.76}$$

Taking the variation of the preceding equation and setting the coefficients of $\delta q_j$ to zero independently yields

$$\frac{\mathrm{d}}{\mathrm{d}t}\left(\frac{\partial \mathcal{L}}{\partial \dot{q}_j}\right) + \frac{\partial \mathcal{D}}{\partial \dot{q}_j} - \frac{\partial \mathcal{L}}{\partial q_j} = Q_j + \sum_{l=1}^{\mathcal{K}} \lambda_l c_{lj}, \qquad Q_j = \sum_{i=1}^{\mathcal{N}} \mathbf{F}_i \cdot \frac{\partial \mathbf{r}_i}{\partial q_j},$$

$$j = 1, 2, 3, \ldots, \mathcal{M} \tag{5.77}$$

$\mathcal{N}$ : Number of particles

$\mathcal{M}$ : Number of generalized coordinates

$\mathcal{K}$ : Number of constraints

Equation (5.77) represents the most general form of Lagrange's equation and can be used to obtain the equations of motion of particles and rigid bodies subject to non-conservative forces and constraints.

**Example 5.10 Simple Pendulum Dynamics**

Consider the dynamics of the pendulum expressed in terms of the generalized coordinates $x$, $y$, and $\theta$ and consider the relationships $x = l\sin\theta$, $y = l\cos\theta$ as geometric constraints. Find the equation of motion of the pendulum.

The goal of this example is to illustrate the process of using Lagrange's equations to deal with dynamical systems subject to constraints. To this end, we first identify the generalized coordinates and constraints. The system dynamics are described using three generalized coordinates ($\mathcal{M} = 3$), namely $x$, $y$, and $\theta$. The system is subject to two constraints ($\mathcal{K} = 2$), namely $x = l\sin\theta$, and $y = l\cos\theta$. As such, the system has a single degree of freedom whose dynamics can be described using a single equation.

Implementing Lagrange's equations requires knowledge of the kinetic and potential energy of the system, the generalized forces, and the constraints. The kinetic and potential energy can be written as

$$\mathcal{T} = \frac{1}{2}m(\dot{x}^2 + \dot{y}^2), \qquad \mathcal{U} = -mgy.$$

Since there are no external or damping forces $Q_j = 0$ and $\mathcal{D} = 0$.

To account for the influence of the constraints on the dynamics, we express them in the form of Equation (5.74); that is $\mathrm{d}x - l\cos\theta\mathrm{d}\theta = 0$ and $\mathrm{d}y + l\sin\theta d\theta = 0$. This yields $c_{11} = 1$, $c_{21} = 0$, $c_{12} = 0$, $c_{22} = 1$, $c_{13} = -l\cos\theta$, and $c_{23} = l\sin\theta$. Note that

- $c_{11}$ is the coefficient of $\mathrm{d}x$ in the first constraint
- $c_{12}$ is the coefficient of $\mathrm{d}y$ in the first constraint
- $c_{13}$ is the coefficient of $\mathrm{d}\theta$ in the first constraint
- $c_{21}$ is the coefficient of $\mathrm{d}x$ in the second constraint
- $c_{22}$ is the coefficient of $\mathrm{d}y$ in the second constraint
- $c_{23}$ is the coefficient of $\mathrm{d}\theta$ in the second constraint.

It follows from Equation (5.77) that:

*$x$-dynamics:*

$$\frac{\mathrm{d}}{\mathrm{d}t}\left(\frac{\partial \mathcal{L}}{\partial \dot{x}}\right) - \frac{\partial \mathcal{L}}{\partial x} = c_{11}\lambda_1 + c_{21}\lambda_2, \qquad m\ddot{x} = \lambda_1,$$

*$y$-dynamics:*

$$\frac{\mathrm{d}}{\mathrm{d}t}\left(\frac{\partial \mathcal{L}}{\partial \dot{y}}\right) - \frac{\partial \mathcal{L}}{\partial y} = c_{12}\lambda_1 + c_{22}\lambda_2, \qquad m\ddot{y} - mg = \lambda_2,$$

*$\theta$-dynamics:*

$$\frac{\mathrm{d}}{\mathrm{d}t}\left(\frac{\partial \mathcal{L}}{\partial \dot{\theta}}\right) - \frac{\partial \mathcal{L}}{\partial \theta} = c_{13}\lambda_1 + c_{23}\lambda_2, \qquad 0 = \lambda_2 l\sin\theta - \lambda_1 l\cos\theta.$$

Substituting the value of $\lambda_1$ and $\lambda_2$ from the first two equations into the third equation and letting $x = l\sin\theta$ and $y = l\sin\theta$, we obtain the well-known equation of motion of the pendulum

$$\ddot{\theta} + \frac{g}{l}\sin\theta = 0. \tag{5.78}$$

**Example 5.11 Constrained Bead**

Consider the motion of a bead constrained to move along the path $y = x^2$, which in turn rotates with a constant angular velocity $\Omega$ (Figure 5.10). Find the equation of motion of the bead.

To describe the motion of the bead, we define the rotating $A$-frame, such that ${}^N\boldsymbol{\omega}^A = \Omega\hat{a}_2$. We then express the velocity of the particle in the rotating frame such that:

$${}^N\mathbf{v}^{P/O} = \dot{x}\hat{a}_1 + \dot{y}\hat{a}_2 - x\Omega\hat{a}_3.$$

The kinetic and potential energy can be written as

$$\mathcal{T} = \frac{1}{2}m(\dot{x}^2 + \dot{y}^2 + x^2\Omega^2), \qquad \mathcal{U} = mgy.$$

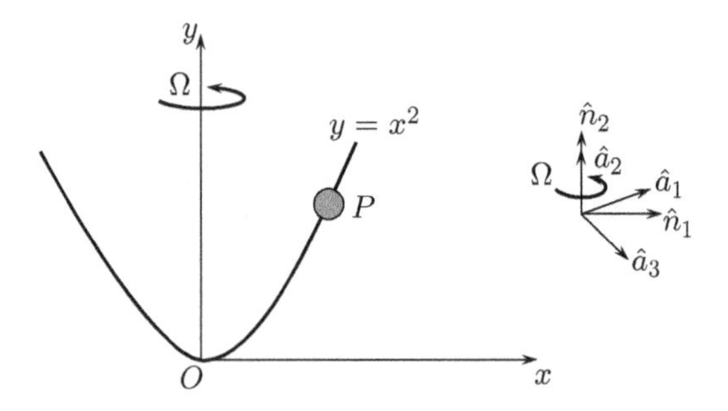


**Figure 5.10** Motion of a constrained bead along the path $y = x^2$.

Using the constraint equation $y = x^2$ or $\mathrm{d}y - 2x\mathrm{d}x = 0$, yields $c_{11} = -2x$ and $c_{12} = 1$. Applying Lagrange's equation along the $x$ and $y$ generalized coordinates, we can write

*x-dynamics:*

$$\frac{\mathrm{d}}{\mathrm{d}t}\left(\frac{\partial \mathcal{L}}{\partial \dot{x}}\right) - \frac{\partial \mathcal{L}}{\partial x} = c_{11}\lambda, \qquad m\ddot{x} - \Omega^2 x = -2x\lambda,$$

$$\frac{\mathrm{d}}{\mathrm{d}t}\left(\frac{\partial \mathcal{L}}{\partial \dot{y}}\right) - \frac{\partial \mathcal{L}}{\partial y} = c_{12}\lambda \qquad m\ddot{y} + mg = \lambda.$$

Eliminating $\lambda$ from the previous equation and rearranging yields

$$(\ddot{x} + 4x^2\ddot{x} + 4x\dot{x}^2 + 2gx) = \Omega^2 x.$$

### 5.4.1 *Physical Interpretation of Lagrange Multipliers*

The Lagrange multipliers are not only mathematical quantities that allow the optimization of functions with constraints. They also have a physical meaning, especially when dealing with dynamical systems through Hamilton's principle and, consequently, Lagrange's equation. The process of accounting for constraints in Lagrange's equations involves adding the variation of the term $\left(\sum_{l=1}^{\mathcal{K}} \lambda_l \sum_{j=1}^{\mathcal{M}} c_{lj}\dot{q}_j + c_l\right)$ to Hamilton's extended principle. To be consistent with other terms in the equation, this term must have units of work/energy. As such, this term represents the virtual work done by the constraints. In essence, when the system attempts to violate the constraints ever so slightly, the constraint forces do work to prevent the prescribed constraints from being violated. Therefore, there is a direct relationship between the Lagrange multipliers and the constraint forces. We attempt to make this relationship clearer through the examples below.

**Example 5.12 Reaction Force and Lagrange Multipliers**

In Figure 5.11, particle $m$ slides on a smooth surface of radius $R$. Find the reaction force $R_N$ exerted by the surface on the particle using Lagrange's approach.

We do not have a direct procedure to obtain the reaction force $R_N$ using Lagrange's equation. This is because the reaction force does not do any work on the particle along

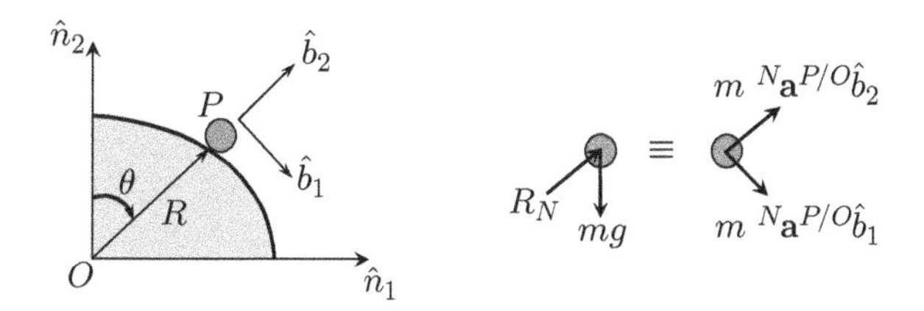


**Figure 5.11** Finding reaction forces using Lagrange's approach.

the virtual displacement, and therefore does not appear naturally in Hamilton's principle. On the other hand, finding the reaction force using Newton's second law of dynamics is straightforward. We start by describing the acceleration of the particle $P$ with respect to point $O$ as

$$^N\mathbf{a}^{P/O} = R\ddot{\theta}\hat{b}_1 - R\dot{\theta}^2\hat{b}_2.$$

Subsequently, we draw the free-body diagram and use it to apply Newton's second law to obtain the equation of motion and the reaction $R_N$ as follows:

*Direction of* $\hat{b}_1$:

$$\mathbf{F}\hat{b}_1 = m\ ^N\mathbf{a}^{P/O}\hat{b}_1,$$

$$\ddot{\theta} - \frac{g}{R}\sin\theta = 0.$$

*Direction of* $\hat{b}_2$:

$$\mathbf{F}\hat{b}_2 = m\ ^N\mathbf{a}^{P/O}\hat{b}_2,$$

$$R_N = m(g\cos\theta - R\dot{\theta}^2).$$

Next, we show how to tackle this problem using Lagrange's equation. To this end, we first assume that the reaction force $R_N$ does not exist. In such a case, the surface cannot constrain the particle to move along the circular surface, and hence, as shown in Figure 5.12, we

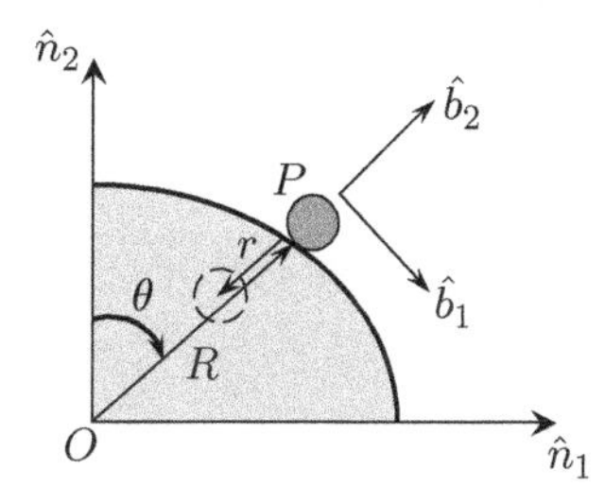


**Figure 5.12** The generalized coordinate $r$ is introduced to release the influence of the constraint force $R_N$.

add another generalized coordinate $r$ to represent the particle moving along the radius of the surface. It follows that the position and velocity of the particle can be written as:

$$\mathbf{OP} = (R-r)\hat{b}_2,$$
$${}^{N}\mathbf{v}^{P/O} = (R-r)\dot{\theta}\hat{b}_1 - \dot{r}\hat{b}_2.$$

The kinetic and potential energy can therefore be written as

$$\mathcal{T} = \frac{1}{2}m((R-r)^2\dot{\theta}^2 + \dot{r}^2),$$
$$\mathcal{U} = mg(R-r)\cos\theta.$$

At this point, we introduce the kinematic constraint, $r = 0$, or $\mathrm{d}r = 0$, to force the particle to move on the surface. Substituting the kinetic and potential energy expressions and the constraint equation into Equation (5.77) yields the following equations:

*$r$-dynamics:*

$$\frac{\mathrm{d}}{\mathrm{d}t}\left(\frac{\partial \mathcal{L}}{\partial \dot{r}}\right) - \frac{\partial \mathcal{L}}{\partial r} = \lambda, \qquad mR\ddot{r} + m(R-r)\dot{\theta}^2 - mg\cos\theta = \lambda,$$

*$\theta$-dynamics:*

$$\frac{\mathrm{d}}{\mathrm{d}t}\left(\frac{\partial \mathcal{L}}{\partial \dot{\theta}}\right) - \frac{\partial \mathcal{L}}{\partial \theta} = 0, \qquad m(R-r)^2\ddot{\theta} - 2m(R-r)\dot{r}\dot{\theta} - mg(R-r)\sin\theta = 0.$$

Upon obtaining the equations of motion in terms of $\theta$ and $r$, we force the constraint $r = 0$ back into the previous equations and obtain

*$r$-dynamics:*

$$m(R\dot{\theta}^2 - g\cos\theta) = \lambda.$$

*$\theta$-dynamics:*

$$\ddot{\theta} - \frac{g}{R}\sin\theta = 0.$$

Comparing the previous equations to those obtained using the Newtonian approach, it becomes evident that $R_N = -\lambda$. As such, the Lagrange multiplier is directly related to the constraint force. In fact, the Lagrange multiplier is nothing but the generalized force resulting from the constraint force; that is $\lambda = Q_r$, where

$$\lambda = Q_r = \sum_{i=1}^{\mathcal{N}} \mathbf{F}_i.\frac{\partial \dot{\mathbf{r}}_i}{\partial \dot{q}_j} = R_N\hat{b}_2.\frac{\partial {}^{N}\mathbf{v}^{P/O}}{\partial \dot{r}} = R_N\hat{b}_2.\frac{\partial}{\partial \dot{r}}((R-r)\dot{\theta}\hat{b}_1 - \dot{r}\hat{b}_2) = -R_N.$$

This implies that

$$R_N = -\lambda = m(g\cos\theta - R\dot{\theta}^2),$$

which is the same answer obtained using the vectorial approach.

**Procedure for Obtaining Constraint and Internal Forces Using Lagrange's Equations**

The following is a general procedure that can be utilized to obtain constraint and internal forces using Lagrange's equations.

1. Remove the constraint forces. For each constraint force removed add a generalized coordinate to fully represent the motion of the system in the absence of the associated constraint force.
2. For each constraint force removed, introduce a constraint equation which could be used to capture the influence of the constraint force.
3. Implement Lagrange's equation along each generalized coordinate while taking into account the constraint equations by introducing a Lagrange multiplier for each constraint equation.
4. Equate each Lagrange multiplier to the generalized force resulting from the constraints force.
5. Solve Lagrange's equations for the constraint force.

**Example 5.13 Obtaining Constraint Forces using Lagrange's Approach**

The system shown in Figure 5.13 consists of two particles of mass $(3m, m)$ and three massless rods $(3l, 2l, d)$. The rods are pin connected at a distance $a$ below the support, as shown in the figure. Find the reaction force in the rod $d$ using Lagrange's approach.

We first notice that, due to the reaction force, the angles of rotation of the two pendula are equal: $\theta = \phi$. As such, to find the constraint force, we release this constraint and allow the two pendula to move independently via two generalized coordinates $\theta$ and $\phi$.

We find the velocity of the pendula by introducing the rotating $B$-frame such that ${}^N\omega^B = -\dot{\theta}$ for the first pendulum and ${}^N\omega^B = -\dot{\phi}$ for the second pendulum. It follows that

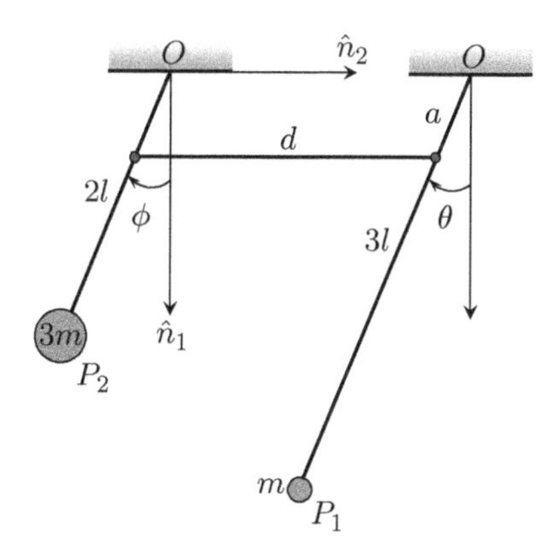


**Figure 5.13** Motion of two constrained pendula.

the velocities can be written as

$$ {}^{N}\mathbf{v}^{P_1/O} = -3l\dot{\theta}\hat{b}_2, \qquad {}^{N}\mathbf{v}^{P_2/O} = -2l\dot{\phi}\hat{b}_2, $$

and that the kinetic and potential energy are

$$ \mathcal{T} = 6ml^2\dot{\phi}^2 + \frac{9}{2}ml^2\dot{\theta}^2, $$
$$ \mathcal{U} = -6mgl\cos\phi - 3mgl\cos\theta. $$

At this point, we re-constrain the system by introducing the constraint $\theta = \phi$ or $\mathrm{d}\theta - \mathrm{d}\phi = 0$. We apply Lagrange's equation and obtain

*$\theta$-dynamics:*

$$ \ddot{\theta} + \frac{1}{3}\frac{g}{l}\sin\theta = \frac{\lambda}{9ml^2}, $$

*$\phi$-dynamics:*

$$ \ddot{\phi} + \frac{1}{2}\frac{g}{l}\sin\phi = -\frac{\lambda}{12ml^2}. $$

As already described, the Lagrange multiplier is nothing but the generalized force associated with the constraint force, $F_c$ as shown in Figure 5.14; that is

$$ \lambda = Q_\theta = -F_c\hat{n}_2 \cdot \frac{\partial}{\partial\dot{\theta}}(-a\dot{\theta}\hat{b}_2) = F_c a\cos\theta, $$
$$ -\lambda = Q_\phi = F_c\hat{n}_2 \cdot \frac{\partial}{\partial\dot{\phi}}(-a\dot{\phi}\hat{b}_2) = -F_c a\cos\phi $$

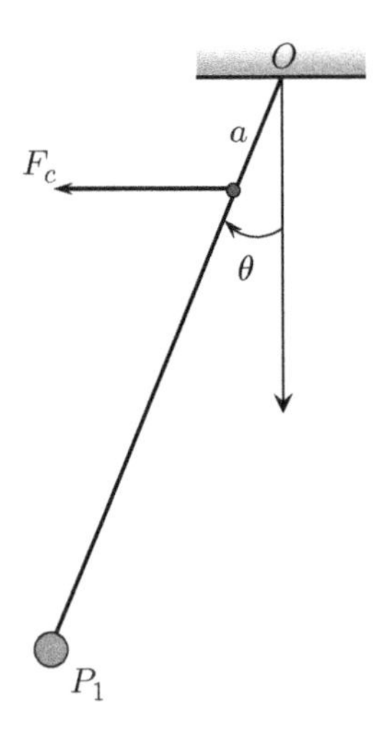


**Figure 5.14** Constraint force, $F_c$.

Letting $\theta = \phi$ and $\lambda = F_c a \cos\theta$ in the equations of motion for $\theta$ and $\phi$, then solving for $F_c$ we obtain

$$F_c = -\frac{6l}{7a} mg \tan\theta.$$

 **Flipped Classroom Exercise 5.5**

The end support of the simple pendulum is forced to move horizontally in a smooth slot, as shown in the figure. Use Lagrange's approach to find the tension in the cable.

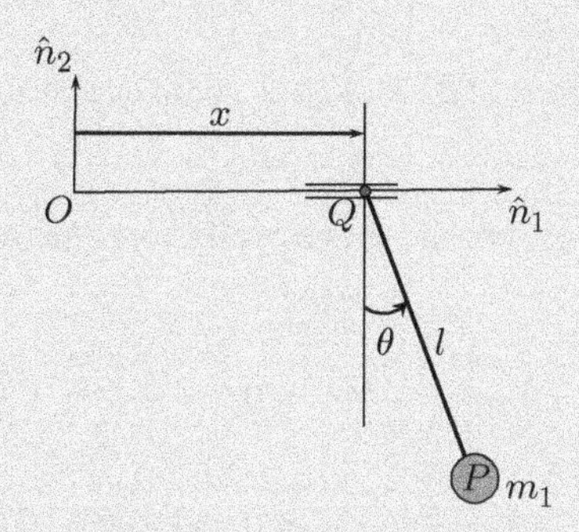


1. Remove the constraint force and let the length of the cable $l$ be time-varying $l(t)$. As such, you need two generalized coordinates $(\theta, l)$ to describe the dynamics.
2. Account for the influence of the constraint force on the dynamics by introducing the constraint equation $l = l_0$, where $l_0$ is a constant.
3. Implement Lagrange's equation along the $l$ and $\theta$ generalized coordinates. Do not forget to include the influence of the constraint using a Lagrange multiplier.
4. Set $l = l_0$ in the resulting equations for $\theta$ and $l$.
5. Equate the Lagrange multiplier to the generalized force associated with the $l$ generalized coordinates; that is, $\lambda = Q_l$.
6. Show that the tension is given by

$$T = m(g\cos\theta + l\dot{\theta}^2 - \ddot{x}\sin\theta).$$

## 5.5 Cyclic Coordinates

A *cyclic* or *ignorable* coordinate is a generalized coordinate for which the Lagrangian depends *only* on the derivative of the generalized coordinate (generalized velocity) but not the coordinate itself. For instance, consider a conservative system for which the Lagrangian takes the form

$$\mathcal{L} = \mathcal{L}(q_1, q_2, q_4, q_6, \ldots, q_{\mathcal{M}}, \dot{q}_1, \dot{q}_2, \dot{q}_3, \dot{q}_4, \dot{q}_5, \dot{q}_6, \ldots, \dot{q}_{\mathcal{M}}), \tag{5.79}$$

where $\mathcal{M}$ is the number of generalized coordinates used to describe the configuration of the system. Since the Lagrangian does not depend on $q_3$ and $q_5$, these generalized coordinates

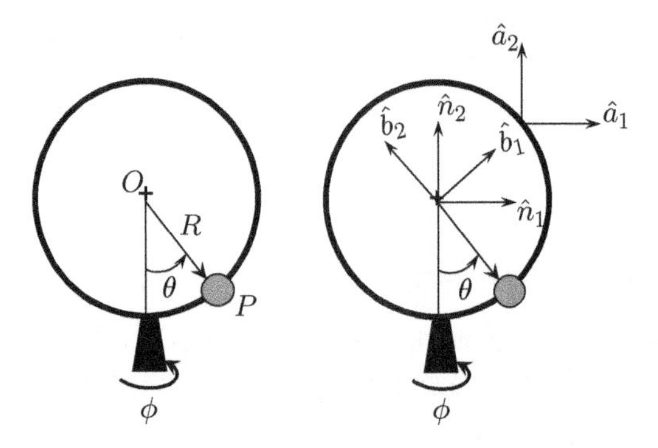


**Figure 5.15** Motion of a bead on a rotating hoop.

are called cyclic coordinates. For these two cyclic coordinates, applying Lagrange's equation yields

$$\frac{\mathrm{d}}{\mathrm{d}t}\left(\frac{\partial \mathcal{L}}{\partial \dot{q}_3}\right) = 0 \Longrightarrow \frac{\partial \mathcal{L}}{\partial \dot{q}_3} = C_3$$

$$\frac{\mathrm{d}}{\mathrm{d}t}\left(\frac{\partial \mathcal{L}}{\partial \dot{q}_5}\right) = 0 \Longrightarrow \frac{\partial \mathcal{L}}{\partial \dot{q}_5} = C_5,$$

where $C_3$ and $C_5$ are constants determined from the initial conditions of the system. In general, for any cyclic coordinate $q_c$, we can write

$$\frac{\mathrm{d}}{\mathrm{d}t}\left(\frac{\partial \mathcal{L}}{\partial \dot{q}_c}\right) = 0 \Longrightarrow \frac{\partial \mathcal{L}}{\partial \dot{q}_c} = C_c.$$

The previous equation can be solved for $\dot{q}_c$ in terms of $C_c$ and the other system parameters. As such, any dependence on $\dot{q}_c$ can be completely eliminated from the dynamics. Therefore, the resulting equations governing the motion of the system will have no dependence on $q_c$, and hence the name *ignorable* coordinate.

### Example 5.14 Cyclic Coordinate

Consider the motion of a bead on a rotating frictionless massless hoop, as shown in Figure 5.15. The ring rotates with angle $\phi$. Find the equations of motion. Identify the cyclic coordinate and use it to reduce the number of equations of motion.

Define your rotating coordinates such that ${}^N\boldsymbol{\omega}^A = \dot{\phi}\hat{a}_2$, and ${}^A\boldsymbol{\omega}^B = \dot{\theta}\hat{b}_3$. Two generalized coordinates, $\theta$ and $\phi$, are used to describe the dynamics.

The velocity of point $P$ with respect to point $O$ can be written as

$${}^N\mathbf{v}^{P/O} = R\dot{\theta}\hat{b}_1 - R\dot{\phi}\sin\theta\hat{b}_3.$$

The kinetic energy is

$$\mathcal{T} = \frac{1}{2}m{}^N\mathbf{v}^{P/O}.{}^N\mathbf{v}^{P/O} = \frac{1}{2}mR^2(\dot{\theta}^2 + \dot{\phi}^2\sin^2\theta).$$

Choosing the datum to be a horizontal line passing through point $O$, the potential energy becomes

$$\mathcal{U} = -mgR\cos\theta.$$

The Lagrangian can therefore be expressed as

$$\mathcal{L} = \frac{1}{2}mR^2(\dot{\theta}^2 + \dot{\phi}^2\sin^2\theta) + mgR\cos\theta.$$

Note that the Lagrangian does not depend explicitly on $\phi$. As such, $\phi$ is an ignorable coordinate. The equations of motion can be written as:

*$\theta$-dynamics:*

$$\frac{\mathrm{d}}{\mathrm{d}t}\left(\frac{\partial\mathcal{L}}{\partial\dot{\theta}}\right) - \frac{\partial\mathcal{L}}{\partial\theta} = 0,$$

$$\ddot{\theta} - \dot{\phi}^2\sin\theta\cos\theta + \frac{g}{R}\sin\theta = 0.$$

*$\phi$-dynamics:*

$$\frac{\mathrm{d}}{\mathrm{d}t}\left(\frac{\partial\mathcal{L}}{\partial\dot{\phi}}\right) = 0,$$

$$\dot{\phi}^2 = \frac{C_\phi}{mR^2\sin^2\theta}.$$

Solving the previous equation for $\dot{\phi}$ and substituting in the first, we obtain

$$\ddot{\theta} - \frac{C_\phi}{mR^2}\cot\theta + \frac{g}{R}\sin\theta = 0.$$

Note that the resulting equation has no dependence on the generalized coordinate $\phi$.

**Flipped Classroom Exercise 5.6**

Consider two masses $(M, m)$, as shown in the figure. Mass $M$ is fixed, while mass $m$ is free to move in space, with its position defined by the generalized coordinates $(r, \theta)$. The two masses interact due to their mutual gravitational forces. Find the equation of motion of the moving mass $m$, in terms of the distance $r$.

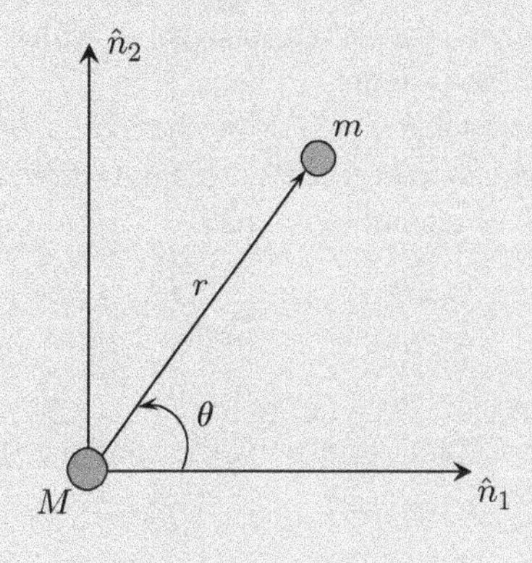

To solve this problem, take the following steps:

1. Use mass $M$ as the inertial point and find the velocity of mass $m$ with respect to the inertial point.
2. Use the velocity of mass $m$ to find the kinetic energy $\mathcal{T}$.
3. Use the gravitational attraction force, $F = G\frac{Mm}{r^2}$, between the particles to determine the potential energy $\mathcal{U} = -\int F dr$.
4. Implement Lagrange's equation along the $\theta$ generalized coordinate. Show that $\theta$ is a cyclic coordinate.
5. Implement Lagrange's equation along the $r$ generalized coordinate. Use the information that $\theta$ is cyclic to reduce the equation of motion into one equation in terms of $r$. Show that the equation takes the form:

$$\ddot{r} - \frac{C_\theta^2}{m^2 r^3} + \frac{GM}{r^2} = 0,$$

where $C_\theta$ is determined from the initial conditions.

## 5.6 Advantages and Disadvantages of the Analytical Approach

Now that we have learned the vectorial and analytical approaches to dynamics, it is worth discussing their advantages and disadvantages. In what follows, we summarize the most obvious advantages and disadvantages of the analytical approach when compared to the vectorial approach.

*Advantages:*

- To obtain the equations of motion using the analytical approach one does not have to create a force balance or calculate the acceleration of a point in space. This is a great advantage when dealing with complex dynamical systems involving a large number of generalized coordinates.
- The analytical approach only deals with scalar quantities. As such, the equations are invariant under any coordinate transformation. Application of the vectorial approach, on the other hand, requires the existence of a Cartesian space in which it is sometimes very difficult to express the dynamics of the systems.
- Internal forces or any forces that do not do work along a virtual displacement do not appear in the derivation of the equation of motion. This is a significant advantage over the vectorial approach when such forces are not of interest.

*Disadvantages:*

- Using the analytical approach does not require much geometrical reasoning. This inhibits any physical intuition about the system dynamics until the end, when the equations of motion are obtained.

- The process of accounting for external forces in the equation of motion requires finding the generalized forces associated with the generalized coordinates. This can sometimes be a difficult task.
- Because internal and constraint forces do not appear naturally when implementing the analytical approach, one has to revert to indirect, and often sophisticated, methodologies to obtain these forces.

## Exercises

5.1 Find the function $y(x)$ that maximizes/minimizes the following functions:

(a) $\int_a^b (y^2(x) - y'^2(x))\mathrm{d}x$.

(b) $\int_a^b (xy'(x) + y'^2(x))\mathrm{d}x$.

(c) $\int_a^b (1+x)y'^2(x)\mathrm{d}x$.

5.2 The pendulum shown in the Figure 5.16 is made by welding two massless rods at a 90° angle to the fixed mass $m$ as shown. Another mass, $\frac{m}{2}$, slides along one of the rods and is connected through a spring of stiffness $k$ to the stationary mass. Assuming that all surfaces are smooth, find the equation of motion of the system in terms of the generalized coordinates $\theta$ and $x$.

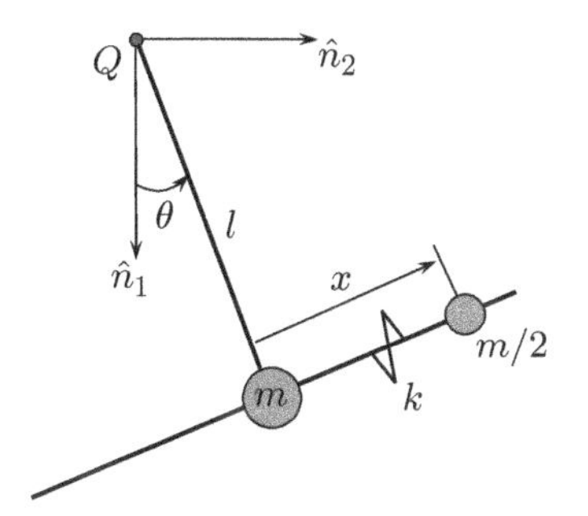


**Figure 5.16** Exercise 5.2

5.3 Mass $m$ is attached to the inner surface of the massless hoop shown in Figure 5.17. The hoop rotates without slipping on the rough horizontal surface. Find the equation of motion of the mass.

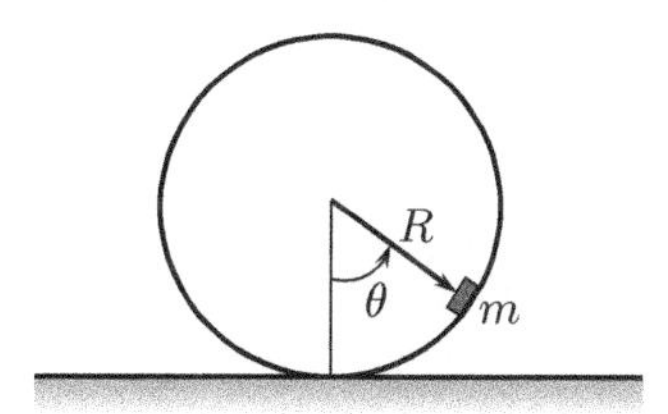


**Figure 5.17** Exercises 5.3, 5.4.

5.4 For the system described in Exercise 5.3, find the reaction force using Lagrange's approach.

5.5 The pendula shown in Figure 5.18 oscillate in the horizontal plane and interact due to their mutual gravitational attraction. Find their equation of motion.

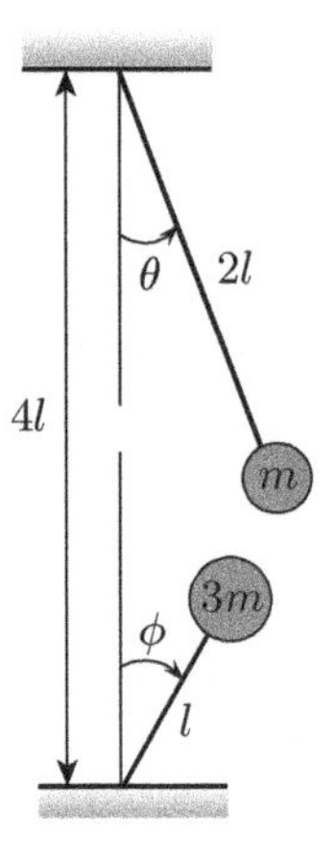


**Figure 5.18** Exercise 5.5.

5.6 The mass shown in Figure 5.19 slides on the stationary surface shown. Use Lagrange's approach to find the reaction force.

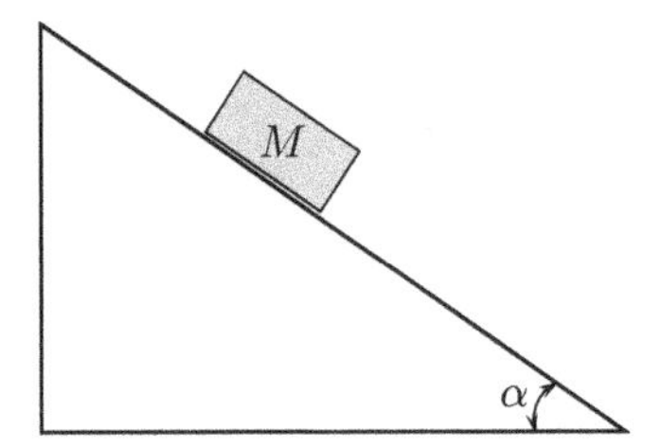


**Figure 5.19** Exercise 5.6.

5.7 The massless ring shown in Figure 5.20 is forced to rotate about point $O$ with angular velocity $\Omega$. A mass $m$ slides along the smooth surface of the ring as the ring rotates. Find the equation of motion of the mass.

5.8 Show that the function $g(y)$ that extremizes the functional

$$J = \int_{x_1}^{x_2} g(y)\sqrt{1 + \left(\frac{\partial y}{\partial x}\right)^2}\, dx$$

must satisfy the differential equation

$$1 + \left(\frac{\partial y}{\partial x}\right)^2 = Cg(y)^2$$

where $C$ is an integration constant.

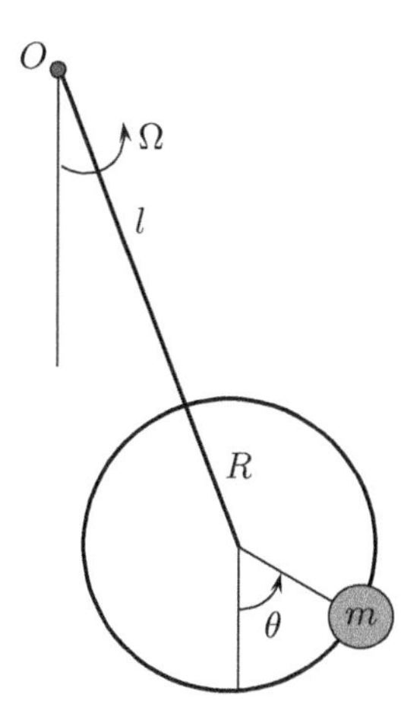


**Figure 5.20** Exercise 5.7.

5.9 The two masses $2m$ and $3m$ are attached to the center of the massless disks shown in Figure 5.21. Find the equations of motion of both masses using Lagrange's approach. Assume that the cable is rigid and that all surfaces are smooth.

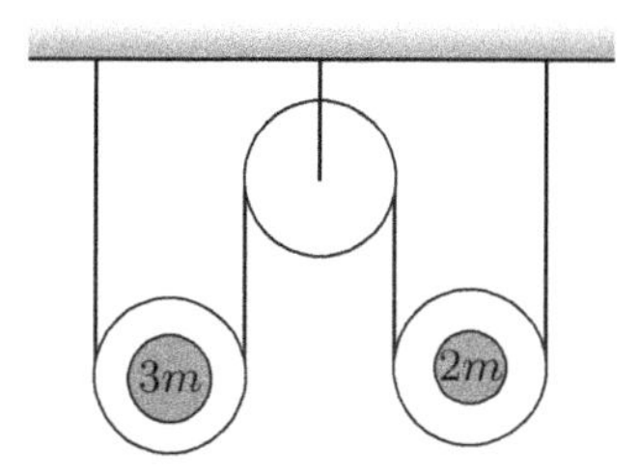


**Figure 5.21** Exercise 5.9.

5.10 Consider the motion of a ball inside a cone of half-angle vertex $\alpha$, as shown in Figure 5.22. Find the equation of motion of the ball. Are there any cyclic coordinates? If so, use this information to reduce the number of equations of motion.

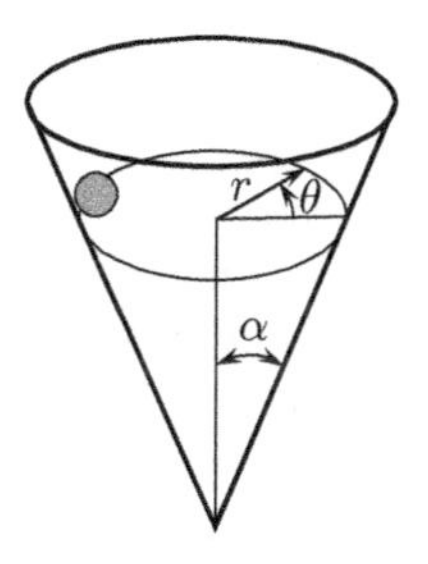


**Figure 5.22** Exercise 5.10.

5.11 Consider the motion of two pendula attached through a rigid string of length $l$, as shown in Figure 5.23. The left-hand pendulum is given an initial condition and allowed to oscillate. Write the equations governing the motion of the system.

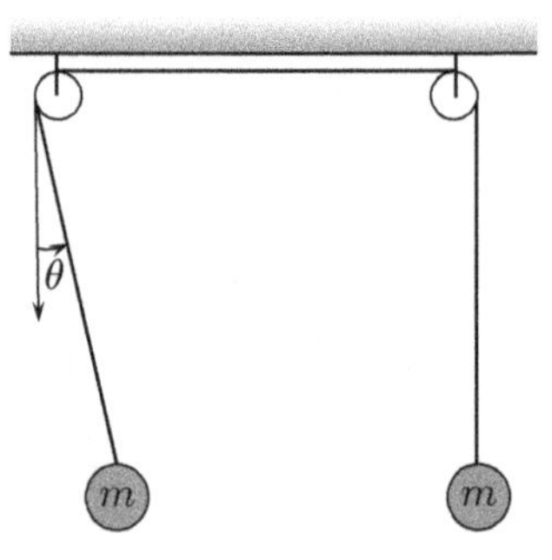


**Figure 5.23** Exercise 5.11.

5.12 Repeat Exercise 5.11 but with an elastic cable of stiffness $k$.

5.13 Mass $m$ is connected on one end to a spring of stiffness $k$ and on the other end to a viscous damper of damping coefficient $c$. As shown in Figure 5.24, the mass slides on the smooth rod which, in turn, is free to rotate about the pivot at point $O$. Find the equations of motion of the system.

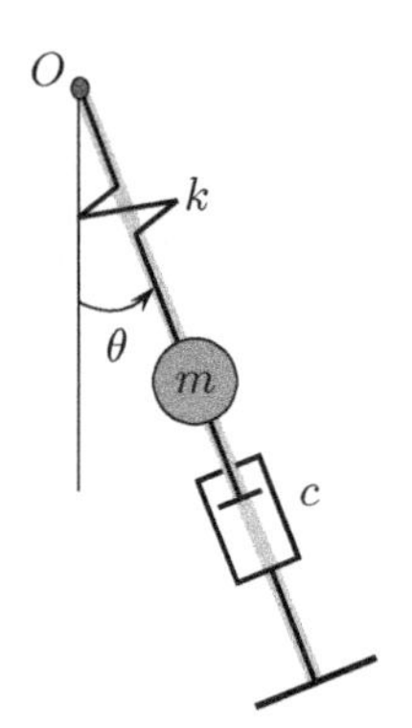


**Figure 5.24** Exercise 5.13.

5.14 A simple pendulum has a natural period which varies with its amplitude. To overcome this problem, Huygens designed a pendulum that varies its length in such a way that it keeps its period constant. As shown in Figure 5.25, he designed the pendulum such that it wraps partially about a cycloidal obstruction as it rotates. The position of any point on the cycloid is described by $\mathbf{r} = R(\theta - \sin\theta)\hat{n}_1 + R(\cos\theta - 1)\hat{n}_2$. Using a pendulum cable of length $4R$, find the equation of motion of the Huygens pendulum in terms of the parametric angle $\theta$. Show that the period of the pendulum is independent of the amplitude of motion.

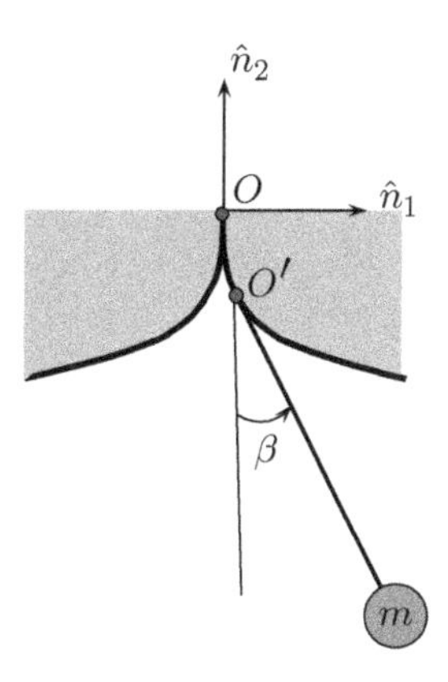


**Figure 5.25** Exercise 5.14.

5.15 Two masses each of mass $m$ are connected through a cable of length $l$ and hang over a smooth massless pulley, as shown in Figure 5.26. One of the masses is released from the position shown. Find the equations of motion of both masses assuming that they do not run into each other during their motion.

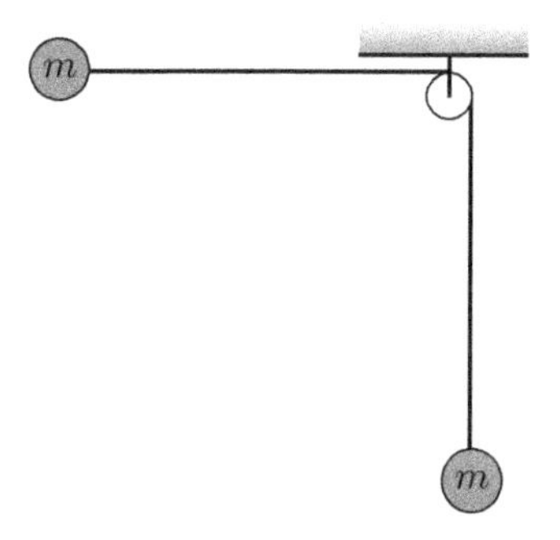


**Figure 5.26** Exercise 5.15.

## References

1. Bernoulli J. (1696) "*Problema novum ad cujus solutionem Mathematici invitantur*". *Acta Eruditorum*, 18, 269.
2. Maupertuis P.L. (1746) "Les loix du mouvement et du repos déduites d'un principe metaphysique". *Histoire de l'Académie Royale des Sciences et des Belles Lettres*, 1746, 267–294.
3. Sommerfeld A. (1956) *Mechanics: Lectures on Theoretical Physics*, vol 1. Academic Press.

**Figure I.1** Aristotle.
Source: https://en.wikipedia.org/wiki/Aristotle.

**Figure I.2** Isaac Newton.
Source: https://en.wikipedia.org/wiki/Isaac_Newton.

*Dynamics of Particles and Rigid Bodies: A Self-Learning Approach*, First Edition. Mohammed F. Daqaq.

Companion website: www.wiley.com/go/daqaq/dynamics

**Figure I.3** Leonhard Euler.
Source: https://en.wikipedia.org
/wiki/Leonhard_Euler.

**Figure I.4** Joseph-Louis Lagrange.
Source: http://ro.math.wikia.com
/wiki/Teorema_lui_Lagrange.

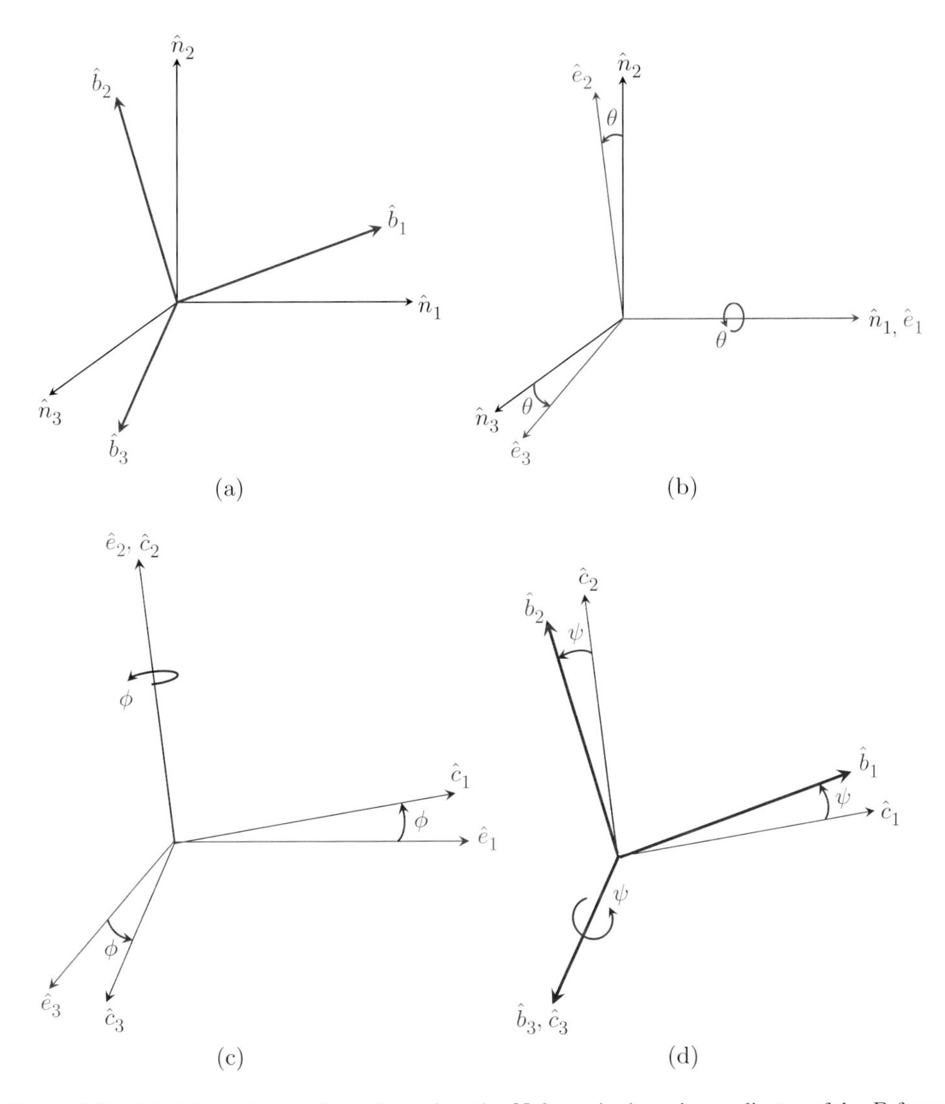


**Figure 1.2** A 1-2-3 rotation performed to orient the $N$-frame in the unit coordinates of the $B$-frame.

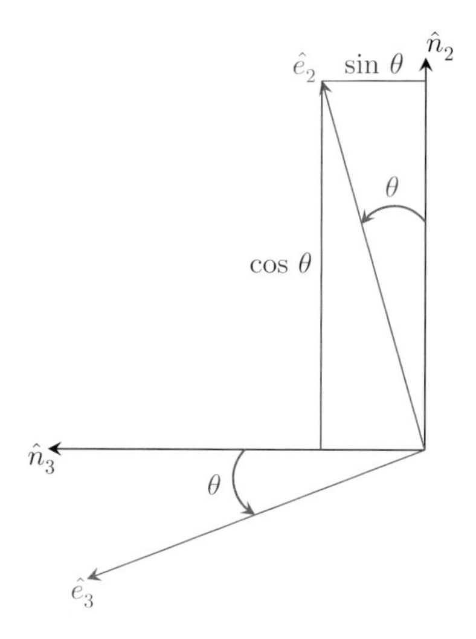


**Figure 1.3** Relationship between the unit vectors of the $E$-frame and the unit vectors of the $N$-frame.

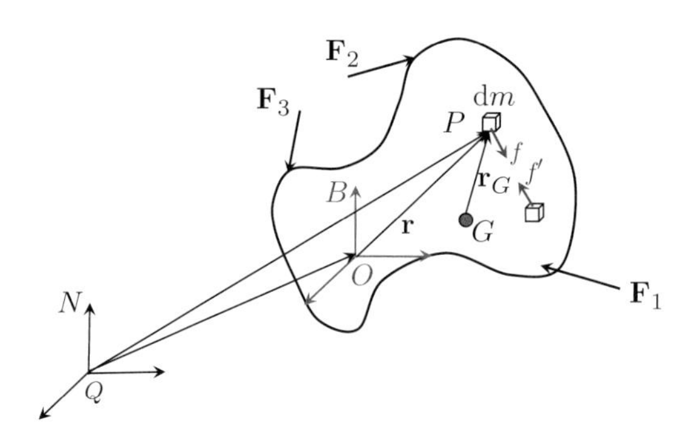


**Figure 3.12** Planar rigid-body dynamics.

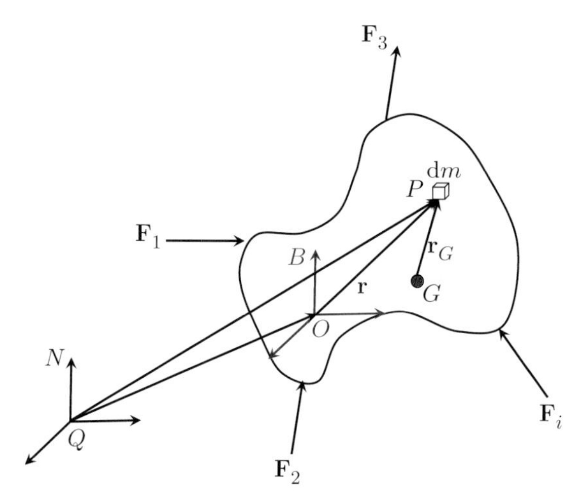


**Figure 7.13** Coordinate system used for the derivation of the angular momentum expression for rigid bodies.

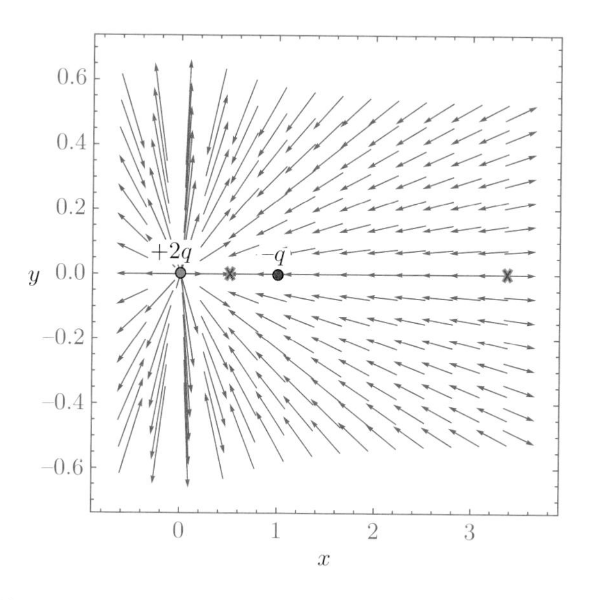


**Figure 8.4** Electric field lines due to the charges $+2q$ and $-q$.

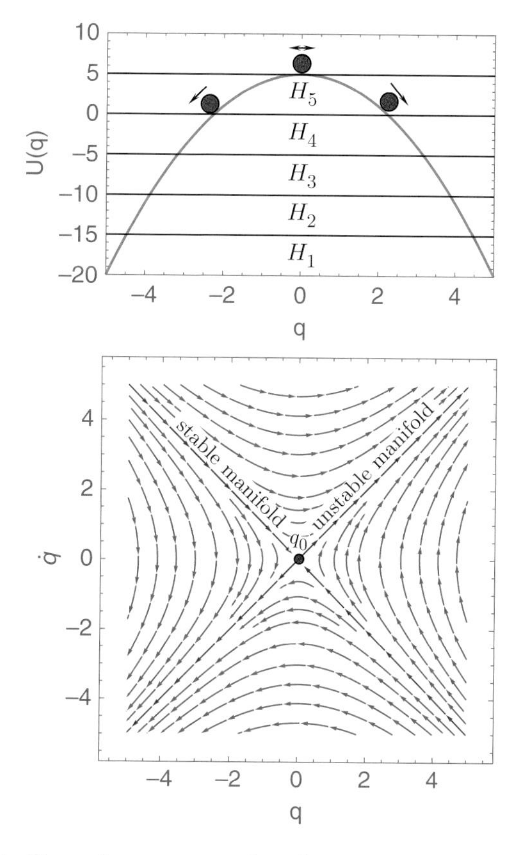


**Figure 9.3** Phase-plane representation of a concave up potential function.

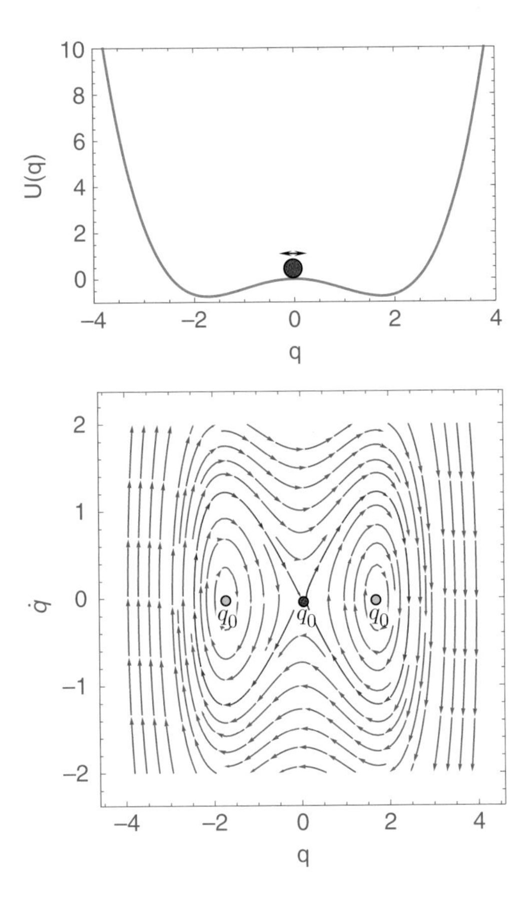


**Figure 9.4** Phase-plane representation of a twin-well potential function.

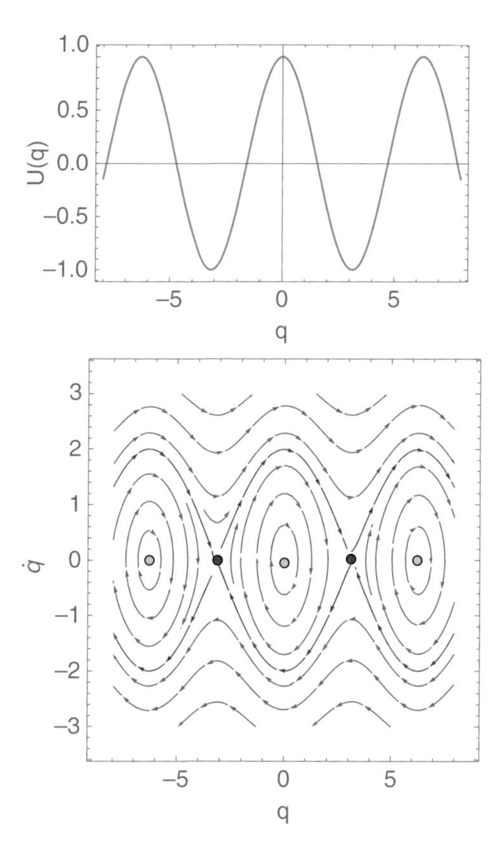


**Figure 9.5** Phase-plane representation of simple pendulum motion.

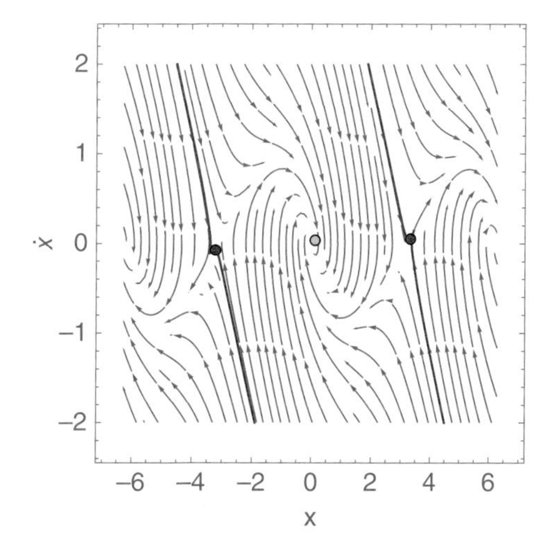


**Figure 9.11** Phase-plane representation of the damped simple pendulum.

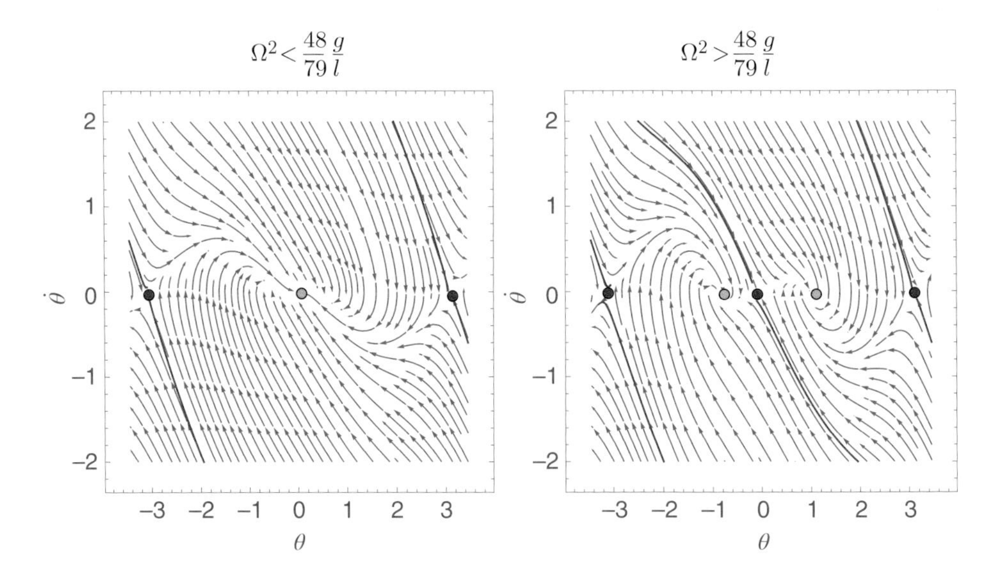


**Figure 9.12** Phase-plane representation of a damped rotating T-shaped rod.

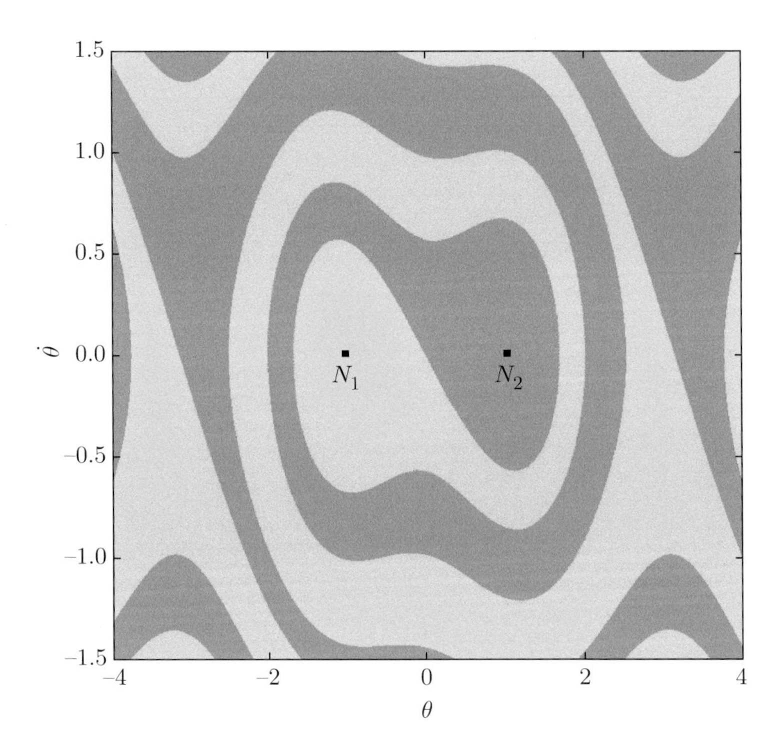


**Figure 9.26** Basins of attraction of the equilibrium solutions of the rotating T-shaped rod for $\Omega^2 > \frac{48g}{79l}$.

# 6

# Dynamics of Rigid Bodies: Analytical Approach

In this chapter, we extend the analytical approach to the motion of rigid bodies. We begin by deriving the kinetic energy expression for rigid bodies and use it in conjunction with Lagrange's equation to find the equations of motion for planar and non-planar rigid bodies.

## 6.1 Kinetic Energy of a Rigid Body

Hamilton's extended principle and Lagrange's equation are still valid in the form derived in Chapter 5 when applied to rigid or even elastic bodies. The only necessary modification lies in the expression of the kinetic energy of the rigid body, which must contain additional terms to account for the rotary inertia of the body.

To derive an expression for the kinetic energy of a rigid body, consider Figure 6.1, which depicts a rigid body of an arbitrary shape. Take a differential element, $dm$, on the body and express its position vector as

$$\mathbf{QP} = \mathbf{QO} + \mathbf{r}. \tag{6.1}$$

It follows that the velocity ${}^N\mathbf{v}^{P/Q}$ of point $P$ with respect to point $Q$, as observed in the inertial, $N$-frame, can be written as

$${}^N\mathbf{v}^{P/Q} = {}^N\mathbf{v}^{O/Q} + {}^N\boldsymbol{\omega}^B \times \mathbf{r}, \tag{6.2}$$

where ${}^N\boldsymbol{\omega}^B$ is the angular velocity of the rigid body described in the $B$-frame. Note that in Equation (6.2), terms involving $\dot{\mathbf{r}}$ vanish because the body is rigid.

The kinetic energy of the differential element can be written as

$$\mathcal{T} = \frac{1}{2}\mathrm{dm}{}^N\mathbf{v}^{P/Q}.{}^N\mathbf{v}^{P/Q}. \tag{6.3}$$

---

*Dynamics of Particles and Rigid Bodies: A Self-Learning Approach*, First Edition. Mohammed F. Daqaq.

Companion website: www.wiley.com/go/daqaq/dynamics

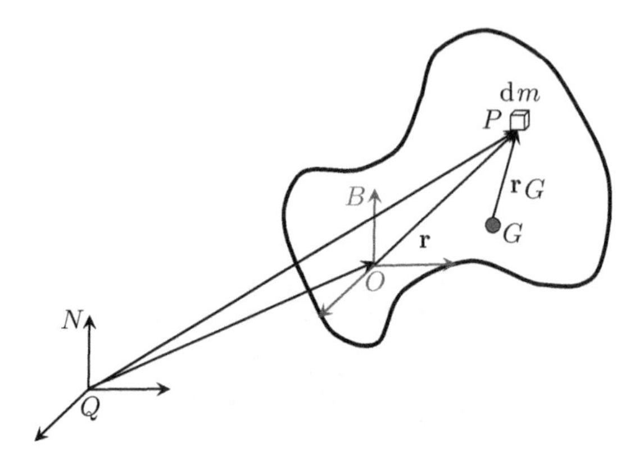


**Figure 6.1** Kinetic energy of a rigid body.

To obtain the kinetic energy of the rigid body, we integrate Equation (6.3) over the volume $\mathcal{V}$, which yields

$$\begin{aligned} \mathcal{T} &= \int_{\mathcal{V}} \frac{1}{2}\, {}^{N}\mathbf{v}^{P/Q}.{}^{N}\mathbf{v}^{P/Q} \mathrm{d}m, \\ &= \int_{\mathcal{V}} \frac{1}{2}({}^{N}\mathbf{v}^{O/Q} + {}^{N}\boldsymbol{\omega}^{B} \times \mathbf{r}).({}^{N}\mathbf{v}^{O/Q} + {}^{N}\boldsymbol{\omega}^{B} \times \mathbf{r})\mathrm{d}m. \end{aligned} \tag{6.4}$$

Since the velocity, ${}^{N}\mathbf{v}^{O/Q}$, does not depend on the location of $\mathrm{d}m$, it can be taken outside the integral, and the previous equation can be reduced to:

$$\begin{aligned} \mathcal{T} = \frac{1}{2}\, {}^{N}\mathbf{v}^{O/Q}.{}^{N}\mathbf{v}^{O/Q} \int_{\mathcal{V}} \mathrm{d}m + {}^{N}\mathbf{v}^{O/Q}.\left({}^{N}\boldsymbol{\omega}^{B} \times \int_{\mathcal{V}} \mathbf{r}\mathrm{d}m\right) \\ + \frac{1}{2}\int_{\mathcal{V}} ({}^{N}\boldsymbol{\omega}^{B} \times \mathbf{r}) \cdot ({}^{N}\boldsymbol{\omega}^{B} \times \mathbf{r})\mathrm{d}m. \end{aligned} \tag{6.5}$$

Using the identity $\mathbf{A}.(\mathbf{B} \times \mathbf{C}) = \mathbf{B}.(\mathbf{C} \times \mathbf{A})$, the last term in the previous equation can be simplified to

$$\frac{1}{2}\int_{\mathcal{V}} ({}^{N}\boldsymbol{\omega}^{B} \times \mathbf{r}).({}^{N}\boldsymbol{\omega}^{B} \times \mathbf{r})\mathrm{d}m = \frac{1}{2}\left({}^{N}\boldsymbol{\omega}^{B}.\int_{\mathcal{V}} \mathbf{r} \times ({}^{N}\boldsymbol{\omega}^{B} \times \mathbf{r})\mathrm{d}m\right). \tag{6.6}$$

Next, we utilize the identity $\mathbf{A} \times (\mathbf{B} \times \mathbf{C}) = \mathbf{B}(\mathbf{A}.\mathbf{C}) - \mathbf{C}(\mathbf{A}.\mathbf{B})$, to completely eliminate the cross products from the previous equation, and obtain:

$$\frac{1}{2}\left({}^{N}\boldsymbol{\omega}^{B}.\int_{\mathcal{V}} \mathbf{r} \times ({}^{N}\boldsymbol{\omega}^{B} \times \mathbf{r})\mathrm{d}m\right) = \frac{1}{2}\, {}^{N}\boldsymbol{\omega}^{B}.\int_{\mathcal{V}} ({}^{N}\boldsymbol{\omega}^{B}(\mathbf{r}.\mathbf{r}) - \mathbf{r}.(\mathbf{r}.{}^{N}\boldsymbol{\omega}^{B})\mathrm{d}m. \tag{6.7}$$

Substituting $\mathbf{r} = x\hat{b}_1 + y\hat{b}_2 + z\hat{b}_3$, ${}^{N}\boldsymbol{\omega}^{B} = \omega_1\hat{b}_1 + \omega_2\hat{b}_2 + \omega_3\hat{b}_3$ into the previous equation and simplifying yields

$$\frac{1}{2}\, {}^{N}\boldsymbol{\omega}^{B}.\int_{\mathcal{V}} ({}^{N}\boldsymbol{\omega}^{B}(\mathbf{r}.\mathbf{r}) - \mathbf{r}.(\mathbf{r}.{}^{N}\boldsymbol{\omega}^{B}))\mathrm{d}m = \frac{1}{2}\, {}^{N}\boldsymbol{\omega}^{B}.(I_O\, {}^{N}\boldsymbol{\omega}^{B}), \tag{6.8}$$

where $I_O$ is the moment of inertia matrix about a set of axes whose origin is located at point $O$. It follows from Equation (6.5) that the kinetic energy can be expressed as

$$\mathcal{T} = \frac{1}{2}m\,{}^N\mathbf{v}^{O/Q}.{}^N\mathbf{v}^{O/Q} + {}^N\mathbf{v}^{O/Q}.\left({}^N\omega^B \times \int_V \mathbf{r}\mathrm{d}m\right) + \frac{1}{2}\,{}^N\omega^B.(I_O\,{}^N\omega^B) \tag{6.9}$$

Equation (6.9) can be simplified further by choosing point $O$ to coincide with the center of mass of the rigid body $G$, for which $\int \mathbf{r}_G \mathrm{d}m = 0$. This yields

$$\mathcal{T} = \frac{1}{2}m\,{}^N\mathbf{v}^{G/Q}.{}^N\mathbf{v}^{G/Q} + \frac{1}{2}\,{}^N\omega^B.(I_G\,{}^N\omega^B) \tag{6.10}$$

where the first term in the kinetic energy expression is due to translational motion, the second term is due to rotational motion, and $I_G$ is the moment of inertia matrix about a set of axes whose origin is located at the center of mass $G$. Equation (6.9) can also be simplified if we choose point $O$ to be a fixed point, $S$, on the body. In such cases, ${}^N\mathbf{v}^{S/Q} = 0$ and Equation (6.9) simplifies to:

$$\mathcal{T} = \frac{1}{2}\,{}^N\omega^B.(I_S\,{}^N\omega^B) \tag{6.11}$$

where $I_S$ is the moment of inertia about a set of axes whose origin is located at the fixed point, $S$.

**Kinetic Energy of a Rigid Body**

The expression of the kinetic energy of a rigid body changes depending on the type of point used to calculate it.

- For an arbitrary point on the body, $O$, it can be expressed as

$$\mathcal{T} = \frac{1}{2}m\,{}^N\mathbf{v}^{O/Q}.{}^N\mathbf{v}^{O/Q} + {}^N\mathbf{v}^{O/Q}.\left({}^N\omega^B \times \int_V \mathbf{r}\mathrm{d}m\right) + \frac{1}{2}\,{}^N\omega^B.(I_O\,{}^N\omega^B),$$

  where $O$ is a moving point located on the rigid body, $Q$ is an inertial point and $I_O$ is the moment of inertia about a set of axes whose origin is located at point $O$.
- For the center of mass, $G$, it can be expressed as

$$\mathcal{T} = \frac{1}{2}m\,{}^N\mathbf{v}^{G/Q}.{}^N\mathbf{v}^{G/Q} + \frac{1}{2}\,{}^N\omega^B.(I_G\,{}^N\omega^B),$$

  where $I_G$ is the moment of inertia about a set of coordinates whose origin is located at point $G$.
- For a fixed point, $S$, located on the body, it can be expressed as

$$\mathcal{T} = \frac{1}{2}\,{}^N\omega^B.(I_S\,{}^N\omega^B),$$

where $I_S$ is the moment of inertia about a set of axes whose origin is located at the fixed point, $S$.

**Example 6.1 Kinetic Energy of a Rotating Rod**

For the system shown in Figure 6.2, find the kinetic energy of the rod, of length $l$ and mass $M$.

Note that there is no fixed point on the rod. As such, the best way to obtain the kinetic energy is to use the expression involving the center of mass; that is

$$\mathcal{T} = \frac{1}{2}m\ {}^N\mathbf{v}^{G/Q}.{}^N\mathbf{v}^{G/Q} + \frac{1}{2}\ {}^N\boldsymbol{\omega}^B.(I_G\ {}^N\boldsymbol{\omega}^B).$$

To this end, we need to find ${}^N\mathbf{v}^{G/Q}$, ${}^N\boldsymbol{\omega}^B$, and $I_G$. To find ${}^N\mathbf{v}^{G/Q}$, we introduce the $A$-frame such that ${}^N\boldsymbol{\omega}^A = \Omega\hat{a}_2$ and the $B$-frame such that ${}^A\boldsymbol{\omega}^B = -\dot{\theta}\hat{b}_3$. Using these frames, the position vector from point $Q$ to point $G$ can be described as

$$QG = L\hat{a}_2 + R\hat{a}_1 + \frac{l}{2}\hat{b}_2.$$

The velocity of point $G$ with respect to point $Q$ is

$${}^N\mathbf{v}^{G/Q} = {}^N\boldsymbol{\omega}^A \times (R\hat{a}_1 + L\hat{a}_2) + {}^N\boldsymbol{\omega}^B \times \frac{l}{2}\hat{b}_2,$$

$${}^N\mathbf{v}^{G/Q} = \frac{l}{2}\dot{\theta}\hat{b}_1 - \left(R\Omega + \frac{l}{2}\sin\theta\Omega\right)\hat{b}_3.$$

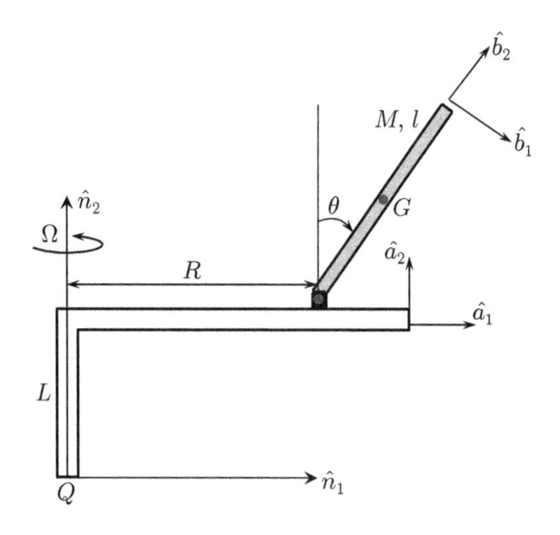


**Figure 6.2** Kinetic energy of a rotating rod.

Now that the velocity of the center of mass has been obtained, we turn our attention to finding the angular velocity vector, which can be written as

$$^N\boldsymbol{\omega}^B = \,^N\boldsymbol{\omega}^A + \,^A\boldsymbol{\omega}^B = \Omega\hat{a}_2 - \dot{\theta}\hat{b}_3 = -\Omega\sin\theta\hat{b}_1 + \Omega\cos\theta\hat{b}_2 - \dot{\theta}\hat{b}_3.$$

Expressing the angular velocity in the matrix form, $^N\boldsymbol{\omega}^B = [\omega_1\ \omega_2\ \omega_3]^T$ yields $\omega_1 = -\Omega\sin\theta$, $\omega_2 = \Omega\cos\theta$, and $\omega_3 = -\dot{\theta}$. It is worth mentioning that it is best to describe the angular velocity vector in the body's rotating frame, in which the moment of inertia is much easier to obtain.

Finally, the moment of inertia matrix $I_G$ is obtained and expressed in the body rotating $B$-frame, as

$$I_{11G} = \frac{1}{12}Ml^2, \qquad I_{22G} = 0\ (\text{thin rod}), \qquad I_{33G} = \frac{1}{12}Ml^2. \tag{6.12}$$

Note that in the $B$-frame, all the products of inertia terms are zero. Therefore, the inertia matrix can be written as

$$I_G = Ml^2\begin{pmatrix}\frac{1}{12} & 0 & 0\\ 0 & 0 & 0\\ 0 & 0 & \frac{1}{12}\end{pmatrix}.$$

Now that all the elements needed to obtain the kinetic energy have been defined, we can calculate the translational component of the kinetic energy as

$$\mathcal{T}_{trans} = \frac{1}{2}M\,^N\mathbf{v}^{G/Q}.\,^N\mathbf{v}^{G/Q} = \frac{1}{2}M\left(R\Omega^2(R + l\sin\theta) + \frac{l^2}{4}\sin^2\theta\Omega^2 + \frac{l^2}{4}\dot{\theta}^2\right).$$

The rotational component can be written as

$$\mathcal{T}_{rot} = \frac{1}{2}Ml^2\begin{bmatrix}\omega_1 & \omega_2 & \omega_3\end{bmatrix}\begin{pmatrix}\frac{1}{12} & 0 & 0\\ 0 & 0 & 0\\ 0 & 0 & \frac{1}{12}\end{pmatrix}\begin{bmatrix}\omega_1\\ \omega_2\\ \omega_3\end{bmatrix} = \frac{1}{24}ml^2(\Omega^2\sin^2\theta + \dot{\theta}^2).$$

It follows that the kinetic energy of the rotating rod is

$$\mathcal{T} = \frac{1}{2}M\left(R^2\Omega^2 + Rl\Omega^2\sin\theta + \frac{l^2}{4}\sin^2\theta\Omega^2 + \frac{l^2}{4}\dot{\theta}^2\right) + \frac{1}{24}ml^2(\Omega^2\sin^2\theta + \dot{\theta}^2).$$

**Flipped Classroom Exercise 6.1**

Consider a cube of mass $m$ and edge $a$ rotating about an axis that passes through its center of gravity and the mid-points of two of its edges, as shown in the figure. Find the kinetic energy of the cube.

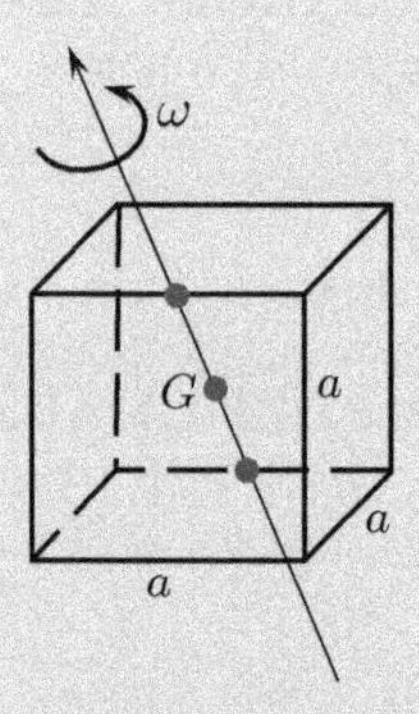


There are two ways to solve this problem: the first, and not so smart approach, is to find the moment of inertia about the axis of rotation defined in the problem. This would require finding the moment of inertia of the cube about a set of axes parallel to its sides then rotating it by using the proper rotation matrix. The shorter, and more intelligent approach, is to project the angular velocity vector in the direction of the cube's principal axes, then use the principal axes of rotation to obtain the kinetic energy of the body about point $G$. In other words, you can follow the following steps:

1. Find the principal moment of inertia matrix of the cube about point $G$.
2. Project the angular velocity vector in the direction of the principal axes.
3. Use the kinetic energy expression to show that

$$\mathcal{T} = \frac{1}{2} m a^2 \omega^2.$$

## 6.2 Lagrange's Equation Applied to Rigid Bodies

As described in the last section, Lagrange's equation in its different forms holds whether we are dealing with particles or rigid bodies. In what follows, we present several examples to demonstrate the implementation of Lagrange's equations on rigid bodies.

### Example 6.2 Rotating T-shaped Structure

The system shown in Figure 6.3 consists of two welded rods, each of mass $m$ and length $l$. The system is pinned at point $O$ to a column that rotates with a constant angular velocity $\Omega$. Find the equation of motion of the rod in terms of the angle $\theta$.

To obtain the equation of motion, we define the following frames: ${}^N\boldsymbol{\omega}^A = \Omega \hat{a}_2$ and ${}^N\boldsymbol{\omega}^B = \dot{\theta}\hat{b}_3$. Using these frames, we find the kinetic energy of the body. Since point $O$ is a fixed point on the body, we can write

$$\mathcal{T}_{\text{rigidbody}} = \frac{1}{2}\, {}^N\boldsymbol{\omega}^B . (I_O\, {}^N\boldsymbol{\omega}^B),$$

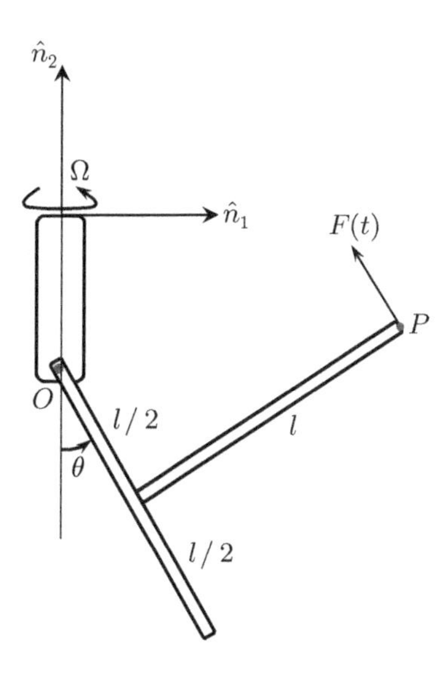


**Figure 6.3** Rotating T-shaped rod.

where ${}^N\boldsymbol{\omega}^B = {}^N\boldsymbol{\omega}^A + {}^A\boldsymbol{\omega}^B = \Omega\hat{a}_2 + \dot{\theta}\hat{b}_3 = \Omega\sin\theta\hat{b}_1 + \Omega\cos\theta\hat{b}_2 + \dot{\theta}\hat{b}_3$, and

$$I_O = ml^2 \begin{pmatrix} \frac{7}{12} & \frac{1}{4} & 0 \\ \frac{1}{4} & \frac{1}{3} & 0 \\ 0 & 0 & \frac{11}{12} \end{pmatrix}.$$

It follows that the kinetic energy is

$$\mathcal{T}_{\text{rigidbody}} = \frac{ml^2\Omega^2}{6}\left(\frac{7}{4}\sin^2\theta + \cos^2\theta\right) + \frac{ml^2\Omega^2}{4}\sin\theta\cos\theta + \frac{11}{24}ml^2\dot{\theta}^2.$$

Choosing the datum to be a horizontal line passing through point $O$ yields

$$\mathcal{U}_{\text{rigidbody}} = mgl\left(\frac{1}{2}\sin\theta - \cos\theta\right).$$

The generalized force due the force $F(t)$ can be written as

$$Q_\theta = F(t)\hat{b}_2.\frac{\partial^N\mathbf{v}^{P/O}}{\partial\dot{\theta}} = F(t)l.$$

Applying Lagrange's equation along the $\theta$ generalized coordinate yields

$$\frac{11}{12}ml^2\ddot{\theta} + mgl\left(\frac{1}{2}\cos\theta + \sin\theta\right) - \frac{ml^2\Omega^2}{4}\left(\cos 2\theta + \frac{1}{2}\sin 2\theta\right) = F(t)l.$$

**Example 6.3 Double Pendulum with Viscous Damping**

For the system shown in Figure 6.4, the rod is thin and homogeneous with a mass $m$ and length $l$. The external force, $F(t)$, is always perpendicular to the link connecting the mass to the rigid rod. Use Lagrange's method to obtain the equations governing the motion of the system.

To obtain the equation of motion, we define the following frames: ${}^N\boldsymbol{\omega}^A = \dot{\theta}\,\hat{a}_3$ and ${}^N\boldsymbol{\omega}^B = \dot{\phi}\,\hat{b}_3$. Since the system consists of a rigid body and a particle, we need to find the kinetic and potential energy of each. For the rigid body, we note that point $O$ is a fixed point on the body; therefore we can write

$$\mathcal{T}_{\text{rigidbody}} = \frac{1}{2}\,{}^N\boldsymbol{\omega}^A.(I_O\,{}^N\boldsymbol{\omega}^A).$$

Since the motion of the rod involves a single rotation ${}^N\boldsymbol{\omega}^A = \dot{\theta}$, the previous equation can be simplified to

$$\mathcal{T}_{\text{rigidbody}} = \frac{1}{2}I_{33O}\,{}^N\boldsymbol{\omega}^A.{}^N\boldsymbol{\omega}^A = \frac{1}{6}ml^2\dot{\theta}^2.$$

The kinetic energy of the particle is given by

$$\mathcal{T}_{\text{particle}} = \frac{1}{2}m\,{}^N\mathbf{v}^{P/O}.\mathbf{v}^{P/O},$$

where you can show that

$${}^N\mathbf{v}^{P/O} = (l\dot{\theta} + l\dot{\phi}\cos(\phi-\theta))\hat{a}_1 + l\dot{\phi}\sin(\phi-\theta)\hat{a}_2.$$

It follows that the total kinetic energy of the system can be written as

$$\mathcal{T} = \frac{2}{3}ml^2\dot{\theta}^2 + \frac{1}{2}ml^2\dot{\phi}^2 + ml^2\dot{\theta}\dot{\phi}\cos(\phi-\theta).$$

The potential energy consists of the potential energy of the rigid body and the particle. Choosing our datum to be a horizontal line passing through point $O$, we can write

$$\mathcal{U}_{\text{rigidbody}} = -mg\frac{l}{2}\cos\theta,$$

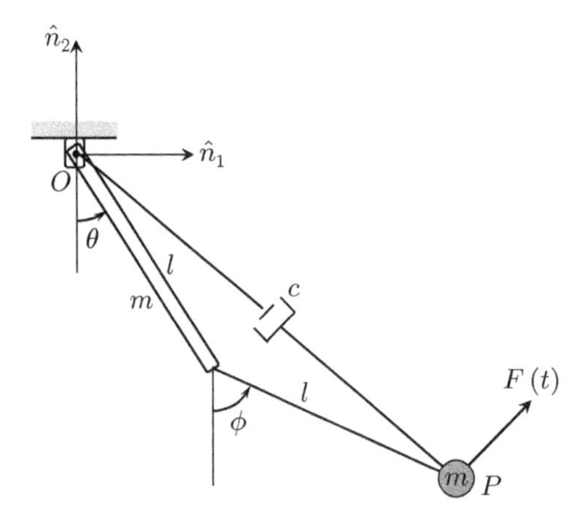


**Figure 6.4** Double compound pendulum with viscous damping.

$$\mathcal{U}_{\text{particle}} = -mgl\cos\theta - mgl\cos\phi,$$

$$\mathcal{U} = -\frac{3}{2}mgl\cos\theta - mgl\cos\phi.$$

The power dissipated by the damper is given by

$$\mathcal{D} = \frac{1}{2}c\dot{p}^2,$$

where $\dot{p}$ is the temporal variation of the damper length $p = l\sqrt{2(1+\cos(\phi-\theta))}$.

The generalized forces due to the force $F(t)$ are given by

$$Q_\theta = F(t)\hat{b}_1.\frac{\partial}{\partial\dot{\theta}}({}^N\mathbf{v}^{P/O}) = F(t)l\cos(\phi-\theta),$$

$$Q_\phi = F(t)\hat{b}_1.\frac{\partial}{\partial\dot{\phi}}({}^N\mathbf{v}^{P/O}) = F(t)l.$$

The equation of motion can be written as

$$\frac{\mathrm{d}}{\mathrm{d}t}\left(\frac{\partial\mathcal{L}}{\partial\dot{\theta}}\right) + \frac{\partial\mathcal{D}}{\partial\dot{\theta}} - \frac{\partial\mathcal{L}}{\partial\theta} = Q_\theta,$$

$$\frac{\mathrm{d}}{\mathrm{d}t}\left(\frac{\partial\mathcal{L}}{\partial\dot{\phi}}\right) + \frac{\partial\mathcal{D}}{\partial\dot{\phi}} - \frac{\partial\mathcal{L}}{\partial\phi} = Q_\phi.$$

 **Flipped Classroom Exercise 6.2**

Consider a thin plate of mass $m$, height $h$ and width $b$, which is pinned to a rotating frame at point $O$, as shown in the figure. The frame rotates at a constant angular velocity $\Omega$. Find the equation of motion for the plate.

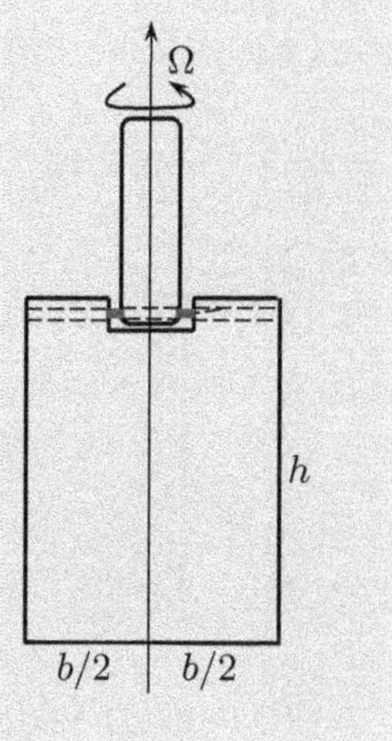


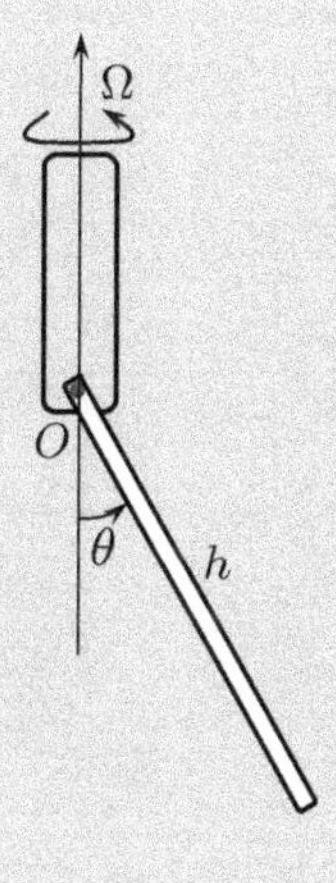

To solve the problem, do the following:

1. Define the following rotating frames: $A$-frame such that ${}^N\boldsymbol{\omega}^A = \Omega\hat{a}_2$ and $B$-frame ${}^A\boldsymbol{\omega}^B = \dot{\theta}\hat{b}_3$.
2. Describe the angular velocity vector, ${}^N\boldsymbol{\omega}^B$ in term of its components $\omega_1, \omega_2$, and $\omega_3$.
3. Find the inertia matrix $I_O$ in the rotating body, $B$-frame, about a set of axes whose origin is located at point $O$.
4. Use the inertia matrix and the angular velocity vector to find the kinetic energy, $\frac{1}{2}{}^N\boldsymbol{\omega}^B.(I_O\ {}^N\boldsymbol{\omega}^B)$, of the plate and show that it can be written as
$$\mathcal{T} = \frac{1}{6}m\left(h^2 + \frac{b^2}{4}\right)\Omega^2\sin^2\theta + \frac{1}{24}mb^2\Omega^2\cos^2\theta + \frac{1}{6}mh^2\dot{\theta}^2.$$
5. Show that the potential energy of the plate can be written as
$$\mathcal{U} = -mg\frac{h}{2}\cos\theta.$$
6. Use Lagrange's equation to write the equation of motion.

**Example 6.4 Rolling Disk**

A disk of mass $m$ and radius $R$ rolls without slipping as it unwinds down the cable shown in Figure 6.5. Find the tension in the cable using Lagrange's approach.

To find the tension force using Lagrange's approach, we remove the tension force and add the generalized coordinate $x$ to fully describe the motion of the disk in the absence of the constraint force. We then replace the tension force by the constraint equation $x = 0$ or $\mathrm{d}x = 0$.

Using the generalized coordinates $x$ and $\phi$, we express the kinetic energy of the disk as
$$\mathcal{T} = \frac{1}{2}m{}^N\mathbf{v}^{G/O}.{}^N\mathbf{v}^{G/O} + \frac{1}{2}I_G\dot{\phi}^2,$$
where ${}^N\mathbf{v}^{G/O} = (\dot{x} + R\dot{\phi})\hat{n}_1$ and $I_G = \frac{1}{2}mR^2$; this yields
$$\mathcal{T} = \frac{1}{2}m\dot{x}^2 + \frac{3}{4}mR^2\dot{\phi}^2 + mR\dot{\phi}\dot{x}.$$

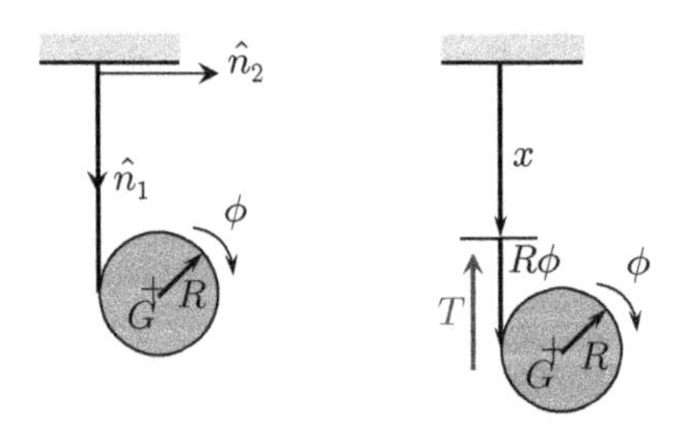


**Figure 6.5** Disk rolling down a cable.

The potential energy is

$$\mathcal{U} = -mg(x + R\phi).$$

Using the expression for the kinetic and potential energy and the constraint equation $\mathrm{d}x = 0$, the equations of motion can be written as

$$\frac{\mathrm{d}}{\mathrm{d}t}\left(\frac{\partial \mathcal{L}}{\partial \dot{x}}\right) - \frac{\partial \mathcal{L}}{\partial x} = \lambda, \qquad m\ddot{x} + mR\ddot{\phi} - mg = \lambda,$$

$$\frac{\mathrm{d}}{\mathrm{d}t}\left(\frac{\partial \mathcal{L}}{\partial \dot{\phi}}\right) - \frac{\partial \mathcal{L}}{\partial \phi} = 0, \qquad \ddot{\phi} + \frac{2}{3R}\ddot{x} - \frac{2g}{3R} = 0.$$

Letting $Q_x = \lambda$, where $Q_x = -T\hat{n}_1 \cdot \frac{\partial \dot{x}}{\partial \dot{x}}\hat{n}_1 = -T$, then setting $x = 0$ and solving the previous equations for $T$, we obtain $T = \frac{mg}{3}$.

### Example 6.5 Shear Force

A rod of length $l$ and mass $m$ is released from the position shown in Figure 6.6. Use Lagrange's approach to find the shear force at point $P$ as a function of the angle, $\theta$.

As described in Chapter 5, to find internal forces using Lagrange's approach, we remove the constraint force and add a generalized coordinate to fully represent the motion of the system in the absence of the constraint force. In this case, removing the shear force at point $P$ allows the lower part of the rod to slide with respect to the upper part in a direction parallel to the shear force. This is represented by an additional generalized coordinate $y$, as shown in Figure 6.7. We replace the shear force by a constraint equation $y = 0$, or $\mathrm{d}y = 0$.

Now we treat the rod as two different rods and find the kinetic and potential energy for each. For the upper part of the rod, the kinetic energy is simply

$$\mathcal{T}_1 = \frac{1}{2} I_O ({}^N\boldsymbol{\omega}^B . {}^N\boldsymbol{\omega}^B),$$

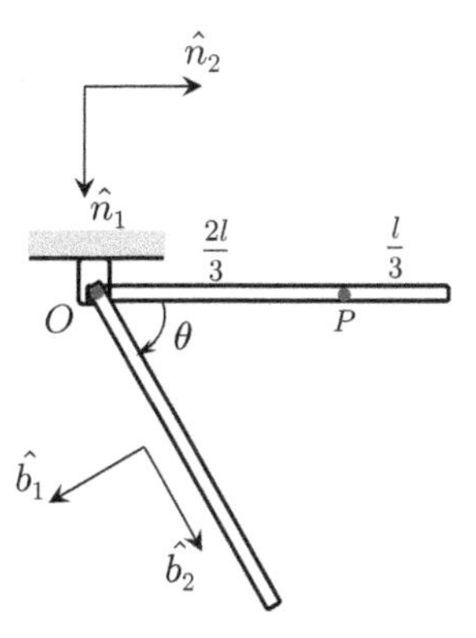


**Figure 6.6** Thin rod released from rest.

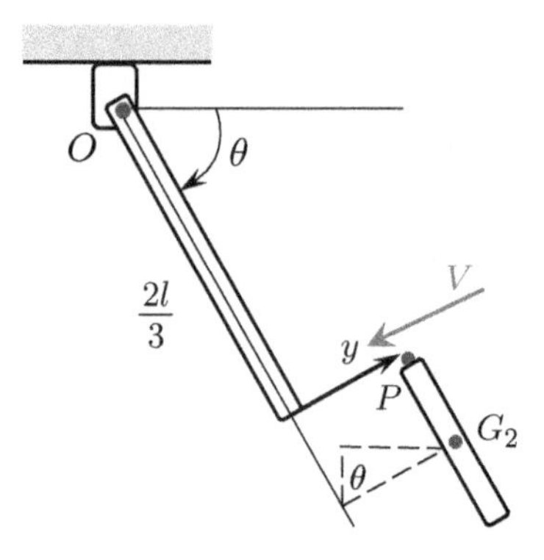


**Figure 6.7** Releasing the shear force.

where $I_O = \frac{1}{3}\left(\frac{2}{3}m\right)\left(\frac{2}{3}l\right)^2 = \frac{8}{81}ml^2$ and ${}^N\omega^B = -\dot{\theta}\hat{b}_3$; this yields

$$\mathcal{T}_1 = \frac{4}{81}ml^2\dot{\theta}^2.$$

For the lower part, the kinetic energy can be obtained using

$$\mathcal{T}_2 = \frac{1}{2}\left(\frac{1}{3}m\right)\ {}^N\mathbf{v}^{G_2/O}\cdot{}^N\mathbf{v}^{G_2/O} + \frac{1}{2}I_{G2}\dot{\theta}^2,$$

where $G_2$ is the center of mass of the lower part. $I_{G2} = \frac{1}{324}ml^2$ and ${}^N\mathbf{v}^{G_2/O}$ can be obtained as follows:

$$\mathbf{OG}_2 = -y\hat{b}_1 + \frac{5}{6}l\hat{b}_2,$$

$${}^N\mathbf{v}^{G_2/O} = \left(-\dot{y} + \frac{5}{6}l\dot{\theta}\right)\hat{b}_1 + y\dot{\theta}\hat{b}_2.$$

The total kinetic energy of the system becomes

$$\mathcal{T} = \mathcal{T}_1 + \mathcal{T}_2 = \frac{1}{6}m\left(l^2\dot{\theta}^2 + \dot{y}^2 + y^2\dot{\theta}^2 - \frac{5}{3}l\dot{y}\dot{\theta}\right).$$

The potential energy of the system consists of the potential energy of the two rods; that is

$$\mathcal{U} = -mg\frac{l}{2}\sin\theta + \frac{1}{3}mgy\cos\theta.$$

Using the kinetic and potential energy expressions as well as the constraint equation, the equations of motion for the $\theta$ and $y$ generalized coordinates become:

$\theta$-*dynamics*:

$$\frac{\mathrm{d}}{\mathrm{d}t}\left(\frac{\partial\mathcal{L}}{\partial\dot{\theta}}\right) - \frac{\partial\mathcal{L}}{\partial\theta} = 0,$$

$$\frac{1}{3}m(l^2+y^2)\ddot{\theta}+\frac{2}{3}my\dot{y}\dot{\theta}-\frac{5}{18}ml\ddot{y}-mg\frac{l}{2}\cos\theta-\frac{1}{3}my\sin\theta=0.$$

*y-dynamics:*

$$\frac{\mathrm{d}}{\mathrm{d}t}\left(\frac{\partial\mathcal{L}}{\partial\dot{y}}\right)-\frac{\partial\mathcal{L}}{\partial y}=\lambda \qquad \frac{1}{3}m\left(\ddot{y}-\frac{5}{6}l\ddot{\theta}-y\dot{\theta}+g\cos\theta\right)=\lambda.$$

Letting $y,\dot{y},\ddot{y}=0$ in the previous equations, and solving for $\lambda$, we obtain

$$\lambda=-\frac{1}{12}mg\cos\theta.$$

To find the shear force $V$, we use the fact that the generalized force $Q_y$ associated with the shear force is equal to the Lagrange multiplier $\lambda$; that is,

$$Q_y=\lambda=V\hat{b}_1.\frac{\partial}{\partial\dot{y}}({}^N\mathbf{v}^{P/O})=V\hat{b}_1.\frac{\partial}{\partial\dot{y}}\left(\left(\dot{y}+\frac{2}{3}l\dot{\theta}\right)\hat{b}_1+y\dot{\theta}\hat{b}_2\right)=V.$$

As such, $V=\lambda=-\frac{1}{12}mg\cos\theta$.

**Example 6.6 Reaction Force**

Particle $P$ of mass $m$ is attached to a ring of mass $M$ and radius $R$. The ring rolls without slipping on a horizontal surface, as shown in Figure 6.8. Find the equation of motion of the system and the reaction force using the Lagrange approach.

We can find the equation of motion and the reaction force simultaneously. To this end, we remove the constraint (reaction) force $R_N$ and introduce a new coordinate $y$ to fully describe motion of the system in the absence of the constraint force (Figure 6.9). We replace the constraint force by a constraint equation $\mathrm{d}y=0$.

Next, we find the kinetic energy of the system

$$\mathcal{T}=\mathcal{T}_{\text{ring}}+\mathcal{T}_{\text{particle}}$$

where

$$\mathcal{T}_{\text{ring}}=\frac{1}{2}M\,{}^N\mathbf{v}^{G/O}.{}^N\mathbf{v}^{G/O}+\frac{1}{2}I_G\dot{\phi}^2,$$

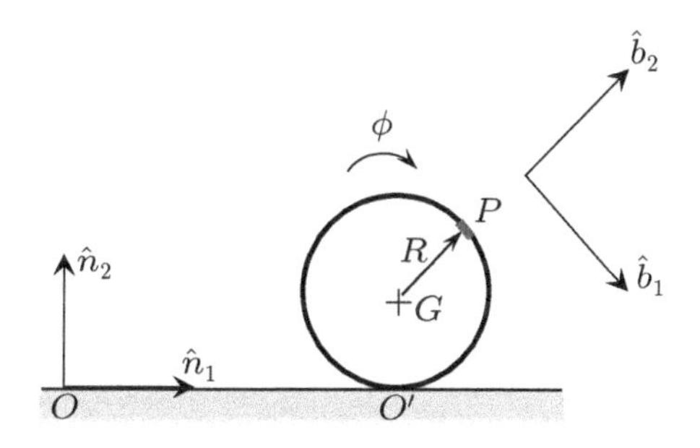


**Figure 6.8** Particle slides over a ring.

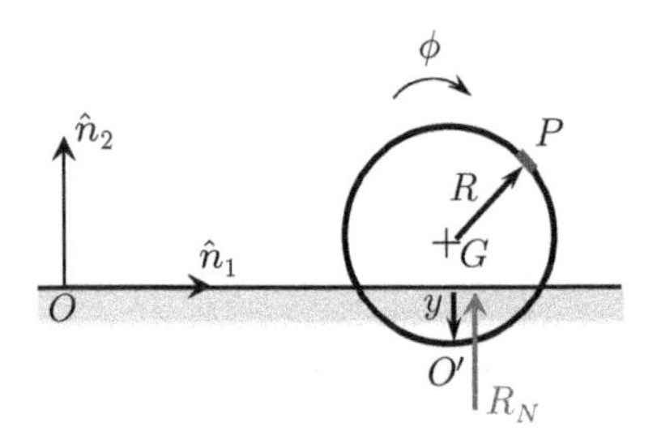


**Figure 6.9** The new coordinate $y$ is introduced to describe motion in the absence of the constraint.

$$\mathbf{OG} = x\hat{n}_1 + (R - y)\hat{n}_2,$$
$${}^N\mathbf{v}^{G/O} = \dot{x}\hat{n}_1 - \dot{y}\hat{n}_2 = R\dot{\phi}\hat{n}_1 - \dot{y}\hat{n}_2.$$

It follows that the kinetic energy of the ring is

$$\mathcal{T}_{\text{ring}} = MR^2\dot{\phi}^2 + \frac{1}{2}M\dot{y}^2 = MR^2\dot{\phi}^2.$$

The kinetic energy of the particle is given by $\mathcal{T} = \frac{1}{2}m^N\mathbf{v}^{P/O}.^N\mathbf{v}^{P/O}$, where

$$\mathbf{v}^{P/O} = R\dot{\phi}(1 + \cos\phi)\hat{n}_1 + (-\dot{y} - R\dot{\phi}\sin\phi)\hat{n}_2.$$

The potential energy of the ring and particle system is given by

$$\mathcal{U} = Mg(R - y) + mg(R - y + R\cos\phi).$$

The equations of motion can be written as

$\phi$*-dynamics*:

$$\frac{\mathrm{d}}{\mathrm{d}t}\left(\frac{\partial\mathcal{L}}{\partial\dot{\phi}}\right) - \frac{\partial\mathcal{L}}{\partial\phi} = 0,$$
$$M\ddot{y} + 2MR^2\ddot{\phi} + 2mR^2\ddot{\phi}(1 + \cos\phi) + mR\ddot{y}\sin\phi - mgR\sin\phi = 0,$$

$y$*-dynamics*:

$$\frac{\mathrm{d}}{\mathrm{d}t}\left(\frac{\partial\mathcal{L}}{\partial\dot{y}}\right) - \frac{\partial\mathcal{L}}{\partial y} = \lambda,$$
$$(m + M)\ddot{y} + mR\sin\phi\ddot{\phi} + mR\cos\phi\dot{\phi}^2 - (m + M)g = -\lambda.$$

where $\lambda = Q_y = R_N\hat{n}_2.\frac{\partial}{\partial\dot{y}}(\dot{x}\hat{n}_1 - \dot{y}\hat{n}_2) = -R_N$. Enforcing the constraint $y = \dot{y} = \ddot{y} = 0$, we obtain

$$2R(M + m(1 + \cos\phi))\ddot{\phi} - mg\sin\phi = 0,$$
$$mR\sin\phi\ddot{\phi} + mR\cos\phi\dot{\phi}^2 + (m + M)g = R_N.$$

The first of these equations is the equation of motion and the second provides the reaction force.

**Flipped Classroom Exercise 6.3**

A rod of mass $m$ and length $l$ is released from rest when $\theta = 0$. Use the Lagrange multiplier approach to obtain an expression for the axial force at point $B$ as a function of $\theta$ only (no $\dot{\theta}$ or $\ddot{\theta}$).

To solve this problem, take the following steps:

1. Remove the constraint force and add a generalized coordinate $x$, to fully describe the motion of the system in the absence of the constraint force. Removing the axial force at point $P$ allows the lower part of the rod to move with respect to the upper part in a direction parallel to the axial force.

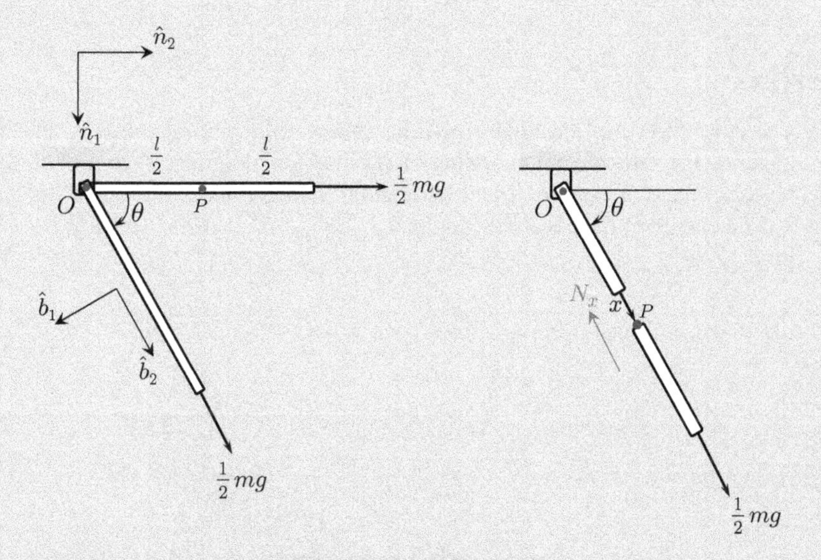


2. Add a constraint equation to replace the constraint force.
3. Show that the kinetic energy of the system can be written as

$$\mathcal{T} = \frac{5}{192} ml^2\dot{\theta}^2 + \frac{1}{4}m\left(\dot{x}^2 + \left(\frac{3}{4}l + x\right)^2 \dot{\theta}^2\right).$$

4. Show that the potential energy of the system can be written as

$$\mathcal{U} = -\frac{1}{8}mgl\sin\theta - \frac{1}{2}mg\left(\frac{3l}{4} + x\right)\sin\theta.$$

5. Using Lagrange's equation, write the equation of motion representing the $\theta$-dynamics. Show that the generalized coordinate $Q_\theta = 0$.

6. Using Lagrange's equation, write the equation of motion representing the $x$-dynamics. Show that the generalized coordinate $Q_x = \frac{1}{2}mg\sin\theta$.
7. Setting $x = \dot{x} = \ddot{x} = 0$, show that

$$\lambda = -\frac{3}{8}ml\dot{\theta}^2 - mg\sin\theta.$$

8. Calculate the generalized force resulting from the axial load and show that it is equal to $-\lambda$.
9. Using the relation $\ddot{\theta}\mathrm{d}\theta = \dot{\theta}\mathrm{d}\dot{\theta}$, show that the axial load is

$$N_x = \frac{17}{8}mg\sin\theta.$$

## Exercises

6.1 The cone shown in the Figure 6.10 has mass $m$, height $h$, radius $r$, and half vertex angle $\alpha$. It rotates without slipping on the horizontal surface shown, such that its vertex $O$ is fixed at one point and its angular velocity about the $\hat{b}_1$ axis is $\omega$. Find the kinetic energy of the cone knowing that $I_{11} = \frac{3}{10}mr^2$, and $I_{22} = \frac{3}{5}m(h^2 + r^2/4)$ about point $O$.

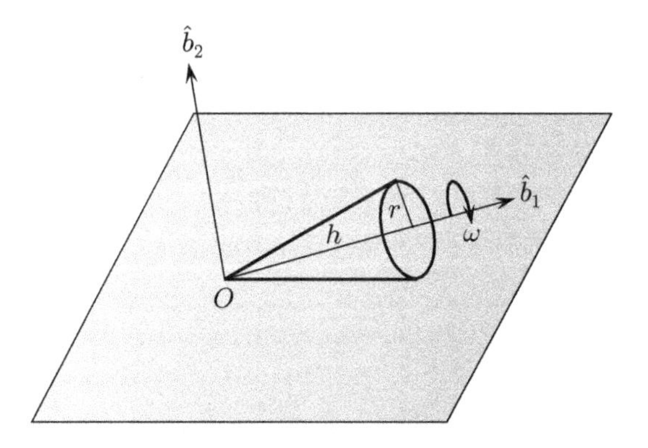


**Figure 6.10** Exercise 6.1.

6.2 A hollow sphere of mass $m$ and radius $r$ rolls without slipping on a surface of radius $R$, as shown in Figure 6.11. The motion takes place in a single plane. Use the Lagrange approach to find the angle $\theta$ at which the sphere loses contact with the surface if it starts from rest with a tiny nudge.

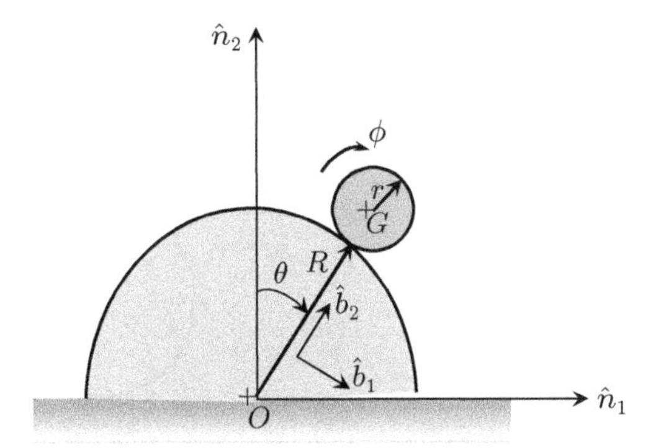


**Figure 6.11** Exercise 6.2.

6.3 A string is wound over a circular cylinder of mass $M$. The string passes over a massless pulley, as shown in Figure 6.12. The other end of the string is attached to a block of mass $m$. Assume that the cylinder rolls without slipping, find the tension in the cable using Lagrange's approach.

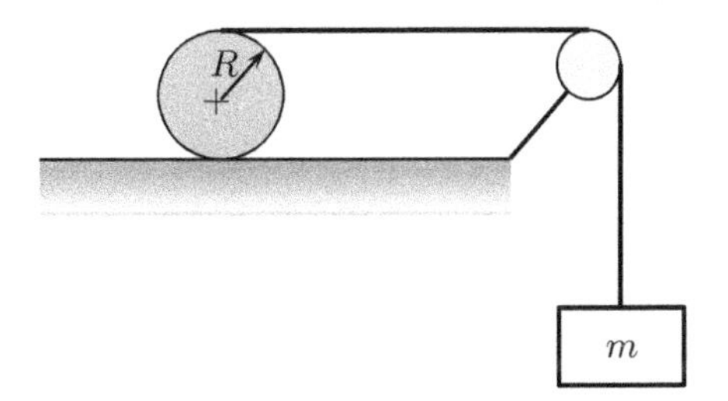


**Figure 6.12** Exercise 6.3.

6.4 Figure 6.13 shows a disk of mass $M$ and radius $R$, which rotates with a constant angular velocity $\Omega$. Find the equation governing the motion of the particle in the frictionless groove. The particle has mass $m$ and the spring has an unstretched length $l_0$.

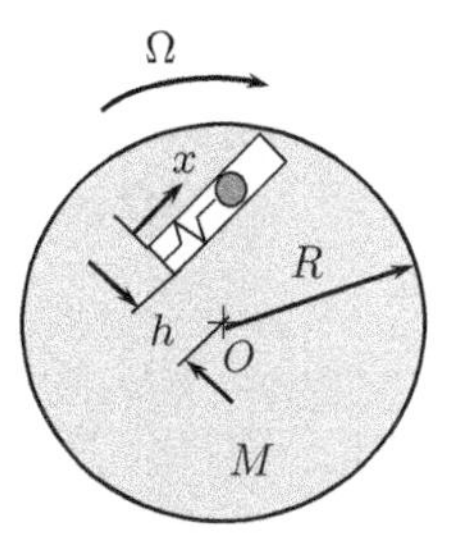


**Figure 6.13** Exercise 6.4.

6.5 A circular hoop of mass $M$ and radius $R$ oscillates in the vertical plane about point $O$, as shown in Figure 6.14. A bead of mass $m$ slides without friction on the hoop. Find the equation of motion of the bead and the hoop.

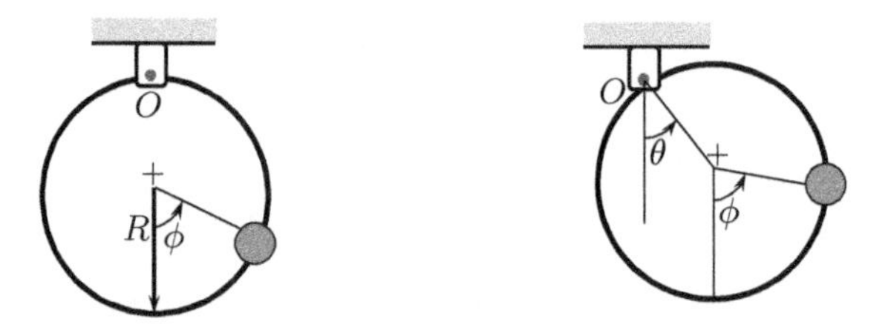


**Figure 6.14** Exercise 6.5.

6.6 Find the equation of motion for the half disk of radius $R$ shown in Figure 6.15. Assume that motion occurs without slipping.

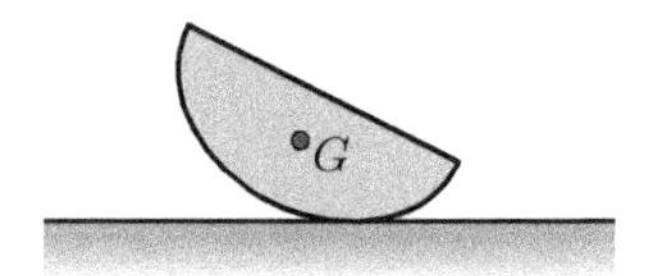


**Figure 6.15** Exercise 6.6.

6.7 An isosceles-triangular thin plate of mass $m$, base $a$, and height $h$ rotates with constant angular velocity $\Omega$, as shown in Figure 6.16. The vertex of the triangle is fixed in its place via a socket-and-ball joint. Determine the angle $\theta$ at which the base of the triangle remains horizontal at all times.

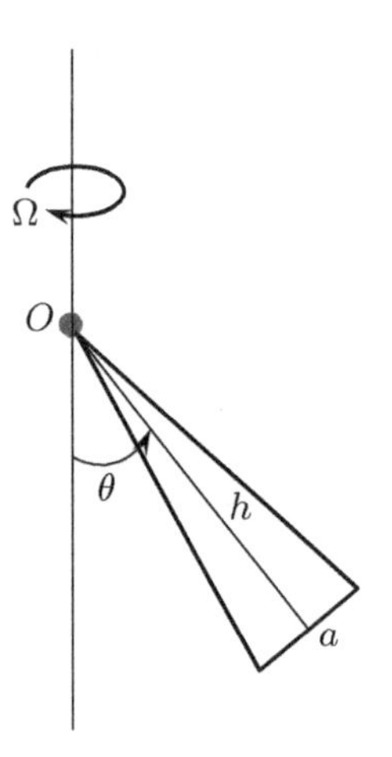


**Figure 6.16** Exercise 6.7.

6.8 A ring of mass $M$ and radius $R$ rolls without slipping on the horizontal surface shown in Figure 6.17. A particle $P$ of mass $m$ slides on the smooth internal surface of the ring. Let $\phi$ be the angle between the vertical and the line from the axis of the cylinder to the particle. If the particle is released from rest at $\phi = \frac{\pi}{2}$, find the position $x$ of the axis of the cylinder as function of $\phi$.

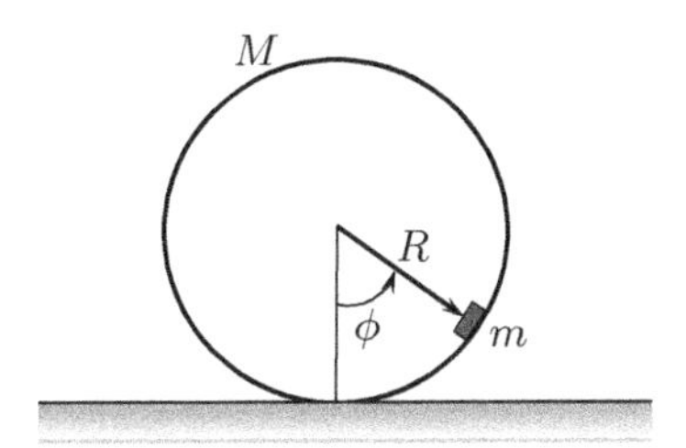


**Figure 6.17** Exercise 6.8.

6.9 A rod of length $l$ and mass $m$ is released from the horizontal position, as shown in Figure 6.18. Use the Lagrange approach to find the internal moment at point $P$ as a function of the angle $\theta$.

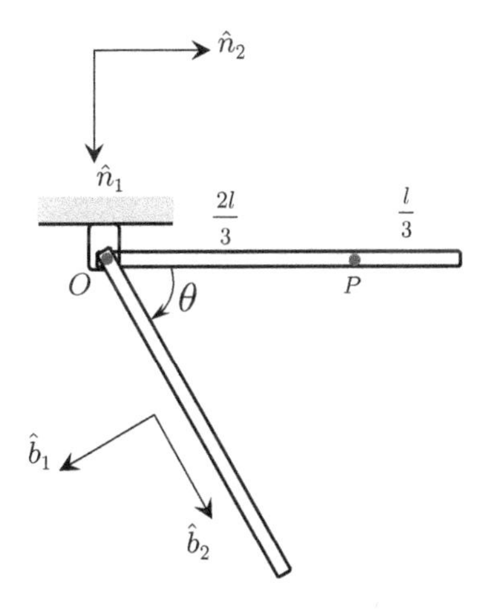


**Figure 6.18** Exercise 6.9.

6.10 A rod of mass $m$ and length $l$ slides without friction on the semi-circular surface shown in Figure 6.19. Determine its equation of motion.

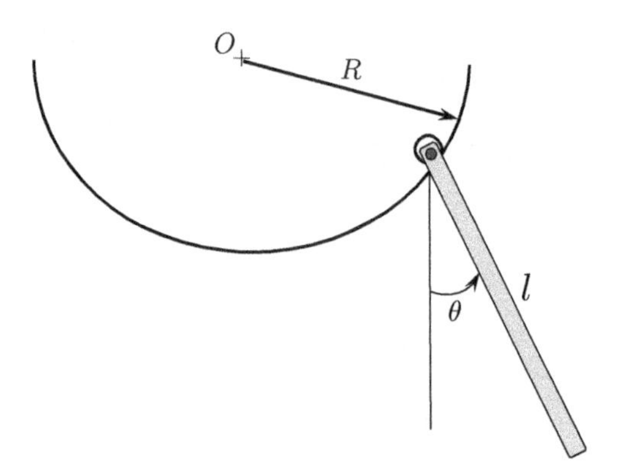


**Figure 6.19** Exercise 6.10.

6.11 A rectangular plate of mass $m$ and sides $2a$ and $2b$ is released from rest in the position shown in Figure 6.20. Assuming that oscillations occur in a single plane, find the equation of motion of the rectangle. What is the angular velocity of the resulting pendulum motion?

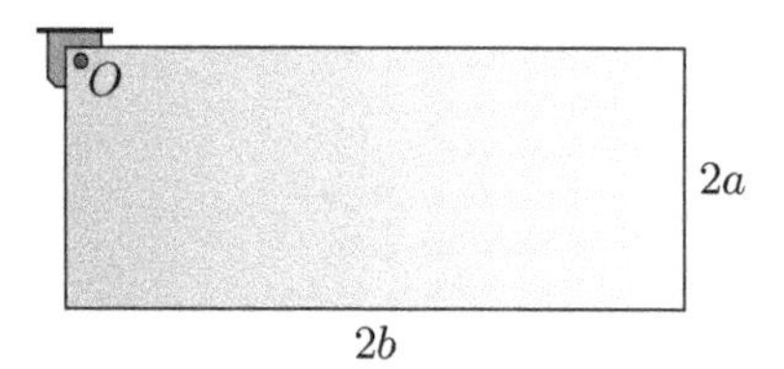


**Figure 6.20** Exercise 6.11.

6.12 A uniform circular disk of radius $R$ and mass $m$ can rotate freely about a fixed horizontal chord $PQ$. The chord is at a distance $\frac{R}{4}$ from the center $O$ of the disk. A particle, also of mass $m$, is fixed at the highest point $A$ on the edge of the disk, as shown in Figure 6.21. The disk is released from the position shown so that it starts rotating about $PQ$. Find the equations of motion and the linear speed of the particle as it reaches its lowest position.

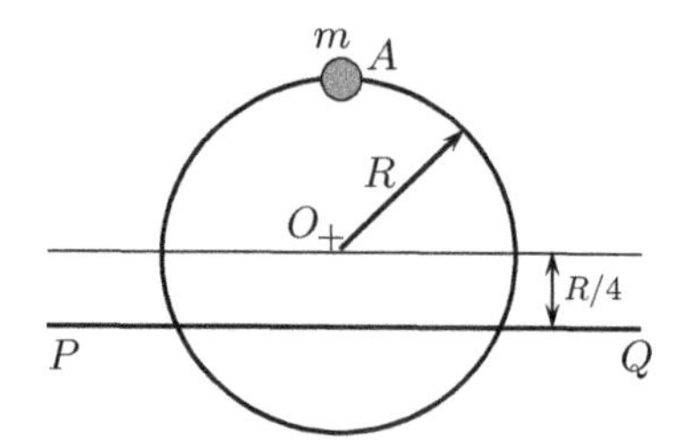


**Figure 6.21** Exercise 6.12.

6.13 A rod of mass $m$ and length $l$ is pinned at its center of mass $G$ to the center of massless rod $AB$, which is of length $2l$. The rod $AB$ is supported by two frictionless bearings and allowed to rotate about its axis with angle $\theta$, as shown in Figure 6.22. The whole setup rotates about the vertical axis with a constant angular velocity $\Omega$. Find the equation of motion of the rod.

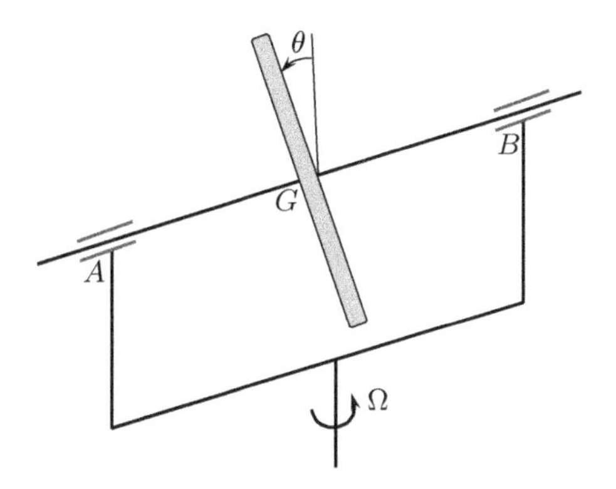


**Figure 6.22** Exercise 6.13.

6.14 A top (Figure 6.23) consists of a homogeneous, circular disk of radius $R_1$, mass $m$, and a massless axis of length $R_1/2$ perpendicular to the disk and passing through its center, $O$. The axis is connected by a joint to point $A$ on the edge of a horizontal circular plate of radius $R_2$ and mass $m$. The plate has a constant angular velocity $\Omega$ about its axis, and the top has a constant angular velocity $\omega$, relative to the plate, about its axis. Find the angle $\alpha$ between the axis of the top and the vertical.

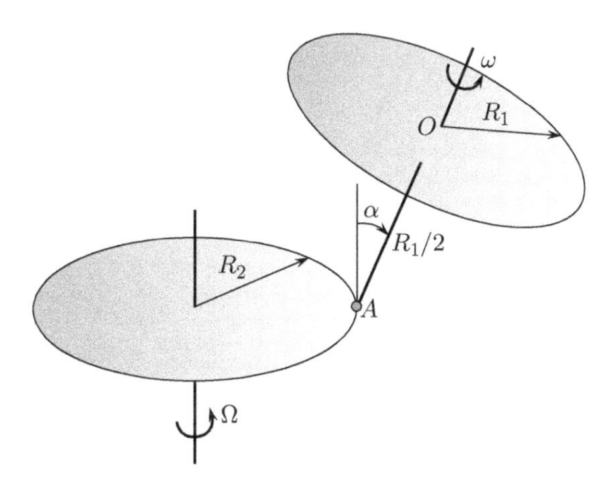


**Figure 6.23** Exercise 6.14.

6.15 The massless link $OQ$ is free to rotate about point $O$ with angle $\theta$, as shown in Figure 6.24. The link is connected to a rod of mass $m$ and length $l$, which rotates at angular velocity $\Omega$ about the link. Find the equation of motion of the rod.

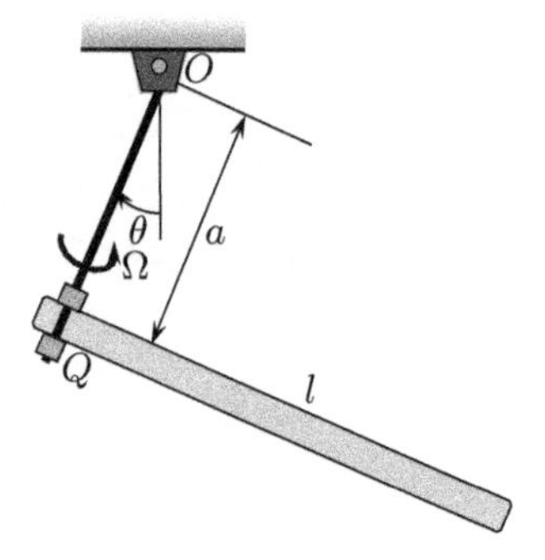


**Figure 6.24** Exercise 6.15.

# 7

# Momentum

This chapter introduces the linear and angular momenta of particles and rigid bodies. It discusses the conditions under which they are conserved and illustrates their relationship with the impulse and angular impulse. The conservation of angular momentum is also used to derive Euler's rotational equations. The concept of generalized momenta is presented and used to derive Hamilton's canonical equations.

## 7.1 Linear Momentum

In Chapter 2, we learned that Newton's second law of dynamics can be expressed mathematically as $\mathbf{F}_i = m_i\ddot{\mathbf{r}}_i$, where $\mathbf{F}_i$ is the net force vector acting on a particle of mass $m_i$ and $\ddot{\mathbf{r}}_i$ is its acceleration measured with respect to an inertial point. In this form, Newton's second law inherently assumes that the mass of the particle does not change with time. However, in general, this might not be the case. When the mass of the particle changes, Newton's second law takes the more general form:

$$\mathbf{F}_i = \frac{\mathrm{d}}{\mathrm{d}t}(m_i\dot{\mathbf{r}}_i) = \frac{\mathrm{d}}{\mathrm{d}t}(\mathbf{p}_i), \tag{7.1}$$

where $m_i\dot{\mathbf{r}}_i = \mathbf{p}_i$ is a vectorial quantity known as the linear momentum of the particle. Thus, in its most general form, Newton's second law states that the net force acing on a particle is equal to the time rate of change of its linear momentum.

Integrating the general form of Newton's equation from some initial time $t_1$ to some final time, $t_2$, yields

$$\mathbf{p}_i(t_2) - \mathbf{p}_i(t_1) = \int_{t_1}^{t_2} \mathbf{F}_i \mathrm{d}t, \tag{7.2}$$

where the integral on the right-hand side is called the *impulse*.

*Dynamics of Particles and Rigid Bodies: A Self-Learning Approach*, First Edition. Mohammed F. Daqaq.

Companion website: www.wiley.com/go/daqaq/dynamics

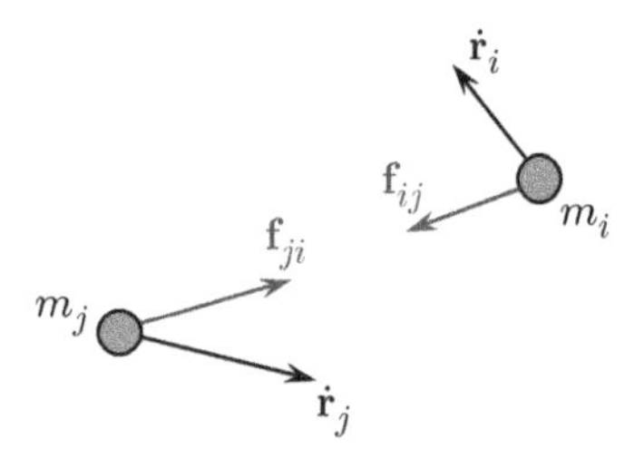


**Figure 7.1** Conservation of linear momentum of two interacting particles.

When Equation (7.2) is applied to two particles $i$ and $j$ that are interacting due to some internal forces $\mathbf{f}_{ij} = -\mathbf{f}_{ji}$, but are not subject to any external forces, as shown in Figure 7.1, we can write:

$$\mathbf{p}_i(t_2) - \mathbf{p}_i(t_1) = \int_{t_1}^{t_2} \mathbf{f}_{ij} dt, \tag{7.3a}$$

$$\mathbf{p}_j(t_2) - \mathbf{p}_j(t_1) = -\int_{t_1}^{t_2} \mathbf{f}_{ij} dt. \tag{7.3b}$$

Adding the previous set of equations, we obtain

$$\mathbf{p}_i(t_1) + \mathbf{p}_j(t_1) = \mathbf{p}_i(t_2) + \mathbf{p}_j(t_1). \tag{7.4}$$

Equation (7.4) states that the sum of the linear momentum of both particles at time $t_1$ is equal to the sum of their linear momentum at time $t_2$. In other words, in the absence of external forces, the linear momentum of the two particles is conserved.

For $\mathcal{N}$ particles, the linear momentum is the sum of the linear momenta of all particles; that is

$$\mathbf{p} = \sum_{i=1}^{\mathcal{N}} m_i \dot{\mathbf{r}}_i. \tag{7.5}$$

As such, conservation of linear momentum can be generalized to a system of $\mathcal{N}$ particles; such that

$$\boxed{\sum_{i=1}^{\mathcal{N}} \mathbf{p}_i(t_1) = \sum_{i=1}^{\mathcal{N}} \mathbf{p}_i(t_2)} \tag{7.6}$$

Note that the previous equation is vectorial in nature; in other words, it holds along the three Cartesian coordinates.

### Example 7.1 Particle Fission

A particle of mass $M$ and velocity $v\,\hat{n}_1$, fissions into two small particles, each of mass $\frac{M}{2}$. As shown in Figure 7.2, one of these masses moves in a direction that makes an angle $\theta_1$ with the $\hat{n}_1$ axis, while the other moves in a direction that makes an angle $\theta_2$ with the $\hat{n}_1$ axis. Determine the velocity of each of the imparting particles.

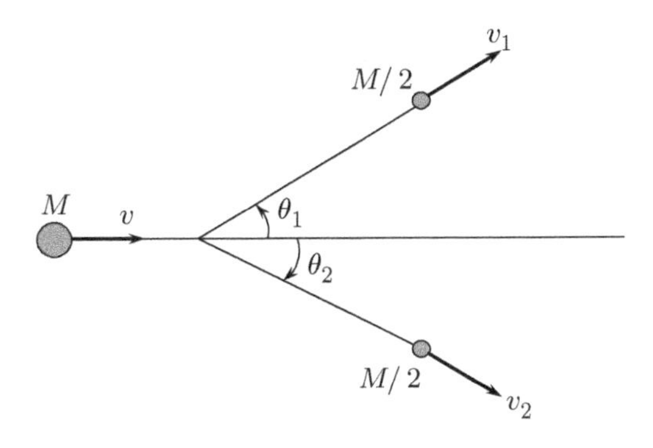


**Figure 7.2** Particle fission.

Since there are no external forces acting on the system, one can use conservation of linear momentum, Equation (7.4), to relate the motion of the particle before and after fission. To this end, we conserve linear momentum along the $\hat{n}_1$ and $\hat{n}_2$ directions and obtain the following equations:

$$p_M(t_1)\hat{n}_1 = p_1(t_2)\hat{n}_1 + p_2(t_2)\hat{n}_1, \qquad Mv = \frac{M}{2}v_1\cos\theta_1 + \frac{M}{2}v_2\cos\theta_2,$$
$$p_M(t_1)\hat{n}_2 = p_1(t_2)\hat{n}_2 + p_2(t_2)\hat{n}_2, \qquad 0 = \frac{M}{2}v_1\sin\theta_1 - \frac{M}{2}v_2\sin\theta_2,$$

where $p_M$ is the linear momentum of the particle before fission, $p_1$ and $p_2$ are the linear momenta of the imparting particles, and $t_1$ and $t_2$ are, respectively, the time immediately before and after fission. Solving the previous set of equations for $v_1$ and $v_2$ yields

$$v_1 = \frac{2v\sin\theta_2}{\sin(\theta_1+\theta_2)},$$
$$v_2 = \frac{2v\sin\theta_1}{\sin(\theta_1+\theta_2)}.$$

**Example 7.2 Throwing a Ball off a Stationary Boat**

A man of mass $m$ standing on a stationary boat of mass $M$ throws a ball of mass $0.1m$ with horizontal velocity, $v_0$. With what speed does the boat move after the man throws the ball? Assume no friction between the man and the boat.

Using conservation of linear momentum, we can write

$$\mathbf{p}_{man}(t_1) + \mathbf{p}_{boat}(t_1) + \mathbf{p}_{ball}(t_1) = \mathbf{p}_{man}(t_2) + \mathbf{p}_{boat}(t_2) + \mathbf{p}_{ball}(t_2),$$
$$m(0) + M(0) + 0.1m(0) = 0.1mv_0 + (m+M)v,$$
$$v = -\frac{0.1mv_0}{m+M}.$$

**Example 7.3 Rocket Propulsion**

Consider a rocket of mass $m_0$ and initial velocity $v = 0$, which is being propelled in the horizontal direction by ejecting burnt fuel at a constant velocity $u$ measured with respect to the

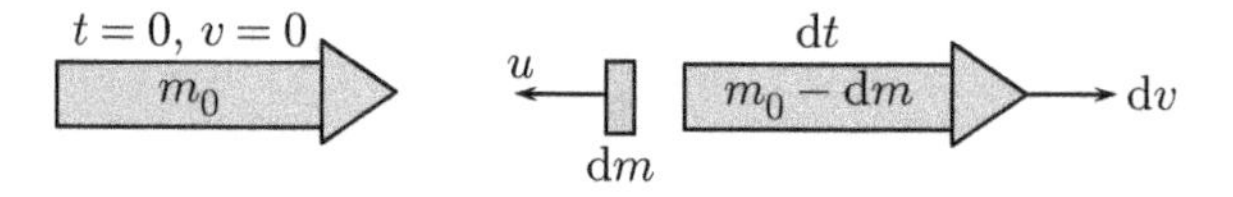


**Figure 7.3** Rocket propulsion.

rocket speed $v$ at a rate $\frac{dm}{dt} = -\alpha$, as shown in Figure 7.3. Find the velocity of the rocket as a function of time.

Initially, the system is stationary and, as a result, the linear momentum is zero. After time $dt$, a mass $dm$ has already been ejected at a velocity equal to $dv - u$. Therefore, the rocket has mass $m(t) = m_0 - dm$ and moves with a velocity $dv$.

Applying conservation of linear momentum on the rocket, we have

$$0 = (m_0 - dm)dv + dm(dv - u), \qquad m(t)dv - udm = 0,$$

$$v(t) = \int_{m_0}^{m(t)} u\frac{dm}{m_0}, \qquad v(t) = u\ln\frac{m(t)}{m_0}.$$

Since $m(t) = m_0 - \alpha t$, we obtain

$$v(t) = u\ln\frac{m_0 - \alpha t}{m_0}.$$

## 7.2 Collision

In general, collisions can be divided into

- *elastic* collisions, in which the kinetic energy is conserved;
- *inelastic* collisions, in which part of the kinetic energy is lost as heat, plastic deformation, and friction.

The amount of energy lost during impact is usually measured via a quantity called the *coefficient of restitution*, $e$, which is defined as

$$e = \sqrt{\frac{\text{Kinetic energy after collision}}{\text{Kinetic energy before collision}}} = \sqrt{\frac{T_{\text{after}}}{T_{\text{before}}}}. \tag{7.7}$$

The coefficient of restitution has typical values in the range $0 \leq e \leq 1$, where $e = 0$ represents a purely inelastic collision (for instance, trying to bounce a ball against a surface covered with a thick layer of sand) and $e = 1$ represents a purely elastic collision in which the kinetic energy is conserved.

For a particle bouncing against a stationary surface, we can use Equation (7.7) to reduce the coefficient of restitution to

$$e = \frac{v_a}{v_b}, \tag{7.8}$$

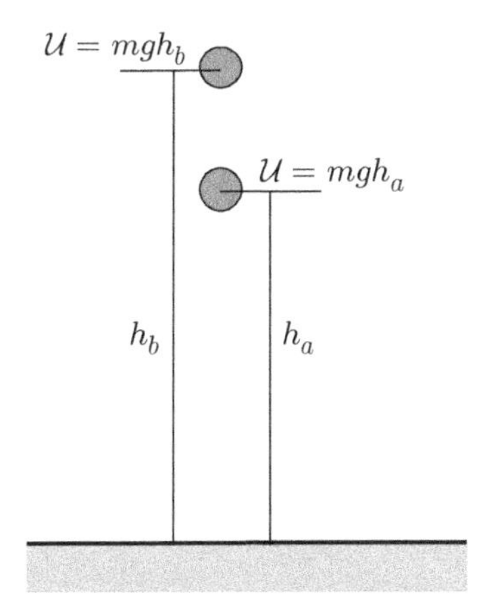


**Figure 7.4** Coefficient of restitution of a ball released from rest at height $h_b$.

where $v_a$ is the velocity after collision and $v_b$ is the velocity before collision. If, as shown in Figure 7.4, the particle is released from rest at a height $h_b$, the kinetic energy just before impact can be related to the potential energy via $\mathcal{T}_{\text{before}} = \mathcal{U}_{\text{before}} = mgh_b$, and the kinetic energy right after impact can be related to the potential energy at the final height $h_a$ via $\mathcal{T}_{\text{after}} = \mathcal{U}_{\text{after}} = mgh_a$. Therefore, using Equation (7.7) we can write:

$$e = \sqrt{\frac{h_a}{h_b}}. \tag{7.9}$$

When the collision involves two particles impacting along a single line (a one-dimensional collision), the coefficient of restitution can be expressed as

$$e = \frac{v_{1a} - v_{2a}}{v_{2b} - v_{1b}}, \tag{7.10}$$

where $v_{ia}$ is the velocity of particle $i$ after collision and $v_{ib}$ is the velocity of particle $i$ before collision.

For two-dimensional collisions of two or more particles, the velocities used for the calculation of the coefficient of restitution in Equation (7.10) are those collinear to the line of impact: the line that is normal to the surfaces that are in contact during impact. Figure 7.5 depicts the line of impact for several impacting bodies.

**Example 7.4 Bouncing against the Ground**

A particle of mass $m$ is thrown horizontally with an initial velocity $v_0$ at a height $h$ above the ground, as shown in Figure 7.6. Determine the time required for the ball to stop bouncing off the ground knowing that the coefficient of restitution between the particle and the ground is $e$.

Prior to the first bounce off the ground, one can calculate the vertical velocity $v_{1y}$ just before impact using Newton's second law and show that it is given by $v_{1y} = \sqrt{2gh}$. Furthermore, the time $t_0$ that elapses before the ball impacts the ground is $t_0 = \sqrt{\frac{2h}{g}}$. Using the definition of

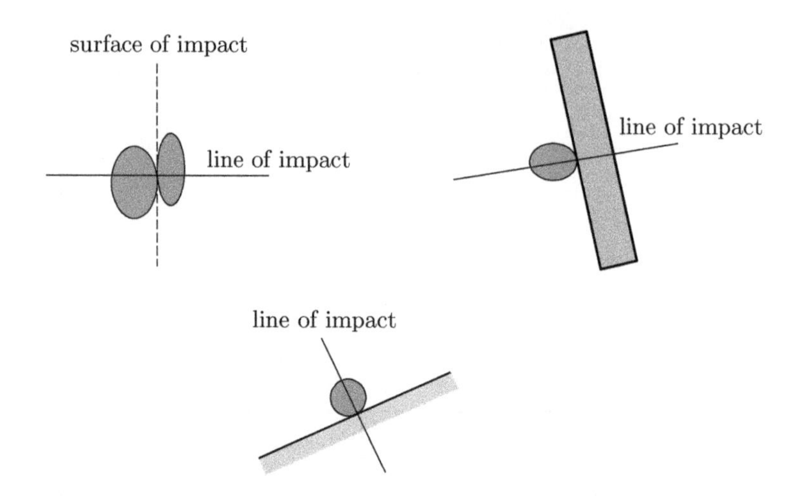


**Figure 7.5** Line of impact for different collisions.

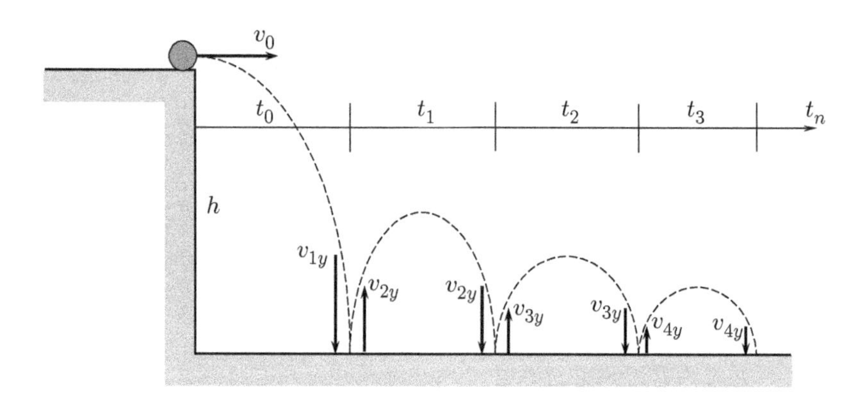


**Figure 7.6** Ball bouncing against the ground.

the coefficient of restitution, we can write

$$
\begin{aligned}
e &= \frac{v_{2y}}{v_{1y}}, \qquad v_{2y} = e\sqrt{2gh}, \\
e &= \frac{v_{3y}}{v_{2y}}, \qquad v_{3y} = e^2\sqrt{2gh}, \\
e &= \frac{v_{4y}}{v_{3y}}, \qquad v_{4y} = e^3\sqrt{2gh}, \\
e &= \frac{v_{(i+1)y}}{v_{iy}}, \qquad v_{(i+1)y} = e^i\sqrt{2gh}.
\end{aligned}
$$

Next, we calculate the time required to complete one cycle, $t_1, t_2, \ldots, t_n$. Applying Newton's second law on the particle during the first cycle, we know that

$$m\ddot{y} = mg, \qquad \dot{y} = gt - v_{2y}, \qquad y = \frac{g}{2}t^2 - v_{2y}t.$$

During a complete cycle, the particle rises from a height $h = 0$ and returns back to a height $h = 0$. Therefore, we can write $t_1 = \frac{2v_{2y}}{g}$, or, in general, $t_i = \frac{2v_{(i+1)y}}{g}$. Substituting $v_{(i+1)y} = e^i\sqrt{2gh}$ into the expression for $t_i$, we obtain

$$t_i = 2e^i\sqrt{\frac{2h}{g}}, \qquad i = 1, 2, \ldots, n.$$

As such, the total time can be written as

$$\begin{aligned} T_{\text{tot}} &= t_0 + t_1 + \ldots + t_n = \sqrt{\frac{2h}{g}} + 2e\sqrt{\frac{2h}{g}} + \ldots + 2e^n\sqrt{\frac{2h}{g}} \\ T_{\text{tot}} &= \sqrt{\frac{2h}{g}}(1 + 2e + 2e^2 + \ldots + 2e^n). \end{aligned}$$

The previous series can be expressed in the form

$$T_{\text{tot}} = \sqrt{\frac{2h}{g}}\left(\frac{1+e}{1-e}\right).$$

Note that the total time is independent of the initial velocity or the mass of the particle. When $e \to 0$, $T_{\text{tot}} \to \sqrt{\frac{2h}{g}}$; on the other hand, when $e \to 1$, $T_{\text{tot}} \to \infty$. Using $h = 1$ m and $e = 0.9$, one can find that $T_{\text{tot}} = 8.58$ s.

**Example 7.5 Elastic Planar Collision**

A billiard ball of mass $M$ and velocity $v$ impacts an identical stationary ball. The balls bounce off each other in such a way that the incoming ball deflects at a known angle $\theta$ from the horizontal, while the stationary ball deflects at an unknown angle $\phi$, as shown in Figure 7.7. Assuming a perfectly elastic collision, find the velocity of both particles after impact and the angle $\phi$.

To solve this problem, we first note that we have three unknowns: the velocity of each particle after impact ($v_1, v_2$) and the angle $\phi$. To find these unknowns, we need three equations. Two equations result from conservation of linear momentum along the horizontal and vertical directions, while the third equation comes from the conservation of kinetic energy, since the collision is a purely elastic one. Using conservation of linear momentum we can write

$$\begin{aligned} Mv &= Mv_1\cos\theta + Mv_2\cos\phi, \\ 0 &= Mv_1\sin\theta - Mv_2\sin\phi. \end{aligned}$$

Using conservation of kinetic energy, we obtain

$$\frac{1}{2}Mv^2 = \frac{1}{2}Mv_1^2 + \frac{1}{2}Mv_2^2.$$

Solving the previous equations yields $v_1 = v\cos\theta$, $v_2 = v\sin\theta$, and $\phi = 90^\circ - \theta$.

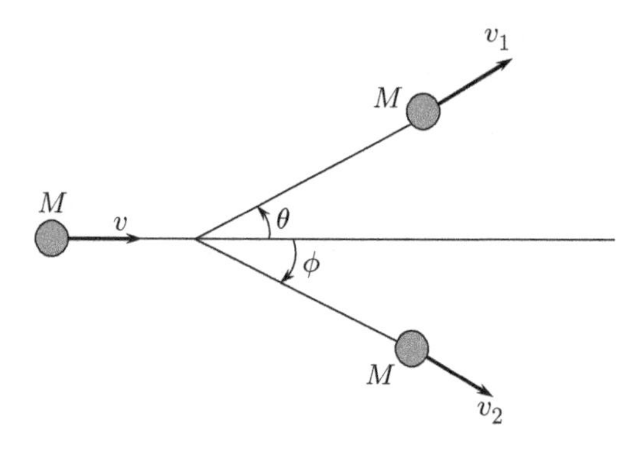


**Figure 7.7** Planar elastic collision.

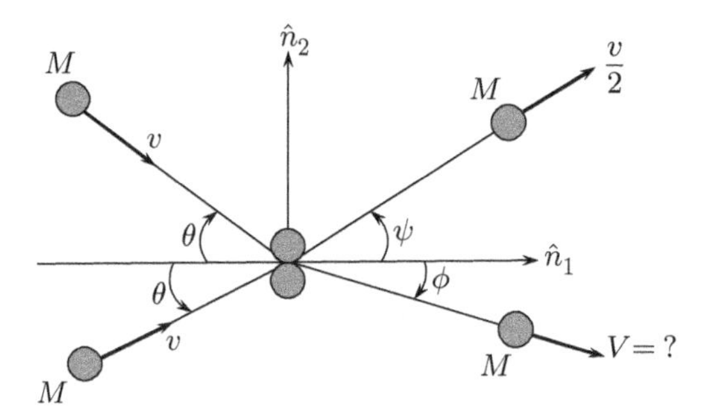


**Figure 7.8** Oblique inelastic impact.

### Example 7.6 Oblique Inelastic Collision

Two identical balls of mass $M$ approach each other at equal velocity $v$ and collide as shown in Figure 7.8. The balls bounce off each other such that one of the balls has a velocity $\frac{v}{2}$ after impact. Assuming a coefficient of restitution $e = 0.5$, find the velocity of the other particle and the angles $\phi$ and $\psi$ that the particles make with the horizontal after impact.

The problem involves three unknowns, $\phi$, $\psi$ and $V$. To obtain these unknowns, we need three equations. Two equations come from conserving linear momentum in the $\hat{n}_1$- and $\hat{n}_2$-directions, while the third equation comes from the relation between the coefficient of restitution and the relative velocities before and after impact.

Conserving linear momentum in the $\hat{n}_1$-direction yields

$$2v\cos\theta = \frac{v}{2}\cos\psi + V\cos\phi.$$

Conserving linear momentum in the $\hat{n}_2$-direction yields

$$\frac{v}{2}\sin\psi = V\sin\phi.$$

Using the definition of the coefficient of restitution, we obtain

$$e = \frac{1}{2} = \frac{V\sin\phi + \frac{v}{2}\sin\psi}{2v\sin\theta}.$$

Solving the previous equations for $\psi$, $\phi$ and $V$, we obtain

$$\begin{aligned}
\psi &= \theta, \\
V &= \frac{v}{2}\sqrt{1+8\cos^2\theta}, \\
\phi &= \arcsin\left(\frac{\sin\theta}{\sqrt{1+8\cos^2\theta}}\right).
\end{aligned}$$

**Flipped Classroom Exercise 7.1**

The figure represents the force exerted by a wall as a tennis ball bounces against it. The ball of mass $m$ approaches the wall with a velocity $v$, and bounces in a purely elastic impact. Find the force $F_{\max}$.

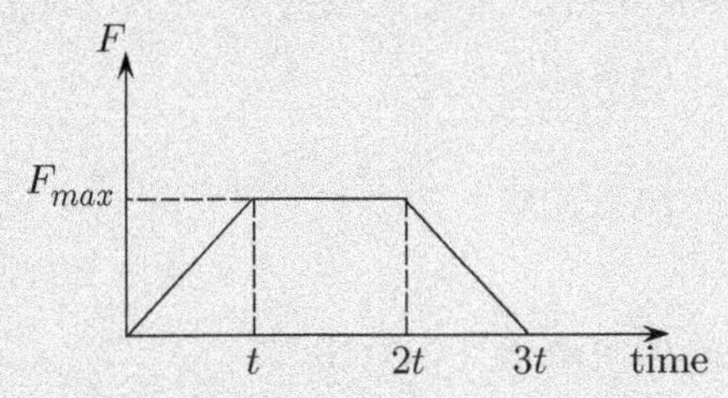


To solve this problem, take the following steps:

1. Calculate the change of linear momentum of the particle.
2. Relate the change of the linear momentum to the impulse expression.
3. Using the given figure, show that $F_{\max} = \frac{mv}{t}$.

**Linear Momentum of Particles**

The linear momentum of a particle is defined as

$$\mathbf{p}_i = m_i \dot{\mathbf{r}}_i$$

where $m_i$ is the mass of the particle and $\dot{\mathbf{r}}_i$ is its velocity with respect to an inertial frame. For $\mathcal{N}$ particles, the linear momentum is the sum of the linear momenta of all particles; that is

$$\mathbf{p} = \sum_{i=1}^{\mathcal{N}} m_i \dot{\mathbf{r}}_i.$$

When a force is applied to a particle for a period of time $[t_1, t_2]$, the resulting force impulse is defined as

$$\mathbf{I} = \int_{t_1}^{t_2} \mathbf{F}_i \mathrm{d}t = \mathbf{p}_i(t_2) - \mathbf{p}_i(t_1).$$

If the net force acting on a system of $\mathcal{N}$ particles is zero, then the impulse vanishes and the previous equation becomes

$$\sum_{i=1}^{\mathcal{N}} \mathbf{p}_i(t_2) = \sum_{i=1}^{\mathcal{N}} \mathbf{p}_i(t_1).$$

The previous equation is known as the conservation of linear momentum.

The time change of the linear momentum of a particle is equal to net forces acting on it; that is,

$$\mathbf{F}_i = \dot{\mathbf{p}}_i.$$

The previous equation is the generalized version of Newton's second law of dynamics, and can be reduced to its more common form, $\mathbf{F}_i = m_i\ddot{\mathbf{r}}_i$ when the mass of the particle is constant.

## 7.3 Angular Momentum of Particles

The angular momentum of a particle is a vector quantity that represents the moment exerted by its linear momentum about a set of coordinates located at some point in space. For a particle of mass $m_i$, located at position $\mathbf{r}_i$ measured from a point $O$, the angular momentum can be defined as

$$\mathbf{L}_{iO} = \mathbf{r}_i \times \mathbf{p}_i = \mathbf{r}_i \times (m_i\dot{\mathbf{r}}_i), \tag{7.11}$$

where, as shown in Figure 7.9, $\dot{\mathbf{r}}_i$ is the velocity of the particle measured with respect to point $O$. For $\mathcal{N}$ particles, the angular momentum is the vectorial sum of the angular momenta of all particles; that is,

$$\mathbf{L}_O = \sum_{i=1}^{\mathcal{N}} \mathbf{r}_i \times \mathbf{p}_i = \sum_{i=1}^{\mathcal{N}} \mathbf{r}_i \times (m\dot{\mathbf{r}}_i). \tag{7.12}$$

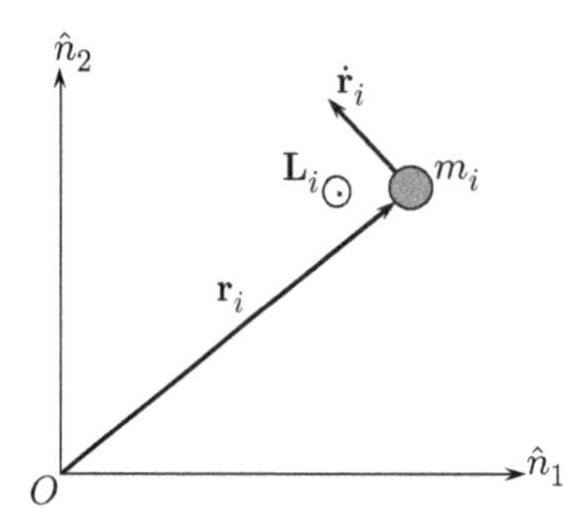


**Figure 7.9** Angular momentum of a particle about point $O$.

Note that the angular momentum is perpendicular to the position vector $\mathbf{r}$ and the velocity $\dot{\mathbf{r}}$. Hence, for planar motion of particles, the angular momentum is *always* perpendicular to the plane. It is also worth noting that, for motion along a line for which $\mathbf{r}_i$ and $\dot{\mathbf{r}}_i$ are parallel, the angular momentum is zero.

The angular momentum can be related to the net moment exerted by the external forces acting on the particle about an inertial point. To find this relationship, we differentiate Equation (7.12) once with respect to time and obtain

$$\frac{\mathrm{d}\mathbf{L}_O}{\mathrm{d}t} = \sum_{i=1}^{\mathcal{N}} \left( \frac{\mathrm{d}\mathbf{r}_i}{\mathrm{d}t} \times \mathbf{p}_i + \mathbf{r}_i \times \frac{\mathrm{d}\mathbf{p}_i}{\mathrm{d}t} \right). \tag{7.13}$$

Noting that $\mathbf{p}_i = m_i \frac{\mathrm{d}\dot{\mathbf{r}}_i}{\mathrm{d}t}$ and substituting into the previous equation, we obtain

$$\frac{\mathrm{d}\mathbf{L}_O}{\mathrm{d}t} = \sum_{i=1}^{\mathcal{N}} \left( \frac{\mathrm{d}\mathbf{r}_i}{\mathrm{d}t} \times m_i \frac{\mathrm{d}\mathbf{r}_i}{\mathrm{d}t} + \mathbf{r}_i \times m_i \ddot{\mathbf{r}}_i \right). \tag{7.14}$$

The first term in the previous equation vanishes. Then, according to Newton's second law, the term $m_i \ddot{\mathbf{r}}_i = \mathbf{F}_i$, where $\mathbf{F}_i$ is the vectorial sum of all external forces acting on particle $m_i$ (internal forces cancel out because they are equal in magnitude and opposite in direction). This yields:

$$\frac{\mathrm{d}\mathbf{L}_O}{\mathrm{d}t} = \sum_{i=1}^{\mathcal{N}} \mathbf{r}_i \times \mathbf{F}_i = \mathbf{M}_O \tag{7.15}$$

where $\mathbf{M}_O$ is the sum of external moments about a set of axes whose origin is located at point $O$.

Equation (7.15) states that the net change of angular momentum of a system of particles about a given inertial point $O$ is equal to the moments exerted by all external forces acting on the system summed about point $O$. It follows from Equation (7.15) that when the net moment exerted by all forces about a given point is zero, the angular momentum about this point is conserved; that is:

$$\frac{\mathrm{d}\mathbf{L}_O}{\mathrm{d}t} = 0, \qquad \mathbf{L}_O(t_1) = \mathbf{L}_O(t_2) \tag{7.16}$$

Equation (7.16) is referred to as the principle of conservation of angular momentum.

**Example 7.7 Spherical Pendulum Revisited**

Consider the spherical pendulum shown in Figure 7.10. The pendulum consists of a mass $m$ and a rigid cable of length $r$. Use the concept of angular momentum to obtain the equation of motion of the pendulum.

As stated in Equation (7.15), time variation of the angular momentum about point $O$ is equal to the moment exerted by all external forces about a set of coordinates located at point $O$; that is

$$\frac{\mathrm{d}\mathbf{L}_O}{\mathrm{d}t} = \mathbf{M}_O,$$

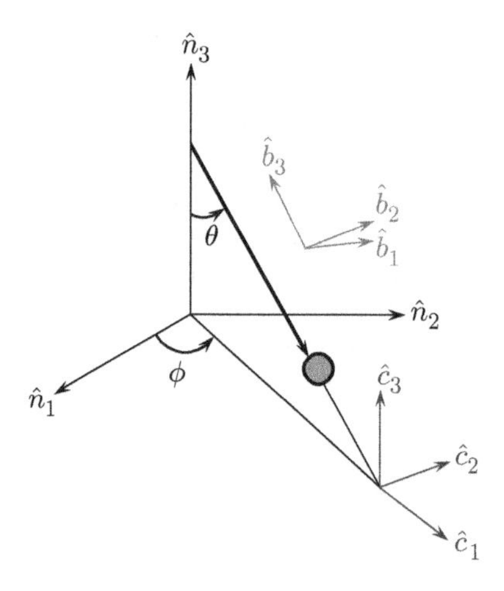


**Figure 7.10** A spherical pendulum with rigid cable.

where

$$\frac{\mathrm{d}\mathbf{L}_O}{\mathrm{d}t} = \frac{\mathrm{d}}{\mathrm{d}t}(\mathbf{OP} \times m^N\mathbf{v}^{P/O}) = \mathbf{OP} \times \frac{\mathrm{d}\mathbf{v}^{P/O}}{\mathrm{d}t} = \mathbf{OP} \times \mathbf{a}^{P/O}.$$

Here $\mathbf{OP} = -r\hat{b}_3$ and

$$\begin{aligned} {}^N\mathbf{a}^{P/O} &= (r\ddot{\theta} - r\dot{\phi}^2 \sin\theta\cos\theta)\hat{b}_1 + (2r\dot{\theta}\dot{\phi}\cos\theta\sin\theta + r\ddot{\phi}\sin^2\theta)\hat{b}_2 \\ &\quad + (r\dot{\theta}^2 + r\dot{\phi}^2\sin^2\theta)\hat{b}_3. \end{aligned}$$

Carrying the cross product $\mathbf{OP} \times {}^N\mathbf{a}^{P/O}$, we obtain

$$\frac{\mathrm{d}\mathbf{L}_O}{\mathrm{d}t} = m(-2r^2\dot{\theta}\dot{\phi}\cos\theta\sin\theta - r^2\ddot{\phi}\sin\theta)\hat{b}_1 + m(-r\ddot{\theta} + r^2\dot{\phi}^2\sin\theta\cos\theta)\hat{b}_2.$$

The pendulum is subject to only one external force, the weight $-mg\ \hat{c}_3$. Therefore, we have

$$\mathbf{M}_O = \mathbf{OP} \times -mg\hat{c}_3 = -mgr\sin\theta\hat{b}_2.$$

Equating $\mathbf{M}_O$ to $\frac{\mathrm{d}\mathbf{L}_O}{\mathrm{d}t}$, we obtain the equations of motion for the spherical pendulum as

$$r\ddot{\theta} + g\sin\theta = r\dot{\phi}^2\sin\theta\cos\theta,$$

$$2r\dot{\theta}\dot{\phi}\cos\theta\sin\theta + r\ddot{\phi}\sin\theta = 0.$$

It is worth mentioning that the moment exerted by external forces about the $\hat{b}_1$ axis located at point $O$ is zero; that is,

$$\frac{\mathrm{d}\mathbf{L}_O}{\mathrm{d}t}\hat{b}_1 = m(-2r^2\dot{\theta}\dot{\phi}\cos\theta\sin\theta - r^2\ddot{\phi}\sin\theta)\hat{b}_1 = 0.$$

The previous equation implies that $\mathbf{L}_O \hat{b}_1 =$ constant, or that the angular momentum about the $\hat{b}_1$ axis is conserved. This can be used to show that

$$\frac{\mathrm{d}}{\mathrm{d}t}(mr\dot{\phi}\sin^2\theta) = 0, \qquad mr\dot{\phi}\sin^2\theta = C_\phi$$

where $C_\phi$ is a constant obtained from the initial conditions.

One can also show that $\phi$ is a cyclic coordinate because the Lagrangian of the system $\mathcal{L} = \mathcal{T} - \mathcal{U}$ does not depend on the generalized coordinate $\phi$ but only depends on the generalized velocity $\dot{\phi}$. Hence, there is a direct relationship between the conservation of angular momentum about the $\hat{b}_1$ axis and the presence of a cyclic coordinate. This relationship will become clearer when we discuss Hamilton's canonical equations in Section 7.6.

 **Flipped Classroom Exercise 7.2**

A satellite of mass $m$ is launched into a geostationary transfer orbit. At its *perigee*, the point where the satellite is closest to the earth, the satellite has an altitude of $r_p$. At *apogee*, the point where the satellite is furthest from the earth, the satellite has altitude $r_a$. Calculate the speed of the satellite at *perigee* $v_p$, and at *apogee*, $v_a$.

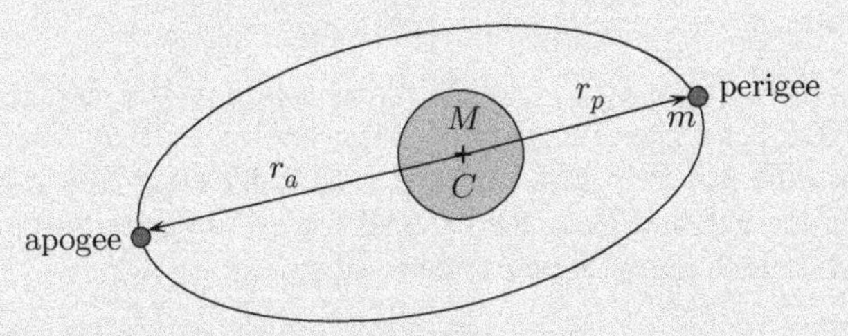


To solve this problem, take the following steps:

1. Treat the earth and the satellites as particles.
2. Use conservation of angular momentum to find a relationship between $v_p$ and $v_a$.
3. Use conservation of energy to find another relationship between $v_p$ and $v_a$. Assume that the kinetic energy of the earth is constant. Note that the satellite is attracted to the center of the earth following Newton's law of attraction.
4. Solve the two relationships for $v_a$ and $v_p$ and show that

$$v_p = \frac{2GMmr_a}{r_p(r_p + r_a)}, \qquad v_a = \frac{2GMmr_p}{r_a(r_p + r_a)},$$

where $G$ is the gravitational constant and $M$ is the effective mass of the earth.

### 7.3.1 *Angular Impulse*

Similar to the force impulse previously defined for linear momentum, there exists an angular impulse associated with angular momentum. To describe the relationship between the angular impulse and angular momentum for one particle, we start by stating that

$$\mathbf{F}_i = \frac{\mathrm{d}\mathbf{p}_i}{\mathrm{d}t}. \tag{7.17}$$

To find the angular momentum about an inertial point $O$, we take the cross product $\mathbf{r}_i \times \mathbf{F}_i$, where $\mathbf{r}_i$ is the position vector between point $O$ and the particle. This yields

$$\mathbf{r}_i \times \mathbf{F}_i = \mathbf{r}_i \times \frac{\mathrm{d}\mathbf{p}_i}{\mathrm{d}t}. \tag{7.18}$$

Integrating the previous equation from $t_1$ to $t_2$, we obtain

$$\int_{t_1}^{t_2} \mathbf{r}_i \times \mathbf{F}_i \mathrm{d}t = \int_{t_1}^{t_2} \mathbf{r}_i \times \mathrm{d}\mathbf{p}_i, \tag{7.19}$$

or

$$\int_{t_1}^{t_2} \mathbf{M}_{iO} \mathrm{d}t = \int_{t_1}^{t_2} \mathrm{d}\mathbf{L}_i = \mathbf{L}_i(t_2) - \mathbf{L}_i(t_1), \tag{7.20}$$

or

$$\int_{t_1}^{t_2} \mathbf{M}_{iO} \mathrm{d}t = \mathbf{L}_i(t_2) - \mathbf{L}_i(t_1) \tag{7.21}$$

where $\mathbf{M}_{iO}$ is the moment exerted by the forces acting on the particle about point $O$. The integral on the left-hand side of Equation (7.23) is known as the *angular impulse*.

Next, we assume that two particles $i$ and $j$ are interacting due to some internal forces $\mathbf{f}_{ij} = -\mathbf{f}_{ji}$, but are not subject to any external forces, as shown in Figure 7.11. We apply Equation (7.23) to each one of these particles and write

$$\int_{t_1}^{t_2} (\mathbf{r}_i \times \mathbf{f}_{ij}) \mathrm{d}t = \mathbf{L}_i(t_2) - \mathbf{L}_i(t_1), \tag{7.22a}$$

$$-\int_{t_1}^{t_2} (\mathbf{r}_j \times \mathbf{f}_{ij}) \mathrm{d}t = \mathbf{L}_j(t_2) - \mathbf{L}_j(t_1). \tag{7.22b}$$

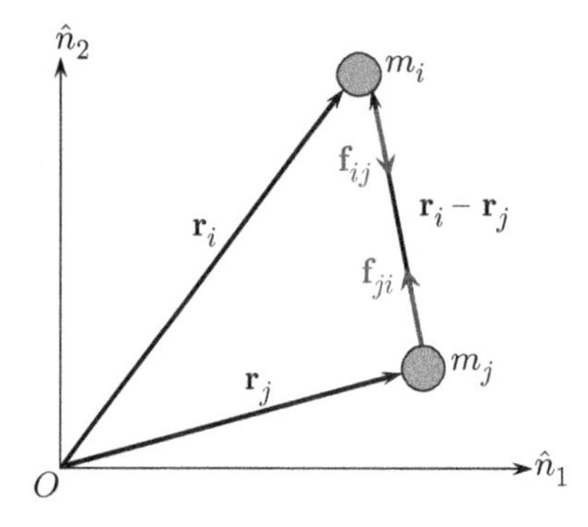


**Figure 7.11** Angular impulse of two interacting particles.

Adding the previous equations we have

$$\int_{t_1}^{t_2} (\mathbf{r}_i - \mathbf{r}_j) \times \mathbf{f}_{ij} dt = \mathbf{L}_i(t_2) - \mathbf{L}_i(t_1) + \mathbf{L}_j(t_2) - \mathbf{L}_j(t_1). \tag{7.23}$$

Note that $(\mathbf{r}_i - \mathbf{r}_j)$ is parallel to $\mathbf{f}_{ij}$. Equation (7.23) can therefore be reduced to

$$\mathbf{L}_i(t_1) + \mathbf{L}_j(t_1) = \mathbf{L}_i(t_1) + \mathbf{L}_j(t_2). \tag{7.24}$$

Therefore, we conclude that, in the absence of external forces, the angular momentum of the particles at time $t_1$ is equal to their angular momentum at time $t_2$. In other words, regardless of the nature of the angular impulse and the associated internal forces causing it, the angular momentum is always conserved.

Equation (7.24) can be extended to $\mathcal{N}$ particles such that

$$\sum_{i=1}^{\mathcal{N}} \mathbf{L}_i(t_1) = \sum_{i=1}^{\mathcal{N}} \mathbf{L}_i(t_2) \tag{7.25}$$

**Example 7.8 Angular Impulse and Momentum**

Two massless rods of length $l$ are welded together at their midpoints, perpendicular to each other, to make the cross shape shown in Figure 7.12. Three particles of mass $m$ are then attached to the end of the rods. The whole system is suspended at point $O$ using a joint that permits rotation about the $\hat{n}_2$-axis only. The right-hand mass is struck so that it receives an impulse of magnitude $-I\hat{n}_3$. Find the angular velocity of the system right after impact.

Using Equation (7.23), we can write

$$\mathbf{r} \times \int_{t_1}^{t_2} \mathbf{F} dt = \mathbf{L}_O(t_2) - \mathbf{L}_O(t_1),$$

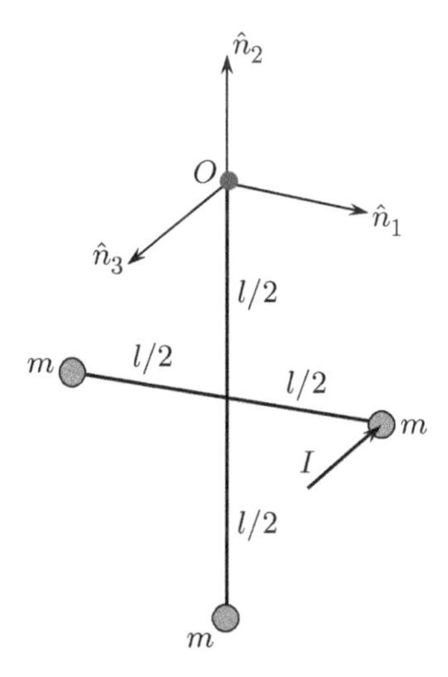


**Figure 7.12** A suspended system subject to an impulse.

where $\mathbf{r}$ is the position vector from point $O$ to the point where the impulse is applied, $\int_{t_1}^{t_2} \mathbf{F}dt = -I\hat{n}_3$ is the impulse force, and $\mathbf{L}_O(t_2)$ and $\mathbf{L}_O(t_1)$ are, respectively, the angular momentum of the system about point $O$ before and after impact.

Noting that $\mathbf{L}_O(t_1) = 0$ since the system is stationary before impact, and applying the previous equation about the $\hat{n}_2$-axis, we obtain

$$\frac{l}{2}\hat{n}_1 \times -I\hat{n}_3 = \frac{l}{2}\hat{n}_1 \times (m\Omega\hat{n}_2 \times \frac{l}{2}\hat{n}_1) - \frac{l}{2}\hat{n}_1 \times ((m\Omega\hat{n}_2 \times -\frac{l}{2}\hat{n}_1)) = m\frac{l^2}{2}\Omega\hat{n}_2,$$

$$\Omega = \frac{I}{ml},$$

where $\Omega$ is the angular velocity of the system about the $\hat{n}_2$-axis right after impact.

**Angular Momentum of Particles**

The angular momentum, $\mathbf{L}$, of a particle of mass $m_i$ about an inertial point $O$ is defined as

$$\mathbf{L}_{iO} = \mathbf{r}_i \times m_i\dot{\mathbf{r}}_i$$

where $\mathbf{r}_i$ is the position vector from point $O$ to the particle, and $\dot{\mathbf{r}}_i$ is the velocity of the particle measured with respect to point $O$. For $\mathcal{N}$ particles, the angular momentum is the sum of the angular momenta of all particles; that is,

$$\mathbf{L}_O = \sum_{i=1}^{\mathcal{N}} \mathbf{L}_{iO}.$$

The moment $\mathbf{M}$ exerted by a net force acting on a particle about point $O$ can be related to the angular momentum of the particle via

$$\int_{t_1}^{t_2} \mathbf{M}_{iO}\mathrm{d}t = \mathbf{L}_{iO}(t_2) - \mathbf{L}_{iO}(t_1),$$

where $\int_{t_1}^{t_2} \mathbf{M}_{iO}\mathrm{d}t$ is the angular impulse. Here, the time interval $[t_1, t_2]$ represents the time for which the moment is applied.

If the net moment acting on a system of $\mathcal{N}$ particles is zero, then the angular impulse vanishes and the previous equation becomes

$$\sum_{i=1}^{\mathcal{N}} \mathbf{L}_{iO}(t_2) = \sum_{i=1}^{\mathcal{N}} \mathbf{L}_{iO}(t_1).$$

The previous equation is known as the conservation of angular momentum.

The time rate of change of the angular momentum about a given point $O$ is equal to the net moments acting on the same point; that is,

$$\frac{\mathrm{d}\mathbf{L}_O}{\mathrm{d}t} = \mathbf{M}_O.$$

## 7.4 Angular Momentum of Rigid Bodies (Planar Motion)

To determine the angular momentum of a rigid body undergoing planar motion, we consider the arbitrarily-shaped rigid body shown in Figure 7.13. The body is rotating with an angular velocity ${}^N\boldsymbol{\omega}^B$ and angular acceleration ${}^N\boldsymbol{\alpha}^B$, where the $B$-frame is a rotating body frame and the $N$-frame is an inertial frame.

Without loss of generality, since we are considering planar rigid-body motion, we can assume that the motion occurs in the $(\hat{b}_1, \hat{b}_2)$-plane and that rotations occur only about the $\hat{b}_3$ axis. As such, we can write ${}^N\boldsymbol{\omega}^B = \omega_3\hat{b}_3$ and ${}^N\boldsymbol{\alpha}^B = \alpha_3\hat{b}_3$, where $\omega_3$ and $\alpha_3$ are, respectively, the magnitudes of the angular velocity and acceleration of the rigid body about the $\hat{b}_3$ axis.

We take a differential element, $\mathrm{d}m$, located at point $P$ on the body, and describe its position and velocity with respect to an inertial point $Q$ as

$$\mathbf{QP} = \mathbf{QO} + \mathbf{r}, \tag{7.26}$$

$${}^N\mathbf{v}^{P/Q} = {}^N\mathbf{v}^{O/Q} + {}^N\boldsymbol{\omega}^B \times \mathbf{r}. \tag{7.27}$$

To obtain the angular momentum of a rigid body, we must specify a point about which it will be calculated. In what follows, we discuss three such points: the first involves finding the angular momentum about an axis passing through the center of mass of the body, the second involves a fixed point on the body, and the third involves finding the angular momentum about an inertial point in space.

### *7.4.1 Angular Momentum about an Axis Passing through the Center of Mass*

The differential angular momentum exerted by a differential element $\mathrm{d}m$ about an axis passing through point $G$ can be written as

$$\mathrm{d}\mathbf{L}_G = \mathbf{r}_G \times {}^N\mathbf{v}^{P/Q}\mathrm{d}m. \tag{7.28}$$

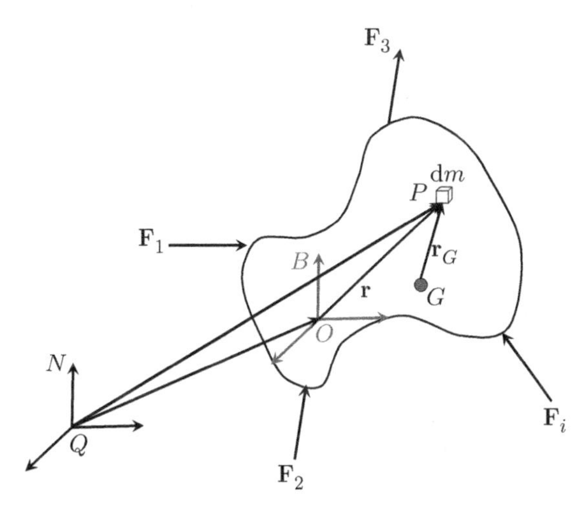


**Figure 7.13** Coordinate system used for the derivation of the angular momentum expression for rigid bodies. (*See color plate section for the color representation of this figure.*)

The total angular momentum of the body about an axis passing through point $G$ can therefore be obtained by integrating the previous equation over the volume to obtain

$$\mathbf{L}_G = \int_{\mathcal{V}} \mathbf{r}_G \times {}^N\mathbf{v}^{P/Q} \mathrm{d}m, \tag{7.29}$$

which upon substituting Equation (7.27) becomes

$$\mathbf{L}_G = \int_{\mathcal{V}} \mathbf{r}_G \mathrm{d}m \times {}^N\mathbf{v}^{G/Q} + \int_{\mathcal{V}} \mathbf{r}_G \times ({}^N\boldsymbol{\omega}^B \times \mathbf{r}_G) \mathrm{d}m. \tag{7.30}$$

Since $\int_{\mathcal{V}} \mathbf{r}_G \mathrm{d}m = 0$, the previous equation reduces to

$$\mathbf{L}_G = \int_{\mathcal{V}} \mathbf{r}_G \times ({}^N\boldsymbol{\omega}^B \times \mathbf{r}_G) \mathrm{d}m. \tag{7.31}$$

For planar motion, we can write $\mathbf{r}_G = x_G \hat{b}_1 + y_G \hat{b}_2$ in the previous equation and obtain

$$\begin{aligned} \mathbf{L}_G &= \int_{\mathcal{V}} (x_G \hat{b}_1 + y_G \hat{b}_2) \times (\omega_3 \hat{b}_3 \times (x_G \hat{b}_1 + y_G \hat{b}_2)) \mathrm{d}m, \\ \mathbf{L}_G &= \omega_3 \int_{\mathcal{V}} (x_G^2 + y_G^2) \hat{b}_3 = I_G \omega_3 \hat{b}_3, \end{aligned} \tag{7.32}$$

where $I_G$ is the moment of inertia of the rigid body about the $\hat{b}_3$ axis, normal to the plane and passing through the center of mass $G$.

Since, for planar motion, the angular momentum is always pointing in the $\hat{b}_3$ direction, we drop the vectorial notation for simplicity, and write

$$\boxed{L_G = I_G\, \omega_3} \tag{7.33}$$

where it is inherently assumed that the vectorial terms in the equation are pointing in the $\hat{b}_3$ direction. The previous equation states that, for planar motion, the angular momentum of a rigid body about an axis passing through its center of mass is equal to its moment of inertia times its angular velocity.

By virtue of Equation (7.15), the moment exerted by all external forces and moments acting on the rigid body about point $G$ can be obtained by differentiating Equation (7.33) with respect to time to obtain

$$\begin{aligned} \frac{\mathrm{d}L_G}{\mathrm{d}t} \hat{b}_3 &= M_G \hat{b}_3 = I_G \dot{\omega}_3 \hat{b}_3 + \omega_3 \hat{b}_3 \times (I_G \omega_3) \hat{b}_3 \\ \frac{\mathrm{d}L_G}{\mathrm{d}t} \hat{b}_3 &= I_G \dot{\omega}_3 \hat{b}_3 \end{aligned} \tag{7.34}$$

where the term $\omega_3 \hat{b}_3 \times (I_G \omega_3) \hat{b}_3$ results from differentiating $\omega_3 \hat{b}_3$, a vector described in the rotating $B$-frame, with respect to an inertial observer located in the $N$-frame. Omitting the vectorial notation from the previous equation, and letting $\dot{\omega}_3 = \alpha_3$, where $\alpha_3$ is the angular acceleration about the $\hat{b}_3$ axis, we obtain

$$\boxed{M_G = I_G \alpha_3} \tag{7.35}$$

### 7.4.2 Angular Momentum about an Axis Passing through a Fixed Point on the Body

A fixed point on the body is another point about which the expression of angular momentum of a rigid body simplifies extensively. In such a case, the differential angular momentum exerted by the differential element $\mathrm{d}m$ about a fixed point $O$ can be written as

$$\mathrm{d}\mathbf{L}_O = \mathbf{r} \times {}^N\mathbf{v}^{P/O}\mathrm{d}m, \tag{7.36}$$

where, as shown in Figure 7.13, ${}^N\mathbf{v}^{P/O} = \omega_3 \times \mathbf{r}$ because point $O$ is fixed. To find the total angular momentum of the body about the fixed point, we integrate the previous equation over the volume to obtain

$$\mathbf{L}_O = \int_{\mathcal{V}} \mathbf{r} \times ({}^N\boldsymbol{\omega}^B \times \mathbf{r})\mathrm{d}m. \tag{7.37}$$

Letting $\mathbf{r} = x\hat{b}_1 + y\hat{b}_2$ and ${}^N\boldsymbol{\omega}^B = \omega_3\hat{b}_3$, it can be shown that the term on the right-hand side of Equation (7.37) is equivalent to $I_O\omega_3$. Here, $I_O$ is the moment of inertia about the $\hat{b}_3$ axis passing through point $O$. It follows that

$$\boxed{L_O = I_O\,\omega_3} \tag{7.38}$$

Again, the moment exerted by all external forces and moments acting on the rigid body about a set of axes located at point $O$ can be obtained by differentiating Equation (7.38) once with respect to time to obtain

$$\boxed{M_O = I_O\,\alpha_3} \tag{7.39}$$

### 7.4.3 Angular Momentum about an Axis Passing through an Arbitrary Inertial Point

By referring to Figure 7.13, we can express the differential angular momentum exerted by the motion of the differential element $\mathrm{d}m$ about an inertial coordinate system located at point $Q$ as

$$\mathrm{d}\mathbf{L}_Q = (\mathbf{QO} + \mathbf{r}) \times {}^N\mathbf{v}^{P/Q}\mathrm{d}m. \tag{7.40}$$

To find the total angular momentum, we integrate the previous equation over the volume and obtain

$$\begin{aligned}
\mathbf{L}_Q &= \int_{\mathcal{V}} (\mathbf{QO} + \mathbf{r}) \times {}^N\mathbf{v}^{P/Q}\mathrm{d}m,\\
\mathbf{L}_Q &= \mathbf{QO} \times {}^N\mathbf{v}^{O/Q}\int_{\mathcal{V}}\mathrm{d}m + \mathbf{QO} \times \left({}^N\boldsymbol{\omega}^B \times \int_{\mathcal{V}}\mathbf{r}\mathrm{d}m\right)\\
&+ \int_{\mathcal{V}}\mathbf{r}\mathrm{d}m \times {}^N\mathbf{v}^{O/Q} + \int_{\mathcal{V}}\mathbf{r} \times ({}^N\boldsymbol{\omega}^B \times \mathbf{r})\mathrm{d}m.
\end{aligned} \tag{7.41}$$

Choosing point $O$ to coincide with the center of gravity $G$, we can write

$$\begin{aligned}
\mathbf{L}_Q &= \mathbf{QG} \times m{}^N\mathbf{v}^{G/Q} + \mathbf{QG} \times \left({}^N\boldsymbol{\omega}^B \times \int_{\mathcal{V}}\mathbf{r}_G\mathrm{d}m\right)\\
&+ \int_{\mathcal{V}}\mathbf{r}_G\mathrm{d}m \times {}^N\mathbf{v}^{G/Q} + \int_{\mathcal{V}}\mathbf{r}_G \times ({}^N\boldsymbol{\omega}^B \times \mathbf{r}_G)\mathrm{d}m.
\end{aligned} \tag{7.42}$$

Since $\int_{\mathcal{V}} \mathbf{r}_G \mathrm{d}m = 0$, the previous equation reduces to

$$\mathbf{L}_Q = \mathbf{QG} \times m^N\mathbf{v}^{G/Q} + \int_{\mathcal{V}} \mathbf{r}_G \times ({}^N\boldsymbol{\omega}^B \times \mathbf{r}_G)\mathrm{d}m. \tag{7.43}$$

Letting $\mathbf{r}_G = x_G\hat{b}_1 + y_G\hat{b}_2$ and ${}^N\boldsymbol{\omega}^B = \omega_3\hat{b}_3$, we can show that the last term in the equation is equivalent to $I_G\,\omega_3$, where $I_G$ is the moment of inertia about the $\hat{b}_3$ axis passing through the center of mass $G$. Therefore, Equation (7.43) can be written as:

$$L_Q\hat{b}_3 = \mathbf{QG} \times m^N\mathbf{v}^{G/Q} + I_G\,\omega_3\hat{b}_3. \tag{7.44}$$

Since the vector $\mathbf{QG} \times m^N\mathbf{v}^{G/Q}$ also points in the $\hat{b}_3$ direction, we omit the vectorial description, and write the previous equation in the following form:

$$\boxed{L_Q = \mathbf{QG} \times m^N\mathbf{v}^{G/Q} + I_G\,\omega_3} \tag{7.45}$$

Hence, for planar motion, the angular momentum of a rigid body about an axis located at an arbitrary inertial point $Q$ is equal to the angular momentum of its center of mass $G$ about $Q$ plus its angular momentum about an axis passing through $G$.

The moment exerted by all external forces and moments acting on the rigid body about point $Q$ can be obtained by differentiating Equation (7.45) to obtain

$$\frac{\mathrm{d}L_Q}{\mathrm{d}t}\hat{b}_3 = M_Q\hat{b}_3 = \mathbf{QG} \times m^N\mathbf{a}^{G/Q} + \omega_3\hat{b}_3 \times I_G\omega_3\hat{b}_3 + I_G\,\dot{\omega}_3\,\hat{b}_3 \tag{7.46}$$

which can be reduced to

$$\boxed{M_Q = \mathbf{QG} \times m^N\mathbf{a}^{G/Q} + I_G\,\alpha_3} \tag{7.47}$$

**Example 7.9 Ball Impacting a Hinged Rod**

A ball of mass $2m$ and velocity $v_0$ impacts a hinged rod of mass $m$ and length $l$ at point $R$, as shown in Figure 7.14. Calculate the angular velocity of the rod and the velocity of the ball just after impact. Assume that the collision is an elastic one and the motion occurs in a horizontal plane.

Note that there are no external forces acting on the system. Therefore, the angular momentum of the system is conserved. As such, we can write

$$\mathbf{L}_p + \mathbf{L}_r(\text{before impact}) = \mathbf{L}_p + \mathbf{L}_r(\text{after impact})$$

where $\mathbf{L}_p$ and $\mathbf{L}_r$ are respectively, the angular momentum of the particle and rod about an axis passing through the fixed point $O$; that is,

$$-\frac{2l}{3}\hat{n}_2 \times 2mv_0\hat{n}_1 = -\frac{2l}{3}\hat{n}_2 \times 2mv_{pa}\hat{n}_1 + I_O\dot{\theta}\hat{n}_3$$

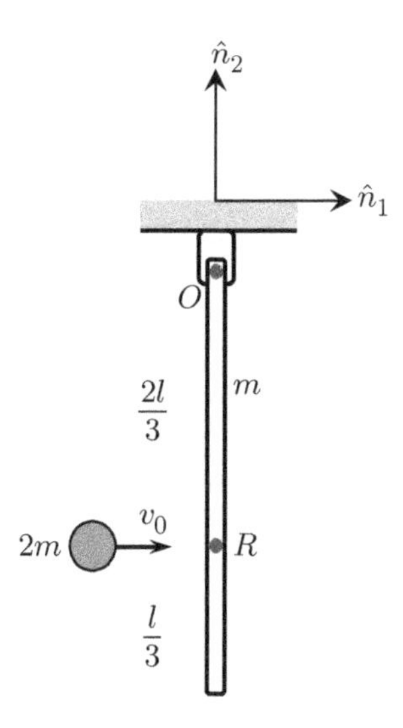


**Figure 7.14** A ball impacting a hinged rod.

where $v_{pa}$ is the velocity of the particle after impact, $\dot{\theta}$ is the angular velocity of the rod after impact, and $I_O = \frac{1}{3}ml^2$ is the moment inertia of the rod about point $O$. Rearranging the previous equation yields

$$v_{pa} = v_0 - \frac{1}{4}l\dot{\theta}.$$

Since the collision is purely elastic, the kinetic energy before and after impact is conserved:

$$\frac{1}{2}(2m)v_0^2 = \frac{1}{2}I_O\dot{\theta}^2 + \frac{1}{2}(2m)v_{pa}^2.$$

Solving the previous two equations together yields

$$v_{pa} = \frac{5}{11}v_0, \qquad \dot{\theta} = \frac{24}{11}\frac{v_0}{l}.$$

### Example 7.10 Sliding to Rolling

Consider a hoop of mass $m$ and radius $R$ that slides over a rough surface with initial velocity $V_0$. Over time, and due to frictional forces, the ring starts to roll (Figure 7.15). Find the velocity at which the ring starts rolling. What happens to the obtained velocity if you replace the ring with a hollow sphere?

This problem can be solved in many different ways. Here we will use conservation of angular momentum. Note that the line of all forces acting on the body passes through the contact point $O$. Therefore, the angular momentum is conserved about a set of coordinates passing through point $O$. As such, we can write

$$\mathbf{L}_O(t_1) = \mathbf{L}_O(t_2).$$

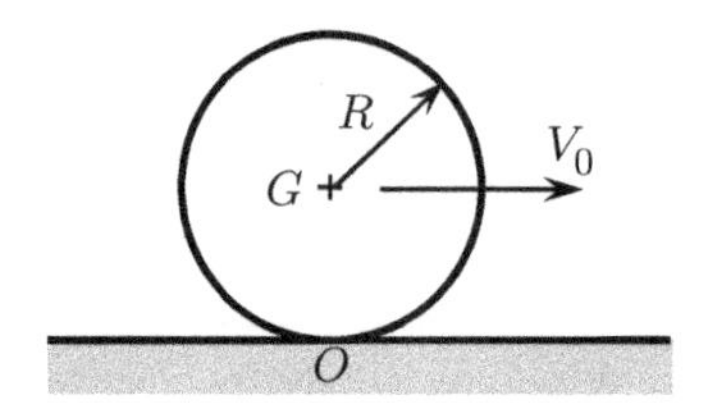


**Figure 7.15** Transition from sliding to rolling.

Initially the ring slides without rolling, and, as a result, the angular momentum of the ring about point $O$ is $\mathbf{L}_O(t_1) = mRV_0$. At the point when the ring only rolls about point $O$, the angular momentum about point $O$ becomes $\mathbf{L}_O(t_2) = I_O\omega$, where $\omega$ can be related to the velocity $V_f$ of the center of mass via $\omega = \frac{V_f}{R}$. Also, for the ring, $I_O = I_G + mR^2 = 2mR^2$. This yields

$$mRV_0 = 2mR^2\frac{V_f}{R}, \qquad V_f = \frac{V_0}{2}.$$

When the ring is replaced with a hollow sphere, the moment of inertia about the center of mass becomes $I_G = \frac{2}{3}mR^2$. Therefore, we can write:

$$mRV_0 = \frac{5}{3}mR^2\frac{V_f}{R}, \qquad V_f = \frac{3}{5}V_0.$$

Thus, the hollow sphere starts rolling at a higher speed.

### Flipped Classroom Exercise 7.3

A ball of mass $2m$ and velocity $v_0$ impacts a free rod of mass $m$ and length $l$ at point $R$, as shown in the figure. Calculate the angular velocity of the rod and the velocity of the ball and the point of impact on the rod just after impact. Assume that the collision is an elastic one and the motion occurs in a horizontal plane.

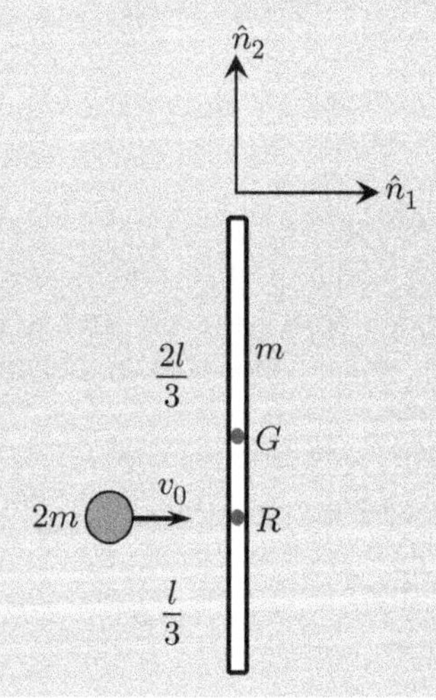

The problem has three unknowns, namely the angular velocity of the rod $\dot{\theta}$, the velocity of the particle after impact $v_{pa}$, and the velocity $v_{ra}$ of the point of impact on the rod after impact. To find these unknowns, take the following steps:

1. Since there are no external forces acting on the system, apply conservation of linear momentum at the point of impact.
2. Apply conservation of angular momentum about the center of mass.
3. Since the impact is elastic, apply conservation of kinetic energy.
4. Solve the previous three equations together and show that

$$\dot{\theta} = \frac{24}{7}\frac{v_0}{l}, \qquad v_{pa} = \frac{4}{7}v_0, \qquad v_{ra} = \frac{6}{7}v_0.$$

## 7.5 Angular Momentum of Rigid Bodies (Non-planar Motion)

The process of deriving the angular momentum expression for non-planar motion of rigid bodies is very similar to that explained in the previous section. The only difference is that the equations are expressed in a general vectorial form, as outlined below.

### 7.5.1 *Angular Momentum about a Set of Axes Located at the Center of Mass*

Referring back to Figure 7.13, the equation governing the angular momentum about an axis passing through the center of mass is

$$\mathbf{L}_G = I_G \, {}^N\boldsymbol{\omega}^B, \tag{7.48}$$

where $I_G$ is the inertia matrix about point $G$. To obtain the moment about a set of axes located at point $G$, we differentiate the previous equation and obtain

$$\frac{\mathrm{d}\mathbf{L}_G}{\mathrm{d}t} = \mathbf{M}_G = {}^N\boldsymbol{\omega}^B \times (I_G{}^N \boldsymbol{\omega}^B) + I_G \, {}^N\boldsymbol{\alpha}^B. \tag{7.49}$$

Note that, unlike planar body motion, the term ${}^N\boldsymbol{\omega}^B \times (I_G \, {}^N\boldsymbol{\omega}^B)$ does not vanish when considering non-planar rigid-body motion. It is also worth noting that the previous equation is nothing but the Euler rotational equation derived in Chapter 3 for non-planar rigid-body motion.

Upon choosing the set of axes whose origin is located at point $G$ to align with the principal axis of rotation (the set of axes about which the product of inertia terms vanish) the previous equation reduces to

$$\begin{aligned} M_{1G} &= I_{11G}\alpha_1 + (I_{33G} - I_{22G})\omega_3\omega_2, \\ M_{2G} &= I_{22G}\alpha_2 + (I_{11G} - I_{33G})\omega_1\omega_3, \\ M_{3G} &= I_{33G}\alpha_3 + (I_{22G} - I_{11G})\omega_1\omega_2. \end{aligned} \tag{7.50}$$

where $M_{iG}$ is the moment exerted by the external forces and moments about the $i$th axis passing through point $G$, $I_{iiG}$ is the mass moment of inertia about the $i$th axis passing through point $G$, $\omega_i$ is the angular velocity about the $i$th axis, and $\alpha_i$ is the angular acceleration about the $i$th axis.

### 7.5.2 Angular Momentum about a Set of Axes Located at a Fixed Point

The equation governing the angular momentum about a set of axes located at a fixed point on the body is given by

$$\mathbf{L}_O = I_O \, {}^N\boldsymbol{\omega}^B, \tag{7.51}$$

where $I_O$ is the inertia matrix about point $O$. To obtain the moment about a set of axes located at point $O$, we differentiate the previous equation and obtain

$$\frac{\mathrm{d}\mathbf{L}_O}{\mathrm{d}t} = \mathbf{M}_O = {}^N\boldsymbol{\omega}^B \times (I_O {}^N\boldsymbol{\omega}^B) + I_O \, {}^N\boldsymbol{\alpha}^B. \tag{7.52}$$

Again, upon choosing the set of axes whose origin is located at point $O$ to align with the principal axis of rotation, the previous equation reduces to

$$\begin{aligned} M_{1O} &= I_{11O}\alpha_1 + (I_{33O} - I_{22O})\omega_3\omega_2, \\ M_{2O} &= I_{22O}\alpha_2 + (I_{11O} - I_{33O})\omega_1\omega_3, \\ M_{3O} &= I_{33O}\alpha_3 + (I_{22O} - I_{11O})\omega_1\omega_2, \end{aligned} \tag{7.53}$$

where $M_{iO}$ is the moment exerted by the external forces and moments about the $i$th axis passing through point $O$, $I_{iiO}$ is the mass moment of inertia about the $i$th axis passing through point $O$, $\omega_i$ is the angular velocity about the $i$th axis, and $\alpha_i$ is the angular acceleration about the $i$th axis.

### 7.5.3 Angular Momentum about a Set of Axes Located at an Arbitrary Inertial Point

The equation governing the angular momentum about a set of axes located at an arbitrary inertial point $Q$ is given by

$$\mathbf{L}_Q = \mathbf{QG} \times m^N\mathbf{v}^{G/Q} + I_G \, {}^N\boldsymbol{\omega}^B. \tag{7.54}$$

The moment exerted by all external forces and moments acting on the rigid body about point $Q$ can be obtained by differentiating Equation (7.54), giving:

$$\frac{\mathrm{d}\mathbf{L}_Q}{\mathrm{d}t} = \mathbf{M}_Q = \mathbf{QG} \times m^N\mathbf{a}^{G/Q} + {}^N\boldsymbol{\omega}^B \times (I_G \, {}^N\boldsymbol{\omega}^B) + I_G \, {}^N\boldsymbol{\alpha}^B. \tag{7.55}$$

### 7.5.4 Conservation of Angular Momentum for Rigid Bodies

When the net moments exerted by forces acting on a rigid body about a set of axes located at an arbitrary point, say $O$, is zero, one can write

$$\frac{\mathrm{d}\mathbf{L}_O}{\mathrm{d}t} = 0. \tag{7.56}$$

Integrating this with respect to time, we can write

$$\mathbf{L}_O = \text{constant} \tag{7.57}$$

or

$$\mathbf{L}_O(t_1) = \mathbf{L}_O(t_2). \tag{7.58}$$

Therefore, the angular momentum of rigid bodies is conserved when there are no external forces or moments acting on them.

**Example 7.11 Rotating Disks**

As shown in Figure 7.16, two disks, each of mass $m$, are rotating around the same axis with equal but opposite angular velocities $\omega_0$. The upper disk has a radius $r$ while the lower disk has a radius $2r$. The upper disk slides down and comes into contact with the lower disk. Due to frictional forces, the two disks start rotating together. What is their angular velocity of rotation? What would the angular velocity of rotation be if the two disks had the same radius?

To solve this problem we first note that there are no external forces or moments acting on the system, so the angular momentum is conserved and we can write

$$\mathbf{L}(t_1) = \mathbf{L}(t_2).$$

Before the two disks come together and start rotating as one rigid body, each disk has a different angular momentum, given by $L_1 = \frac{1}{2}mr^2\omega_0$ for disk 1 and $L_2 = -\frac{1}{2}m(2r)^2\omega_0$ for disk 2. After they come together, they have one angular momentum given by $L_T = \left(\frac{1}{2}mr^2 + \frac{1}{2}m(2r)^2\right)\omega_f$; where $\omega_f$ is the angular velocity of the combined system. Thus, we can write

$$\frac{1}{2}mr^2\omega_0 - \frac{1}{2}m(2r)^2\omega_0 = \left(\frac{1}{2}mr^2 + \frac{1}{2}m(2r)^2\right)\omega_f.$$

Solving the previous equation yields $\omega_f = -\frac{3}{5}\omega_0$. If the two disks have similar radius and mass, then their angular momentum cancels out and $\omega_f = 0$.

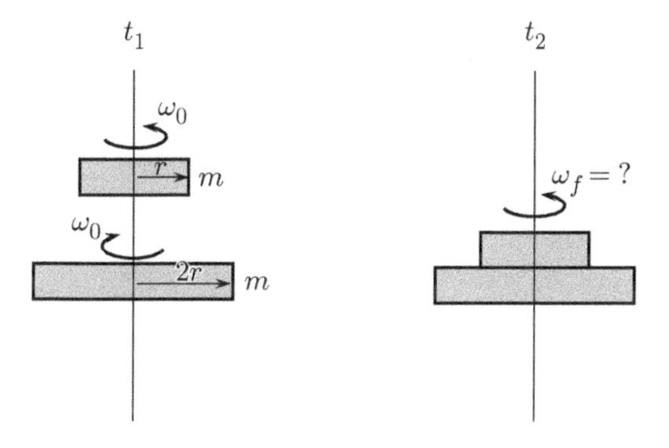


**Figure 7.16** Angular momentum of rotating disks.

### Example 7.12 Sliding Lollipop

A lollipop of mass $m$ and radius $r$ is radially pierced by a massless stick, as shown in Figure 7.17. The free end of the stick is pivoted on the smooth ground. The lollipop slides without rolling on the surface shown, such that the center of the lollipop is always at a distance $R$ from point $Q$ and rotates at an angular frequency $\Omega$. Find the reaction force between the ground and the lollipop.

We can solve this problem in two different ways; either using Newton's translational equations or using Euler's rotational equations. We will describe both approaches.

***Newton's second law:*** To apply Newton's second law, we first find the acceleration of the center of mass $G$ with respect to point $Q$. To this end, we define a new frame, the $A$-frame, such that ${}^N\boldsymbol{\omega}^A = \Omega\hat{a}_2$. Using this definition, the acceleration of point $G$ with respect to point $Q$ can be written as:

$$\mathbf{a}^{G/Q} = -R\Omega^2\hat{a}_1$$

To apply Newton's equations, we draw a free-body diagram, as shown in Figure 7.18. Summing forces in the $\hat{a}_1$ and $\hat{a}_2$ direction, we can write

$$\begin{aligned} Q_x &= -mR\Omega^2, \qquad Q\cos\theta = -mR\Omega^2, \qquad Q = -m\sqrt{R^2+r^2}\Omega^2, \\ Q_y &+ R_N - mg = 0, \qquad R_N = mg - Q\frac{r}{\sqrt{R^2+r^2}}. \end{aligned}$$

Solving the previous two equation for $R_N$ yields

$$R_N = mg + mr\Omega^2.$$

***Euler's rotational equation:*** We sum moments about an axis passing through point $Q$ and write

$$\mathbf{M}_Q = \mathbf{QG} \times m{}^N\mathbf{a}^{G/Q} + {}^N\boldsymbol{\omega}^B \times I_G{}^N\boldsymbol{\omega}^B + I_G\,{}^N\boldsymbol{\alpha}^B.$$

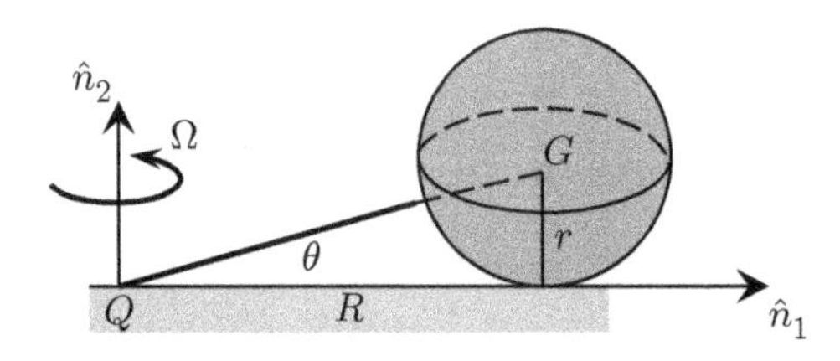


**Figure 7.17** Sliding lollipop.

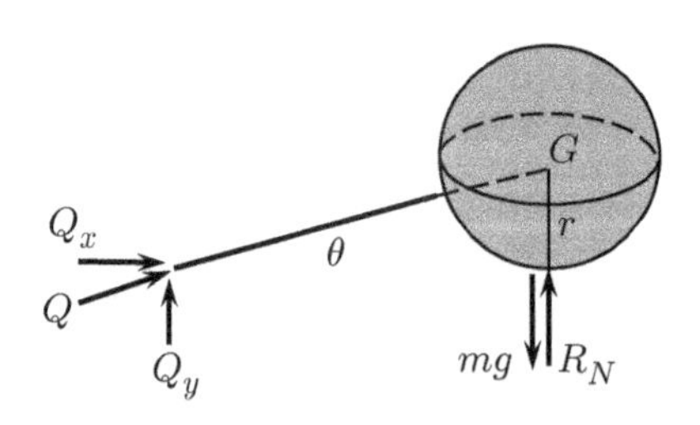


**Figure 7.18** Free-body diagram of a sliding lollipop.

Noting that ${}^N\boldsymbol{\alpha}^B = 0$ and ${}^N\boldsymbol{\omega}^B \times I_G{}^N\boldsymbol{\omega}^B = 0$, we can write

$$(R_N R - mgR)\hat{a}_3 = (R\hat{a}_1 + r\hat{a}_2) \times (-mR\Omega^2\hat{a}_1), \qquad R_N = mg + mr\Omega^2.$$

**Example 7.13 Solid Parallelepiped Impacts an Obstacle**

The corner $P$, of a solid parallelepiped of mass $m$ and dimensions $(l, 2l, l)$ moves with a constant velocity $v\ \hat{n}_1$ before it hits an obstacle, as shown in Figure 7.19. The impact results in an impulse $-I\hat{n}_1$. Find the velocity of the center of mass and the angular velocity of the parallelepiped right after impact.

To solve this problem, we first apply the translational impulse equation to the parallelepiped; that is

$$\mathbf{p}(t_2) - \mathbf{p}(t_1) = \int_{t_1}^{t_2} \mathbf{F}\mathrm{d}t.$$

It follows that

$$m\mathbf{v}_G - mv\hat{n}_1 = -I\hat{n}_1, \qquad \mathbf{v}_G = \left(v - \frac{I}{m}\right)\hat{n}_1.$$

To obtain the angular velocity after impact, we apply the angular impulse equation about the center of mass as follows:

$$\mathbf{L}_G(t_2) - \mathbf{L}_G(t_1) = \int_{t_1}^{t_2} \mathbf{M}_G\mathrm{d}t,$$

which yields

$$I_{G11}\omega_1\hat{n}_1 + I_{G22}\omega_2\hat{n}_2 + I_{G33}\omega_3\hat{n}_3 = \mathbf{GP} \times (-I\hat{n}_1),$$

where $\mathbf{GP} = \frac{1}{2}(l\hat{n}_1 + 2l\hat{n}_2 + l\hat{n}_3)$. Upon rearranging and solving this for the angular velocity components, we obtain

$$\omega_1 = 0, \qquad \omega_2 = \frac{-2lI}{2I_{G22}}, \qquad \omega_3 = \frac{lI}{2I_{G33}},$$

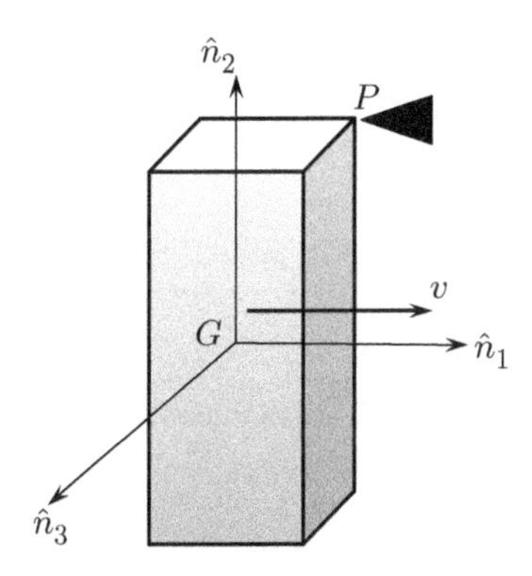


**Figure 7.19** A solid parallelepiped impacting an obstacle.

where $I_{G22} = \frac{5}{12}ml^2$ and $I_{G33} = \frac{1}{3}ml^2$. Thus we can write

$$\boldsymbol{\omega} = \frac{12I}{5ml}\hat{n}_1 + \frac{3I}{ml}\hat{n}_2.$$

### Example 7.14 Ball Impacting a Door

A ball of mass $m$ and velocity $-v_0\hat{n}_3$ impacts the corner of the door of mass $\alpha m$, as shown in Figure 7.20. Find the angular velocity of the door right after impact. The coefficient of restitution between the ball and the door is $e = \frac{3}{4}$.

Since there are no external forces acting on the body, we can use conservation of angular momentum to solve the problem. Before impact, the ball is the only moving body and it has an angular momentum of $-mav_0\hat{n}_2$ about the hinge. Right after impact, we have:

$$\begin{aligned}
\mathbf{L}_{\text{ball}}\hat{n}_2 &= mav_b\hat{n}_2, \\
\mathbf{L}_{\text{door}}\hat{n}_2 &= -I_{\text{door}}\omega = -\tfrac{1}{3}\alpha ma^2\omega\hat{n}_2,
\end{aligned}$$

where $v_b$ is the velocity of the ball right after impact and, $I_{\text{door}}$ is the moment of inertia of the door about the $\hat{n}_2$-axis.

Applying conservation of angular momentum, we obtain

$$v_o = -v_b + \frac{1}{3}\alpha a\omega.$$

This involves two unknowns, namely $v_b$ and $\omega$, so a second equation is necessary. This equation is obtained using the coefficient of restitution, which is defined as

$$e = \frac{v_d - v_b}{-v_0},$$

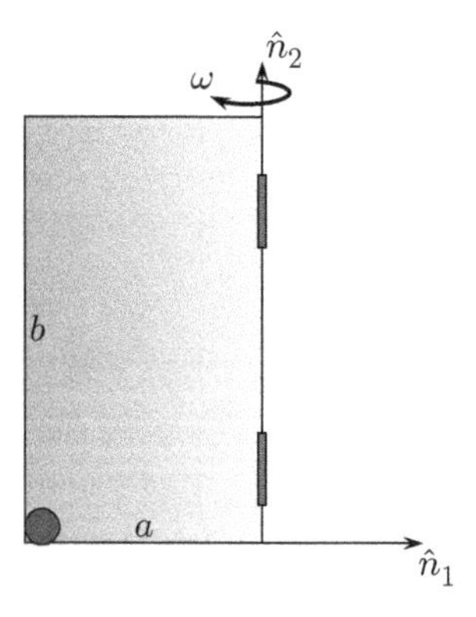


**Figure 7.20** Ball impacting a door.

where $v_d = -a\omega\hat{n}_3$ is the velocity of the door right after impact. It follows that

$$v_b = \frac{3}{4}v_0 - a\omega.$$

Using the two equations for $v_b$ and solving for $\omega$ yields

$$\omega = \frac{7}{4}\frac{v_0}{a(1+\frac{1}{3}\alpha)}.$$

**Flipped Classroom Exercise 7.4**

A homogeneous thin plate in the form of a right-angled triangle of mass $m$ can rotate freely about a fixed ball and socket joint at the right angle $O$. The perpendicular edges have lengths $l$ and $2l$. Initially, point $A$ is stationary, and the plate rotates with angular velocity $\Omega_0$ about $OA$. Corner $A$ is suddenly released, while corner $B$ is simultaneously fixed. Find the new angular velocity $\Omega_1$ about $OB$.

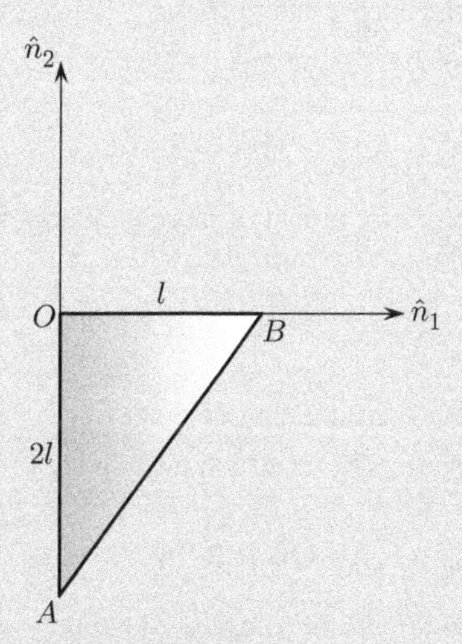


To solve this problem, take the followings steps:

1. Check if there are any external forces acting on the system. If not, use conservation of angular momentum.
2. Find the angular momentum of the plate before point $A$ is released. Do not forget about the contribution of the product of inertia terms.
3. Find the angular momentum of the plate after point $A$ is fixed. Do not forget about the contribution of the product of inertia terms.
4. Apply conservation of angular momentum to show that $\Omega_1 = -\frac{1}{4}\Omega_0$.

**Angular Momentum of Rigid Bodies**

The angular momentum of rigid bodies can be obtained about a set of axes located at any point in space. However, it takes a very simple form when summed about one of the following points:

1. The center of mass $G$, for which the angular momentum, $\mathbf{L}_G$, takes the form:

$$\mathbf{L}_G = I_G\, {}^N\boldsymbol{\omega}^B,$$

where $I_G$ is the moment of inertia matrix about point $G$ and ${}^N\boldsymbol{\omega}^B$ is the angular velocity vector calculated in the body rotating frame. Differentiating the previous equation with respect to time yields:

$$\mathbf{M}_G = {}^N\boldsymbol{\omega}^B \times (I_G\, {}^N\boldsymbol{\omega}^B) + I_G\, {}^N\boldsymbol{\alpha}^B,$$

where $\mathbf{M}_G$ is the net moment acting on point $G$ and ${}^N\boldsymbol{\alpha}^B$ is the angular acceleration vector measured in the body rotating frame.

2. A fixed point on the body, $O$, for which the angular momentum, $\mathbf{L}_O$, takes the form:

$$\mathbf{L}_O = I_O\, {}^N\boldsymbol{\omega}^B,$$

where $I_O$ is the matrix of moment of inertia about point $O$. Differentiating the previous equation with respect to time yields:

$$\mathbf{M}_O = {}^N\boldsymbol{\omega}^B \times (I_O\, {}^N\boldsymbol{\omega}^B) + I_O\, {}^N\boldsymbol{\alpha}^B,$$

where $\mathbf{M}_O$ is the net moment acting on point $O$.

3. An inertial point in space, $Q$ , for which the angular momentum, $\mathbf{L}_Q$, takes the form:

$$\mathbf{L}_Q = \mathbf{QG} \times m {}^N\mathbf{v}^{G/Q} + I_G\, {}^N\boldsymbol{\omega}^B,$$

where $\mathbf{QG}$ is the position vector from point $Q$ to the center of mass $G$, and ${}^N\mathbf{v}^{G/Q}$ is the velocity of point $G$ with respect to point $Q$. The moment exerted by all external forces acting on the rigid body about point $Q$ can be obtained by differentiating the previous equation, to obtain

$$\mathbf{M}_Q = \mathbf{QG} \times m {}^N\mathbf{a}^{G/Q} + {}^N\boldsymbol{\omega}^B \times (I_G\, {}^N\boldsymbol{\omega}^B) + I_G\, {}^N\boldsymbol{\alpha}^B,$$

where ${}^N\mathbf{a}^{G/Q}$ is the acceleration of point $G$ with respect to point $Q$.

## 7.6 Generalized Momenta

Similar to the generalized coordinates and generalized forces we used in Chapter 5 to describe Lagrange's equations, there also exist generalized momenta $p_j$, which can be defined as:

$$p_j = \frac{\mathcal{L}}{\dot{q}_j} \tag{7.59}$$

where $\mathcal{L}$ is the system's Lagrangian and $q_j$ is the generalized coordinate associated with the generalized momentum. As an example, consider the horizontal motion of a particle along a straight line for which the generalized coordinate is $q = x$ and for which the Lagrangian is $\mathcal{L} = \frac{1}{2}m\dot{x}^2$. The generalized momentum of this particle is given by $p_x = \frac{\mathcal{L}}{\dot{x}} = m\dot{x}$, which is nothing but the linear momentum of the particle. Similarly, the Lagrangian of a particle of mass $m$ in circular motion of radius $R$ in a horizontal plane can be expressed as $\mathcal{L} = \frac{1}{2}mR^2\dot{\theta}^2$. Using $\theta$ as the generalized coordinate to describe the motion of the particle, one finds that the generalized momentum of the particle is $p_\theta = mR^2\dot{\theta}$, which is equivalent to the angular momentum of the particle.

In 1835, Hamilton introduced a reformulation of Lagrange's equation based on the use of generalized momenta. The main difference between the Hamilton equation and Lagrange's is that the latter yields a set of differential equations, usually second order, described in terms of the position, velocity, and acceleration of the system; the former yields a set of first-order differential equations described in terms of the position, velocity, and *generalized momenta* of the system.

To express the dynamics of the system in Hamilton's form, we first introduce the following quantity:

$$\mathcal{H}(q_j, p_j, t) = \sum_{j=1}^{\mathcal{M}} p_j\dot{q}_j - \mathcal{L}(q_j, \dot{q}_j, t), \tag{7.60}$$

where $\mathcal{H}$ is called the Hamiltonian, $q_j$ are the generalized coordinates, and $p_j$ are the generalized momenta. Differentiating Equation (7.60) once with respect to $q_j$, we obtain

$$\frac{\partial \mathcal{H}}{\partial q_j} = -\frac{\partial \mathcal{L}}{\partial q_j}. \tag{7.61}$$

Using Lagrange's equation for a conservative system, the term on the right-hand side of the previous equation can be written as

$$\frac{\partial \mathcal{L}}{\partial q_j} = \frac{\mathrm{d}}{\mathrm{d}t}\left(\frac{\partial \mathcal{L}}{\partial \dot{q}_j}\right). \tag{7.62}$$

By virtue of Equation (7.59), the previous equation becomes

$$\frac{\partial \mathcal{L}}{\partial q_j} = \frac{\mathrm{d}}{\mathrm{d}t}\left(\frac{\partial \mathcal{L}}{\partial \dot{q}_j}\right) = \dot{p}_j. \tag{7.63}$$

Substituting the previous equation back into Equation (7.61) yields

$$\frac{\partial \mathcal{H}}{\partial q_j} = -\dot{p}_j \tag{7.64}$$

Taking the partial derivative of Equation (7.60) with respect to $p_i$ yields

$$\frac{\partial \mathcal{H}}{\partial p_j} = \dot{q}_j \tag{7.65}$$

Equations (7.64) and (7.65) are known as Hamilton's canonical equations and can be used directly to describe the motion of a system in terms of its position, velocity, and generalized momenta. In addition to the previous equations, it is also interesting to see that taking a partial derivative of the Hamiltonian described by Equation (7.60) with respect to time, yields the interesting result

$$\frac{\partial \mathcal{H}}{\partial t} = -\frac{\partial \mathcal{L}}{\partial t}. \tag{7.66}$$

It turns out that, under certain conditions that will be described next, the Hamiltonian $\mathcal{H}$ is equal to the total energy of the system; in other words, $\mathcal{H} = \mathcal{T} + \mathcal{U}$, where $\mathcal{T}$ and $\mathcal{U}$ are, respectively, the kinetic and potential energy of the system. To find these conditions, consider the kinetic energy of a system of $\mathcal{N}$ particles each of mass $m_i$, described in terms of their position vector $\mathbf{r}_i$ measured with respect to an inertial point

$$\mathcal{T} = \frac{1}{2} \sum_{i=1}^{\mathcal{N}} m_i \dot{\mathbf{r}}_i . \dot{\mathbf{r}}_i . \tag{7.67}$$

Next, let the position vector $\mathbf{r}_i$ be described in terms of $\mathcal{M}$ generalized coordinates; that is, $\mathbf{r} = \mathbf{r}(q_1, q_2, \dots, q_{\mathcal{M}})$ without dependence on time $t$. The velocity of the $i$th particle can therefore be written as

$$\dot{\mathbf{r}}_i = \sum_{j=1}^{\mathcal{M}} \frac{\partial \mathbf{r}_i}{\partial q_j} \dot{q}_j . \tag{7.68}$$

It follows that

$$\mathcal{T} = \frac{1}{2} \sum_{i=1}^{\mathcal{N}} \sum_{j=1}^{\mathcal{M}} m_i \left( \frac{\partial \mathbf{r}_i}{\partial q_j} \dot{q}_j \right) \cdot \left( \frac{\partial \mathbf{r}_i}{\partial q_j} \dot{q}_j \right) . \tag{7.69}$$

Differentiating the previous equation with respect to $\dot{q}_j$ and multiplying by $\dot{q}_j$, yields

$$\sum_{j=1}^{\mathcal{M}} \frac{\partial \mathcal{T}}{\partial \dot{q}_j} \dot{q}_j = \sum_{i=1}^{\mathcal{N}} \sum_{j=1}^{\mathcal{M}} m_i \left( \frac{\partial \mathbf{r}_i}{\partial q_j} \dot{q}_j \right) \cdot \left( \frac{\partial \mathbf{r}_i}{\partial q_j} \dot{q}_j \right) = 2\mathcal{T}. \tag{7.70}$$

Assuming that the potential energy $\mathcal{U}$ is only a function of the generalized coordinates, $q_j$ but not their derivatives, $\dot{q}_j$ – i.e. $\mathcal{U} = \mathcal{U}(q_j)$ – then using the definition of the Hamiltonian

$$\mathcal{H}(q_j, p_j, t) = \sum_{j=1}^{\mathcal{M}} \frac{\partial \mathcal{L}}{\partial \dot{q}_j}\dot{q}_j - \mathcal{L}(q_j, \dot{q}_j, t), \tag{7.71}$$

we can write

$$\mathcal{H}(q_j, p_j, t) = \sum_{j=1}^{\mathcal{M}} \frac{\partial \mathcal{T}}{\partial \dot{q}_j}\dot{q}_j - \mathcal{T} + \mathcal{U}. \tag{7.72}$$

Substituting Equation (7.70) into the Equation (7.72), we obtain

$$\mathcal{H}(q_j, p_j, t) = 2\mathcal{T} - \mathcal{T} + \mathcal{U} = \mathcal{T} + \mathcal{U}. \tag{7.73}$$

Note that the conclusion that the Hamiltonian is the total energy of the system is valid only when:

- the position of the system of particles does not depend explicitly on time; that is, $\mathbf{r} = \mathbf{r}(q_1, q_2, \dots, q_{\mathcal{M}})$; otherwise, Equation (7.68) does not hold;
- the potential energy $\mathcal{U}$ is only a function of the generalized coordinated; otherwise, $\frac{\partial \mathcal{L}}{\partial \dot{q}_j} \neq \frac{\partial \mathcal{T}}{\partial \dot{q}_j}$;
- the system is conservative; otherwise the total energy of the system is not conserved.

**Example 7.15 Accelerating Particle**

A particle of mass $m$ is constrained to move on the massless rod shown in Figure 7.21. The rod is accelerating vertically against gravity with a constant acceleration $a\,\hat{n}_2$ measured with respect to the inertial frame. Find the Hamiltonian of the system. Is it equal to the total energy? If so why? If not, why not?

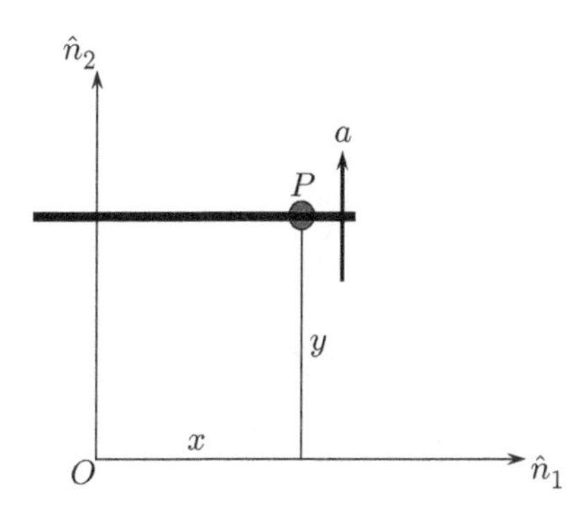


**Figure 7.21** Motion of a particle on a vertically accelerating rod.

First, we obtain the Lagrangian of the system $\mathcal{L}$, which is given by

$$\mathcal{L} = \frac{1}{2} m {}^N v^{P/O} . {}^N v^{P/O} - mgy,$$

where $v^{P/O} = \dot{x}\hat{n}_1 + \dot{y}\hat{n}_2$, $\ddot{y} = a$, $\dot{y} = at$, and $y = at^2/2$ (assuming zero initial conditions). It follows that

$$\mathcal{L} = \frac{1}{2} m(\dot{x}^2 + a^2 t^2) - mga\frac{t^2}{2}.$$

The Hamiltonian of the system is

$$\mathcal{H} = \frac{m}{2}(\dot{x}^2 - a^2 t^2) + mga\frac{t^2}{2}.$$

Note that this is not equal to the total energy $\mathcal{T} + \mathcal{U} = \frac{1}{2} m(\dot{x}^2 + a^2 t^2) + mga\frac{t^2}{2}$. This is because the position of the particle depends explicitly on time: $\mathbf{OP} = x\hat{n}_1 + at^2/2\hat{n}_2$.

**Example 7.16 Simple Pendulum**

Write the Lagrangian and Hamiltonian of the motion of a simple pendulum (Figure 7.22). Show that the Hamiltonian is the total energy of the system. Describe the equations of motion in Hamilton's canonical form.

For the simple pendulum, the kinetic and potential energies can be written as

$$\mathcal{T} = \frac{1}{2} ml^2\dot{\theta}^2, \qquad \mathcal{U} = -mgl\cos\theta.$$

It follows that the Lagrangian can be written as

$$\mathcal{L} = \mathcal{T} - \mathcal{U} = \frac{1}{2} ml^2\dot{\theta}^2 + mgl\cos\theta,$$

and that the Hamiltonian is

$$\mathcal{H} = \frac{\partial \mathcal{L}}{\partial \dot{\theta}}\dot{\theta} - \mathcal{L} = p_\theta\dot{\theta} - mgl\cos\theta - \frac{1}{2} ml^2\dot{\theta}^2.$$

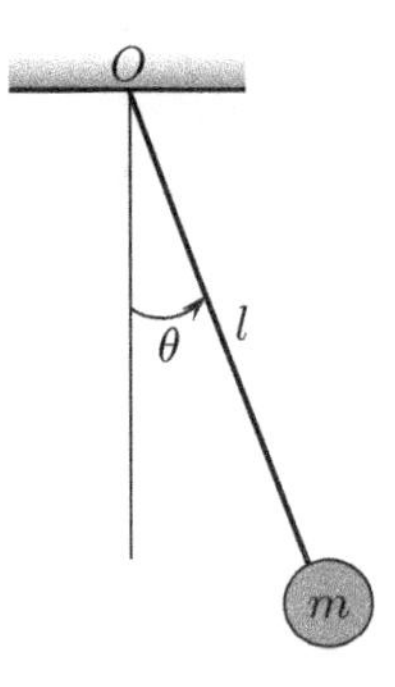


**Figure 7.22** Motion of a simple pendulum.

Since $p_\theta = ml^2\dot{\theta}$, the previous equation reduces to

$$\mathcal{H} = -mgl\cos\theta + \frac{1}{2}ml^2\dot{\theta}^2,$$

which is the total energy of the pendulum. To obtain the equations of motion in the canonical form, we eliminate the dependence on $\dot{\theta}$ from the Hamiltonian and obtain

$$\mathcal{H}(\theta, p_\theta) = -mgl\cos\theta + \frac{p_\theta^2}{2ml^2}.$$

Using the previous equation, the equation of motion can be written as

$$\begin{aligned}
\frac{\partial\mathcal{H}}{\partial\theta} &= -\dot{p}_\theta, \qquad \dot{p}_\theta = -mgl\sin\theta,\\
\frac{\partial\mathcal{H}}{\partial p_\theta} &= \dot{\theta}, \qquad p_\theta = ml^2\dot{\theta}.
\end{aligned}$$

A very interesting result that directly emanates from the application of Hamilton's equation to a system involving a cyclic or ignorable coordinate is that the generalized momentum associated with that coordinate is conserved. Recall that, for an ignorable coordinate $q_j$, the Lagrangian depends only on $\dot{q}_j$ but not on $q_j$ itself. By the definition of the Hamiltonian, it follows that the Hamiltonian also depends only on $\dot{q}_j$ but not on $q_j$ itself. Thus, when applying Hamilton's canonical equation to an ignorable coordinate, we obtain

$$\dot{p}_j = 0 \qquad \text{or} \qquad p_j = \text{constant}. \tag{7.74}$$

**Example 7.17 Cyclic Coordinate**

Consider the motion of a particle of mass $m$ interacting with a point $O$ through a central force such that its potential energy is given by $\mathcal{U}(r)$ (Figure 7.23). Using polar coordinates $(r, \theta)$ to describe the motion of the particle, obtain the equations of motion in Hamilton's canonical form. Are there any cyclic coordinates?

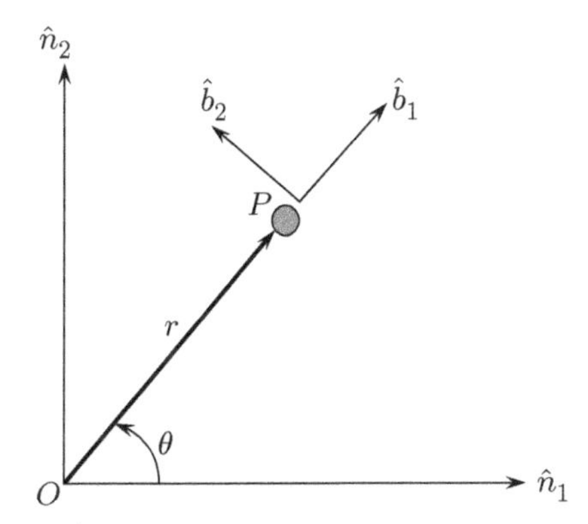


**Figure 7.23** Motion of a particle in a central field.

The velocity of the particle can be written as ${}^N\mathbf{v}^{P/O} = \dot{r}\hat{b}_1 + r\dot{\theta}\hat{b}_2$. Thus, the Lagrangian is

$$\mathcal{L} = \frac{1}{2}m(\dot{r}^2 + r^2\dot{\theta}^2) - \mathcal{U}(r).$$

It follows that the Hamiltonian can be written as

$$\mathcal{H} = \frac{1}{2}m(\dot{r}^2 + r^2\dot{\theta}^2) + \mathcal{U}(r).$$

Eliminating the dependence on $\dot{r}$ and $\dot{\theta}$ from the Hamiltonian yields

$$\mathcal{H} = \frac{1}{2m}p_r^2 + \frac{1}{2mr^2}p_\theta^2 + \mathcal{U}_r.$$

Using Hamilton's canonical equations, we obtain

$$\begin{aligned}
\frac{\partial \mathcal{H}}{\partial r} &= -\dot{p}_r, \qquad \dot{p}_r = \frac{1}{6mr^3}p_\theta^2 - \frac{\partial \mathcal{U}}{\partial r},\\
\frac{\partial \mathcal{H}}{\partial p_r} &= \dot{r}, \qquad p_r = m\dot{r},\\
\frac{\partial \mathcal{H}}{\partial \theta} &= -\dot{p}_\theta, \qquad \dot{p}_\theta = 0, \qquad p_\theta = \text{constant},\\
\frac{\partial \mathcal{H}}{\partial p_\theta} &= \dot{\theta}, \qquad p_\theta = mr^2\dot{\theta}.
\end{aligned}$$

It is obvious that $\theta$ is a cyclic coordinate for which the generalized momentum, here the angular momentum of the particle $m$, is conserved; that is, $mr^2\dot{\theta} = c$.

### Flipped Classroom Exercise 7.5

For the gantry crane shown in the figure, assume that $m_1 = m_2$, then find the equation of motion of the system using Hamilton's canonical equations.

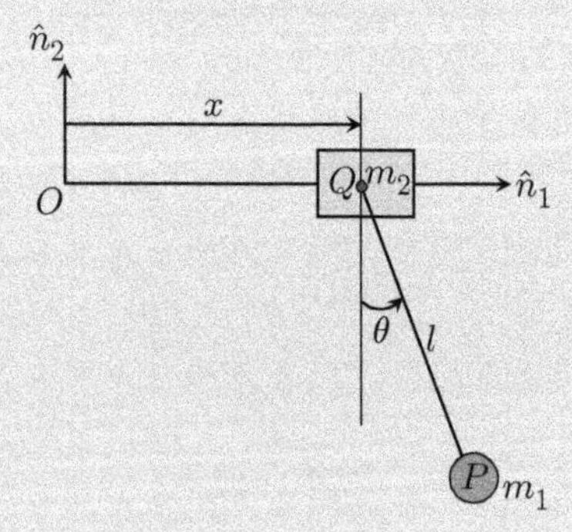


To solve this problem, take the following steps:

1. Find the kinetic and potential energy of the system.
2. Find the Lagrangian of the system.

3. Using the Lagrangian, find the Hamiltonian and check whether it is equal to the total energy of the system.
4. Find the generalized angular momenta, $p_x$ and $p_\theta$.
5. Eliminate the dependence of the Hamiltonian on $\dot{x}$ and $\dot{\theta}$ by expressing them in terms of $p_x$ and $p_\theta$.
6. Apply Hamilton's canonical equation on the Hamiltonian obtained in the previous step.

 **Hamilton's Canonical Equations**

Using Hamilton's canonical equations, the dynamics of a system of particles or rigid bodies can be analyzed using a set of first-order differential equations described in terms of the generalized coordinates $q_j$, their derivatives $\dot{q}_j$, and generalized momenta $p_j$. These equations are given by

$$\begin{aligned}\frac{\partial \mathcal{H}}{\partial q_j} &= -\dot{p}_j,\\ \frac{\partial \mathcal{H}}{\partial p_j} &= \dot{q}_j,\end{aligned} \tag{7.75}$$

where $\mathcal{H}$ is known as the Hamiltonian and is defined as

$$\mathcal{H}(q_j, p_j, t) = \sum_{j=1}^{\mathcal{M}} p_j \dot{q}_j - \mathcal{L}(q_j, \dot{q}_j, t). \tag{7.76}$$

The Hamiltonian reduces to the total energy of the system $\mathcal{H} = \mathcal{T} + \mathcal{U}$ when

- the position of the system of particles does not depend explicitly on time; that is, $\mathbf{r} = \mathbf{r}(q_1, q_2, \dots, q_{\mathcal{M}})$;
- the potential energy $\mathcal{U}$ is only a function of the generalized coordinates;
- the system is conservative.

## Exercises

7.1 A uniform disk of mass $M$ and radius $R$ rotates with angular velocity $\omega$ (Figure 7.24). At one point, a piece of mass $0.1M$ breaks off the disk and flies vertically upward. Find (a) the maximum height of the flying piece, and (b) the angular velocity of the disk after the piece breaks off.

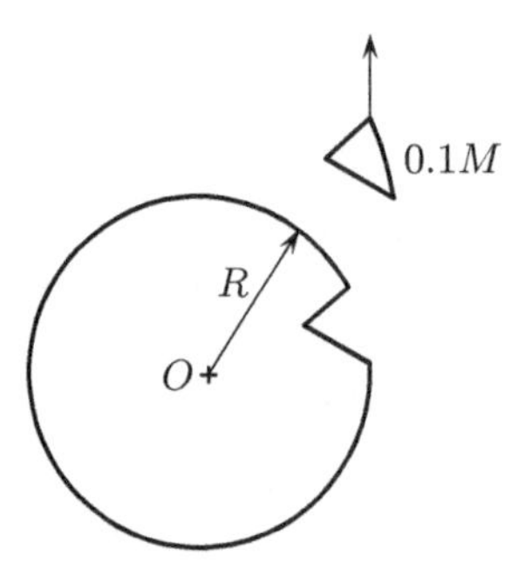


**Figure 7.24** Exercise 7.1.

7.2 A centrifuge consists of four cylindrical containers. Each container has mass $m$ and radius $a$ and is placed at a distance $l$ from the axis of the centrifuge, as shown in Figure 7.25. A torque $\tau$ is applied about the vertical axis of the centrifuge. Find the time required to increase the angular velocity of the system from zero to $\omega$.

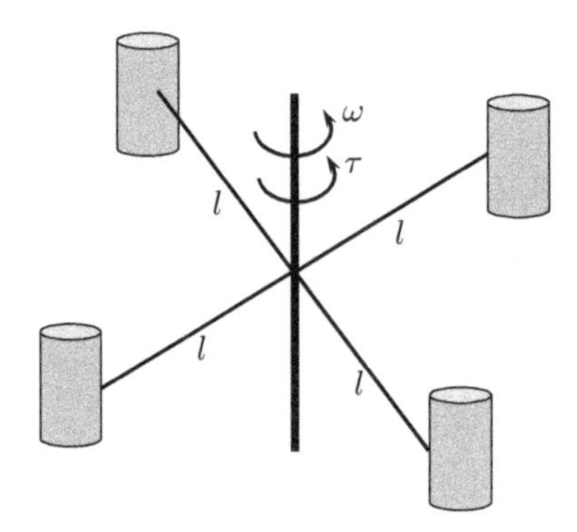


**Figure 7.25** Exercise 7.2.

7.3 A constant force $F$ is applied horizontally at the top of the disk of mass $M$ and radius $R$, as shown in Figure 7.26. Show that the angular momentum of the particle about the point of contact varies linearly with time.

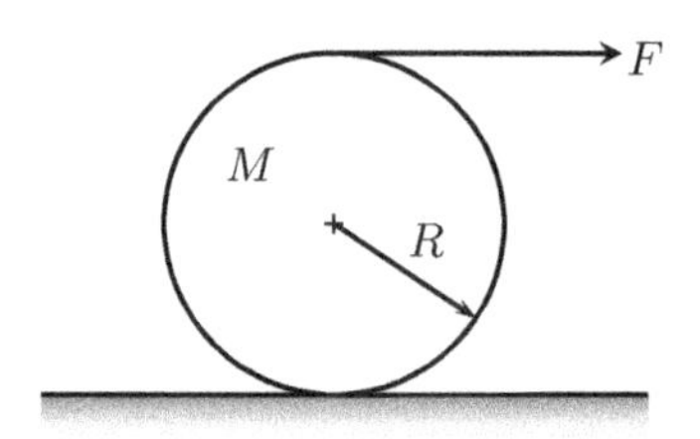


**Figure 7.26** Exercise 7.3.

7.4 A string is wrapped several times around a cylinder of mass $M$ and radius $R$. A block of mass $m$ is tied to the string and placed on a support such that the tension in the string is still zero (Figure 7.27). The block is raised to a height $h$ and the support is removed. The block is then released. Find the velocity of the block $m$, just after the string has become taut.

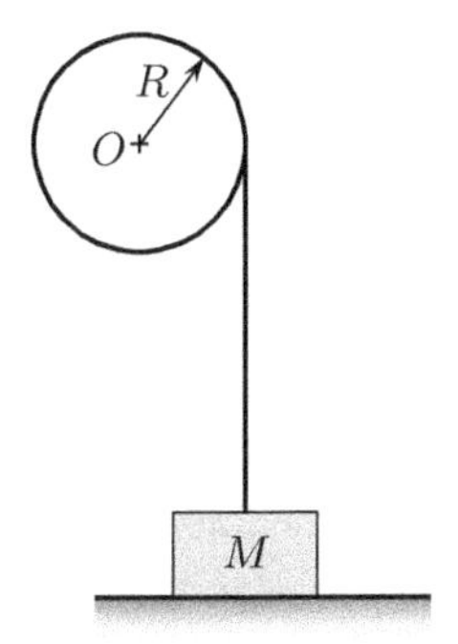


**Figure 7.27** Exercise 7.4.

7.5 A uniform disk of mass $m$ and radius $R$ is free to rotate about point $O$. A particle of mass $m$ strikes it at point $P$ after falling vertically down from a distance $h$ above the center of the wheel. The mass sticks to the wheel at the position shown in Figure 7.28. Find the maximum angular velocity of the wheel.

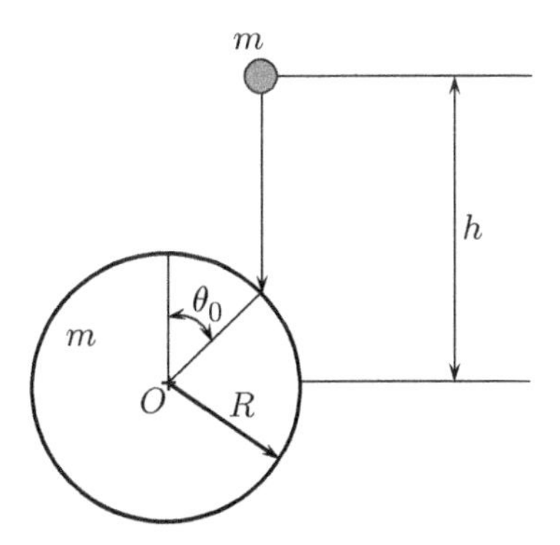


**Figure 7.28** Exercise 7.5.

7.6 A meteorite of mass $m$ strikes the Earth at the equator. Assume that the Earth is a uniform sphere of mass $M$ and radius $R$. If the length of the day was $T$ s before the meteorite struck, by how much does it change after the strike occurs?

7.7 A square block of mass $m$ and side $b$ slides with a constant velocity $v$ on a smooth horizontal plane, as shown Figure 7.29. It hits a ridge at point $O$. What is the angular speed of the block after it hits the ridge?

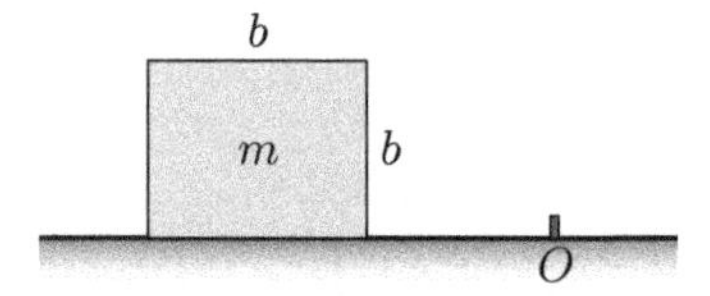


**Figure 7.29** Exercise 7.7.

7.8 A square gate $ABCD$ with sides of length $a$ can rotate freely about line $AB$ (Figure 7.30). The gate is initially standing vertically, but is slightly perturbed and falls. When $AD$ becomes horizontal, the corner $D$ hits a fixed support $O$ and gets stuck. Assuming that corner $B$ loses contact simultaneously (without impact forces), and that the gate starts to rotate about $AO$ with some angular velocity, $\omega$. Find $\omega$.

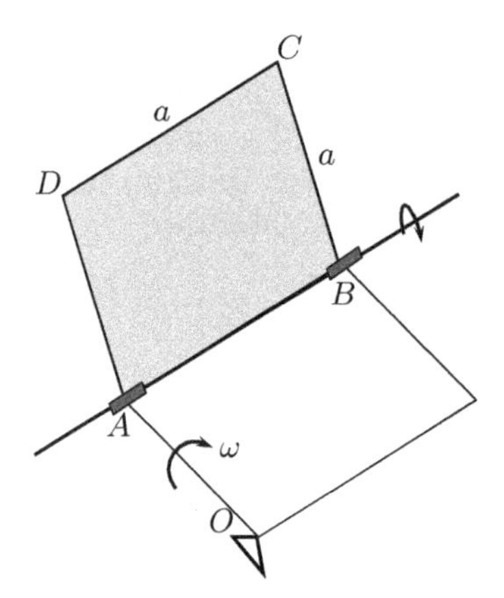


**Figure 7.30** Exercise 7.8.

7.9 A disk of mass $m$ and radius $R$ is hanging via a universal joint at point $A$, as shown in Figure 7.31. The disk is struck at point $P$ (the angle between line $AO$ and $OP$ is a right angle) by a ball of mass $m$ and velocity $v_0$. The ball sticks to the disk and they move together. Determine the angular rotation of motion and the reaction impulse at point $A$.

7.10 An electric motor of mass $M$ rests on two narrow strips that are attached to a horizontal table that turns with a constant angular velocity $\Omega$, as shown in Figure 7.32. The strips are perpendicular to the plane shown in the figure and are at distances $a$ and $b$ from a

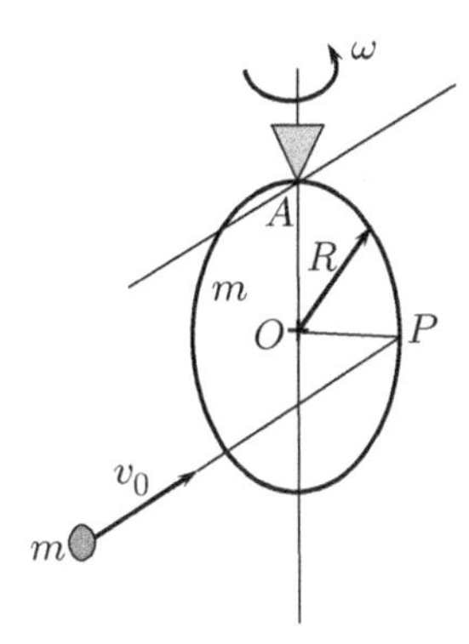


**Figure 7.31** Exercise 7.9.

vertical line passing through the center of mass of the motor. The rotor turns relative to the stator with a constant angular velocity $\omega_r$ counter-clockwise. The rotor is assumed to have equal moments of inertia, $J_r$, about the horizontal and vertical lines passing through its center of mass. Find the maximum allowed $\Omega$ before the rotor starts lifting off the strips.

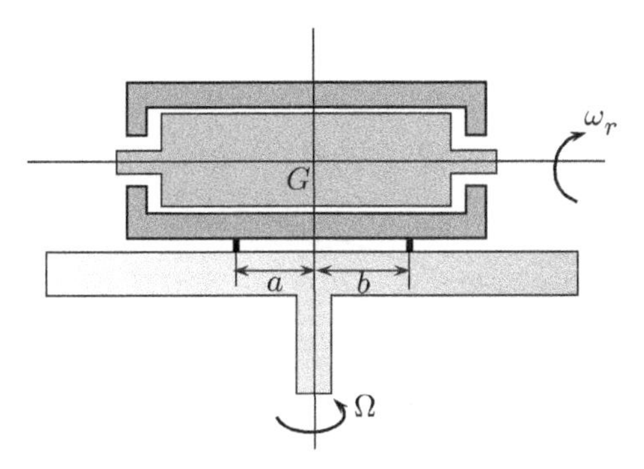


**Figure 7.32** Exercise 7.10.

7.11 A massless rod $AB$ of length $2a$ can rotate freely about a vertical axis passing through point $O$ and remains horizontal at all times (Figure 7.33). A particle of mass $m$ is attached at end $B$, while, at the other end, a disk of mass $m$ and radius $2a$ is attached. The wheel rotates with negligible friction about the axis $AB$, which is normal to the disk and passes through point $O$. The disk remains in contact with a flat rough horizontal floor. Initially, the system is set into motion by giving the disk an initial angular velocity $4\omega_0$ around the axis $AB$, and by giving the axis $AB$ an initial angular velocity $\omega_0$. At the beginning, the wheel slips, but eventually friction forces make it roll without slipping. Find the angular velocities of the wheel and of the axis when the disk starts rolling without slipping.

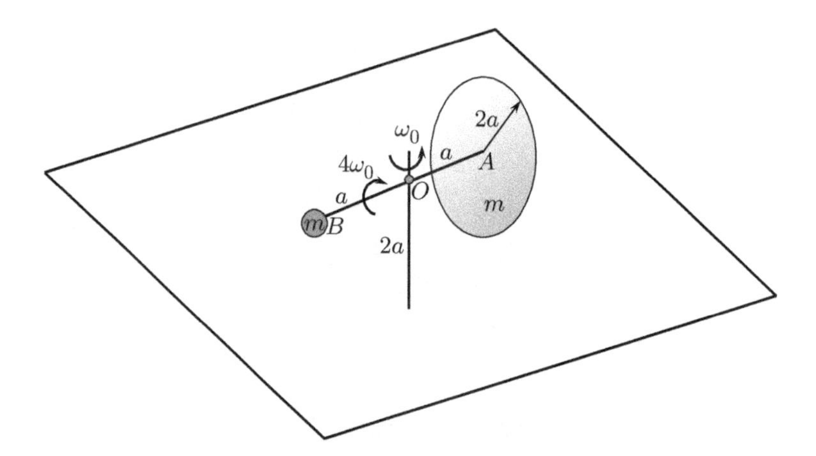


**Figure 7.33** Exercise 7.11.

7.12 A rod of mass $m$ and length $l$ is pinned at its center of mass $G$ to the massless rod $AB$. The rod $AB$ is supported by two frictionless bearings and allowed to rotate about its axis with a constant angular velocity $\omega$, as shown in Figure 7.34. The whole setup rotates about the vertical axis with a constant angular velocity $\Omega$. Find the moments acting on the rod as a result of this motion.

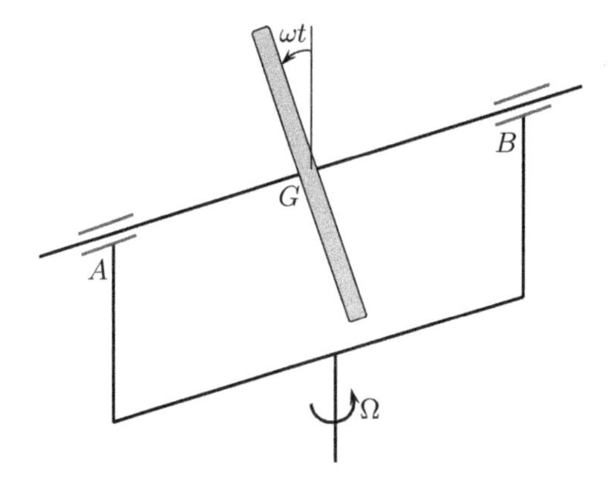


**Figure 7.34** Exercise 7.12.

7.13 The rod of mass $m$ and length $l$ has its end $A$ moving at a constant velocity $v_0$ along the horizontal axis (Figure 7.35). Assuming that there is no friction between the rod and any of the surfaces, find the reaction forces at $A$ and $B$ as a function of the angle $\theta$.

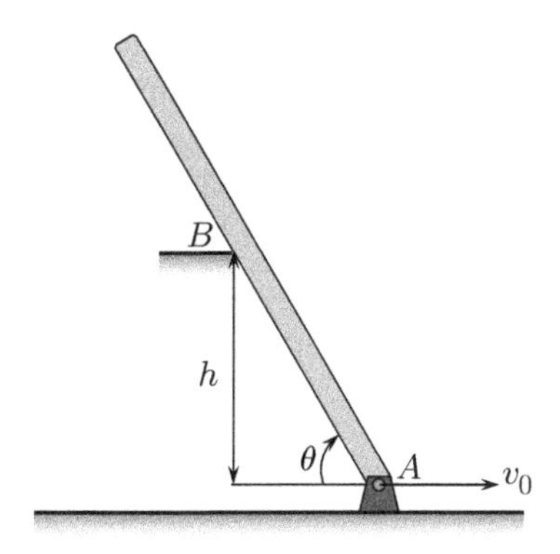


**Figure 7.35** Exercise 7.13.

7.14 The mixer shown in Figure 7.36 makes an angle $\theta_0$ with the vertical, has mass $m$ and moments of inertia $I_{11G}$, $I_{22G}$, and $I_{33G}$ around its center of mass $G$, as calculated in the body rotating $B$-frame. The mixer rotates with constant angular velocity $\Omega$ about the $\hat{b}_2$ coordinate and with varying angular velocity $\alpha(t)$ about the $\hat{b}_1$ axis. Find the reactions at point $A$ and point $B$.

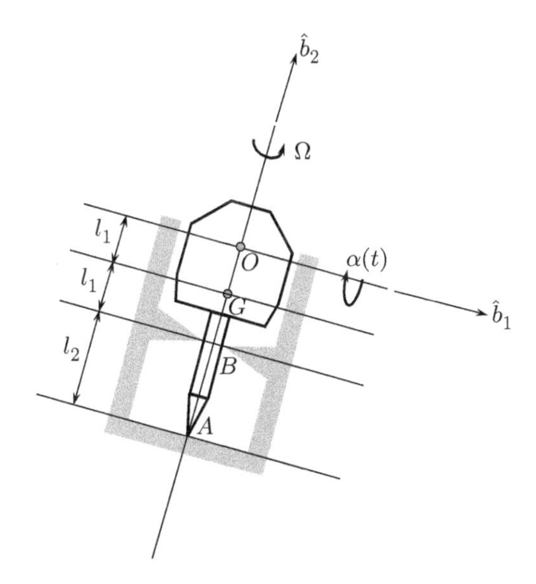


**Figure 7.36** Exercise 7.14.

7.15 The gear wheel shown in Figure 7.37 is being retracted as it rotates with angular velocity $\Omega$ about its rotating axis. Find the reactions at the bearing $A$ at the instant when the landing gear makes an angle $\theta$ about the vertical. Assume that the wheel can be modeled as a disk of mass $m$, radius $R$, and thickness $w$.

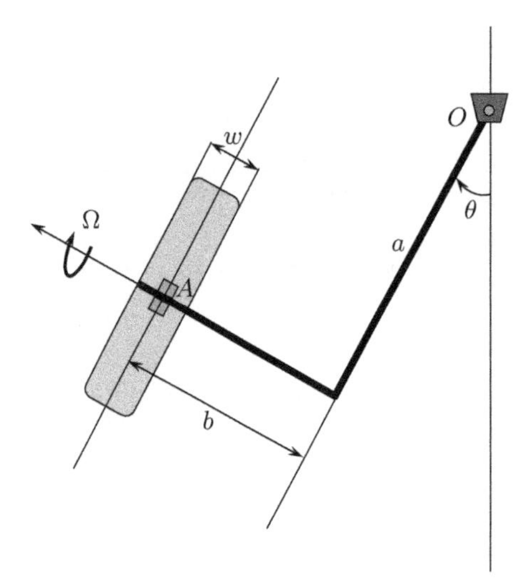


**Figure 7.37** Exercise 7.15.

7.16 Consider a bead which slides without friction on a hoop, as shown in Figure 7.38. The hoop is constrained to rotate at a constant angular velocity $\Omega$. Derive the equation of motion for the bead using Hamilton's canonical equations.

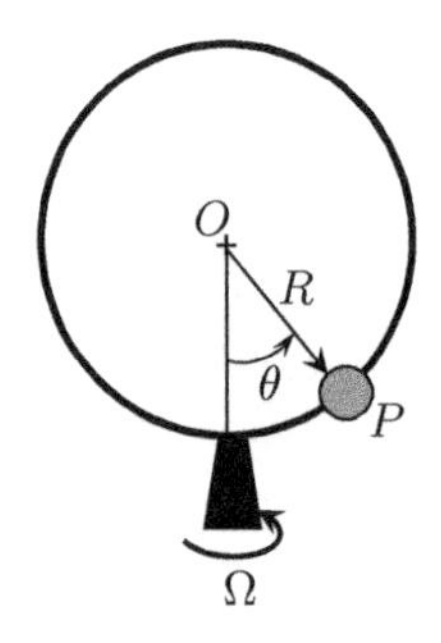


**Figure 7.38** Exercise 7.17.

# 8

# Motion of Charged Bodies in an Electric Field

This chapter focuses on modeling the motion of charged particles and rigid bodies under the influence of electrostatic and electromagnetic forces. It first introduces the fundamental concepts of electrostatics and electromagnetism – force, field, potential, flux and so on – then uses these concepts to derive the equations governing the motion of charged particles and rigid bodies for different scenarios. Using a number of solved examples, this chapter also aims at introducing the various important applications where modeling the interaction of particles and rigid bodies with electrostatic and electromagnetic forces can be used in actuation, sensing, and energy generation.

## 8.1 Electrostatics

### *8.1.1 Electrostatic Forces*

During elementary physics classes, you probably learned about electrostatic forces through a commonly performed experiment in which the teacher rubs a glass rod against a piece of silk, then brings it close to a lightweight conductor. Magically, the rod attracts the conductor. Such attraction forces occur because the glass rod becomes positively charged as it loses electrons when rubbed against the silk sheet. As a result, when the rod is brought close to the conductor, the free electrons in the conductor are attracted to the glass while the positive charges are repelled by it. This process polarizes the material, resulting in an attraction force known as the *Coulomb force of attraction*.

Such attraction forces also occur, but to a lesser extent, if the rod is brought close to a non-conducting material, which does not carry any free electrons. This is because, in the presence of the positive charges on the glass rod, orbital electrons around the nucleus tend to slow down and spend a longer time in the vicinity of the positive charges, resulting in an electric induction in the non-conducting material.

*Dynamics of Particles and Rigid Bodies: A Self-Learning Approach*, First Edition. Mohammed F. Daqaq.

Companion website: www.wiley.com/go/daqaq/dynamics

In 1784, French physicist Charles Augustin de Coulomb published his law of attraction, which predicts the attraction force between two charges of different polarity. He demonstrated that the attraction force $\mathbf{F}_{12}$ can be described by the following equation [1]:

$$\mathbf{F}_{12} = \frac{1}{4\pi\varepsilon}\frac{q_1 q_2}{\|\mathbf{r}\|^2}\hat{r} = \frac{1}{4\pi\varepsilon}\frac{q_1 q_2}{\|\mathbf{r}\|^3}\mathbf{r} \tag{8.1}$$

where, as shown in Figure 8.1a, $q_1$ and $q_2$ represent the charges of the interacting particles, $\mathbf{r}$ and $\|\mathbf{r}\|$ are, respectively, the position vector and the distance from $q_1$ to $q_2$, $\hat{r}$ is the unit vector in the direction of $\mathbf{r}$, and $\varepsilon$ is the electric permittivity of the space enclosing the charged particles. Upon changing the sign of one of the charges in Equation (8.1), the direction of the force is reversed.

In Equation (8.1), the charges are measured in Coulombs (C), which is an extremely large quantity of charge. To put it in perspective, the charge of an electron is only $1.6 \times 10^{-19}$ C.

For more than two charges, it is possible to apply the principle of superposition to find the net force on a given charge. As shown in Figure 8.1b, the force on the $i$th charge due to $\mathcal{N}$ charges can be written as

$$\mathbf{F} = \sum_{j=1}^{\mathcal{N}} \mathbf{F}_{ij}. \tag{8.2}$$

It is worth noting that Coulomb law of attraction looks very similar to Newton's law of gravitation as described in Chapter 2, with the masses of the particles replacing their charges and the gravitational constant replacing the permittivity term $\frac{1}{4\pi\varepsilon}$.

**Example 8.1 Ratio between Electrostatic and Gravitational Forces**

Find the ratio between the electrostatic and gravitational forces for two protons separated by a distance $d$.

The repulsive electrostatic force between the two protons placed at a distance $d$ is given by

$$F_{el} = \frac{1}{4\pi\varepsilon}\frac{q^2}{d^2},$$

where $\varepsilon = 8.854 \times 10^{-12}$ $\mathrm{C^2/(N.m^2)}$ for free space and $q = 1.6 \times 10^{-19}$ C. Substituting the numerical values of $q$ and $\varepsilon$ in the previous equation yields

$$F_{el} = \frac{2.3 \times 10^{-28}}{d^2}\mathrm{N}.$$

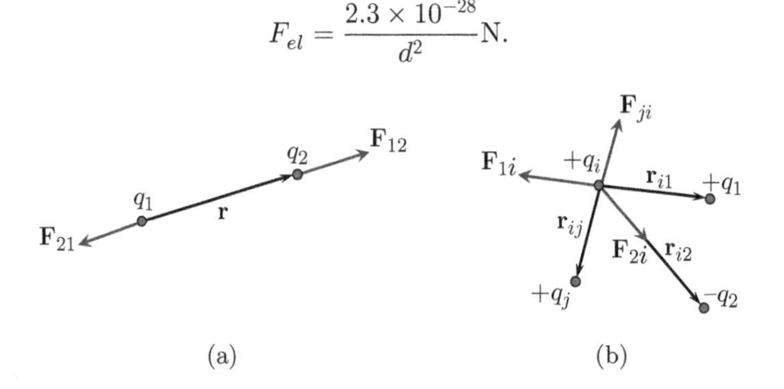


**Figure 8.1** Coulomb forces between charged particles (refer to text for details).

The gravitational force on the other hand is given by

$$F_m = G\frac{m^2}{d^2},$$

where $m = 1.7 \times 10^{-27}$ kg and the gravitational constant $G = 6.7 \times 10^{-11}\text{m}^3/(\text{kg.s}^2)$. Substituting the numerical values of $m$ and $G$ in the previous equation yields

$$F_m = \frac{1.9 \times 10^{-64}}{d^2}\text{N}.$$

Thus, $F_{el}/F_m \approx 10^{36}$. This clearly implies that, compared to the electric force, the gravitational force is very weak at the atomic scale.

### 8.1.2 Electric Field

The electric field $\mathbf{E}_P$ at a point $P$ due to a charge $Q$ can be obtained by assuming that there is a positive test charge $+q$ at $P$, calculating the Coulomb force exerted by $Q$ on $q$, then dividing the Coulomb force by the test charge $q$; that is,

$$\mathbf{E}_P = \frac{\mathbf{F}_{Qq}}{q} = \frac{1}{4\pi\varepsilon}\frac{Q}{\|\mathbf{r}\|^2}\hat{r} \tag{8.3}$$

In other words, as shown in Figure 8.2, the electric field can be defined as the electric force per unit charge. The direction of the electric field is always in the direction of the Coulomb force. The electric field at point $P$ due to a collection of $\mathcal{N}$ charges is the vectorial sum of the electric field at point $P$ due to each charge.

**Example 8.2 Zero Electric Field**

Consider two charges $+2q$ and $-q$ placed on a line and separated by a distance $d$ as shown in Figure 8.3. Find the point along the line where the electric field is zero.

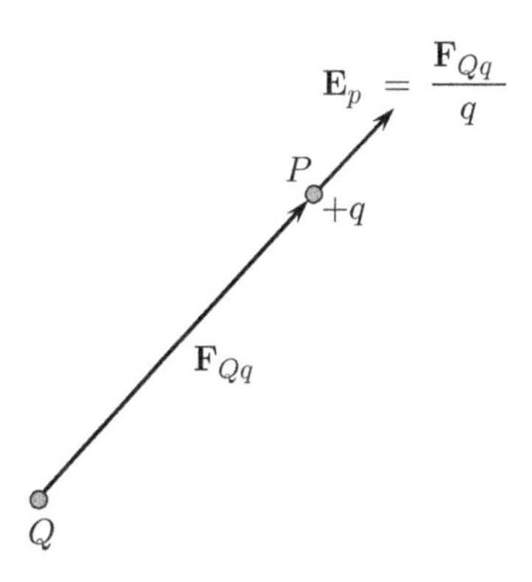


**Figure 8.2** Electric field at point $P$ due to a positive charge $Q$.

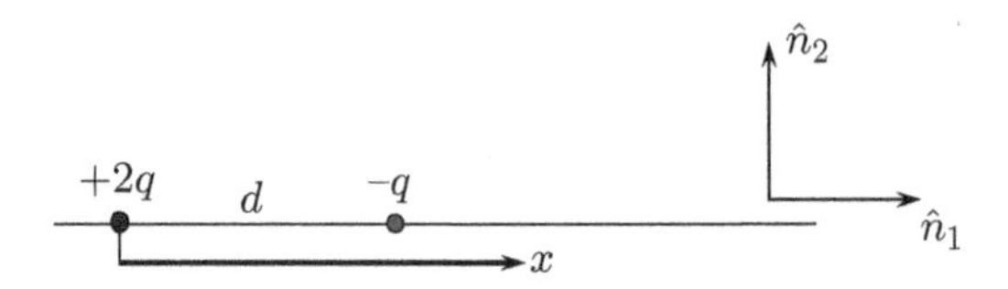


**Figure 8.3** Zero electric field.

To find the point/points along the line where the electric field is zero, we first find the electric field at a distance $x$ from the first charge, which is given by

$$\mathbf{E}_{+2q} = \frac{1}{4\pi\varepsilon}\frac{2q}{x^2}\hat{n}_1$$

Second, we find the electric field at the same point but resulting from the second charge. This can be written as

$$\mathbf{E}_{-q} = -\frac{1}{4\pi\varepsilon}\frac{q}{(x-d)^2}\hat{n}_1.$$

At the point where the electric field vanishes, we have

$$\mathbf{E}_{+2q} + \mathbf{E}_{-q} = \frac{2}{x^2} - \frac{1}{(x-d)^2} = 0.$$

The previous equation yields two solutions for $x$, namely $x_0 = (2 \pm \sqrt{2})d$. As such, there are two points on the line at which the electric field is zero. These points are depicted by $\times$ in Figure 8.4, which illustrates the electric field emanating from the two charges, $+2q$ and $-q$.

**Example 8.3 Electric Dipole**

An electric dipole consists of two charges equal in magnitude and opposite in direction placed at a distance $d$. Find the electric field at a point $p$ located a distance $y$ from the center of the dipole, as shown in Figure 8.5. What is the electric field when $y \gg d$?

To find the electric field, we apply the principle of superposition and add the electric fields resulting from each one of the charges forming the dipole; that is

$$\mathbf{E}^+ = \frac{1}{4\pi\varepsilon}\frac{q}{(y^2+\frac{d^2}{4})^{3/2}}\left(\frac{d}{2}\hat{n}_1 + y\hat{n}_2\right),$$

$$\mathbf{E}^- = -\frac{1}{4\pi\varepsilon}\frac{q}{(y^2+\frac{d^2}{4})^{3/2}}\left(-\frac{d}{2}\hat{n}_1 + y\hat{n}_2\right).$$

Adding the previous equations yields

$$\mathbf{E}_p = \frac{1}{4\pi\varepsilon}\frac{qd}{\left(y^2+\frac{d^2}{4}\right)^{3/2}}\hat{n}_1,$$

It is interesting to note that the electric field at point $p$ is only in the $\hat{n}_1$ direction.

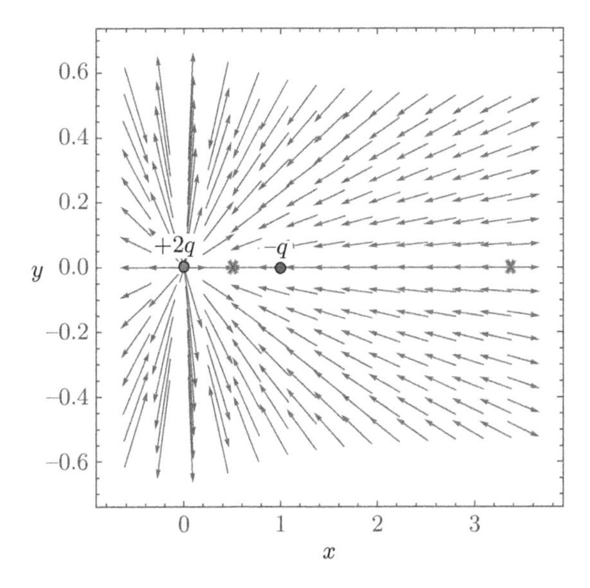


**Figure 8.4** Electric field lines due to the charges $+2q$ and $-q$. (*See color plate section for the color representation of this figure.*)

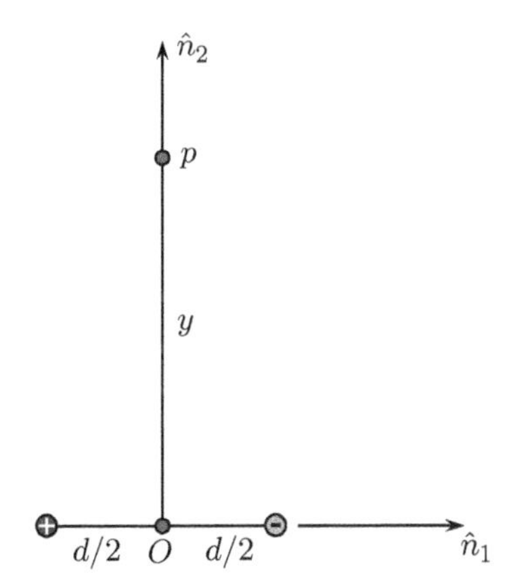


**Figure 8.5** Electric field of an electric dipole.

When $y \gg d$, the previous equation reduces to

$$\mathbf{E}_p = \frac{1}{4\pi\varepsilon} \frac{qd}{y^3} \hat{n}_1.$$

This implies that, far away from the electric dipole, the electric field decays proportional to $1/y^3$. This is much faster than the decay rate associated with a single charge where the electric field decays as $1/y^2$, as given by Equation (8.3).

### 8.1.3 Electric Flux

The electric flux is a measure of the flow of the electric field through a unit area. It is a scalar quantity defined as

$$\Phi = \iint_{\sigma(t)} \mathbf{E}(\mathbf{r}, t).\mathrm{d}\mathbf{A} \tag{8.4}$$

where $\mathbf{E}(\mathbf{r}, t)$ is the electric field at some position $\mathbf{r}$ and time $t$, $\sigma(t)$ is the surface through which the electric lines flow, which can be time varying, and $\mathrm{d}\mathbf{A}$ is a differential element of area on the surface $\sigma(t)$. Note that the dot product implies that only electric field lines parallel to the normal to the differential element $\mathrm{d}\mathbf{A}$ contribute to the electric flux. The unit of the electric flux is $\mathrm{Nm}^2/\mathrm{C}$.

As shown in Figure 8.6, for a uniform, time-invariant electric field and a stationary surface $\sigma(t) = \sigma$, the previous equation reduces to

$$\Phi = \mathbf{E}.\mathbf{A} = EA\cos\theta, \tag{8.5}$$

where $\theta$ is the angle between the electric field and the normal to the surface.

**Example 8.4 Electric Flux of a Unit Charge**

Find the electric flux generated by a unit charge $+q$.

To find the electric flux resulting from a charge, we need to calculate the flow of the electric field lines through a surface enclosing the charge. To this end, we draw a sphere of radius $R$ around the charge and calculate the flux through the sphere. We note that the electric field lines of the charge are always normal to the enclosing spherical surface, as shown in Figure 8.7, so the electric field is always parallel to the normal to the area.

Since, as shown in Figure 8.7, for a given $R$ the electric field, $\mathbf{E} = \frac{Q}{4\pi\varepsilon R^2}\hat{r}$, is constant, the integral in Equation (8.4) reduces to $\Phi = EA$, where $A = 4\pi R^2$ is the surface area of the sphere. This yields

$$\Phi = \frac{Q}{\varepsilon},$$

which means that the electric flux is independent of the radius $R$. This is expected since the number of electric fields lines passing through the spherical surface is independent of the radius of the chosen surface.

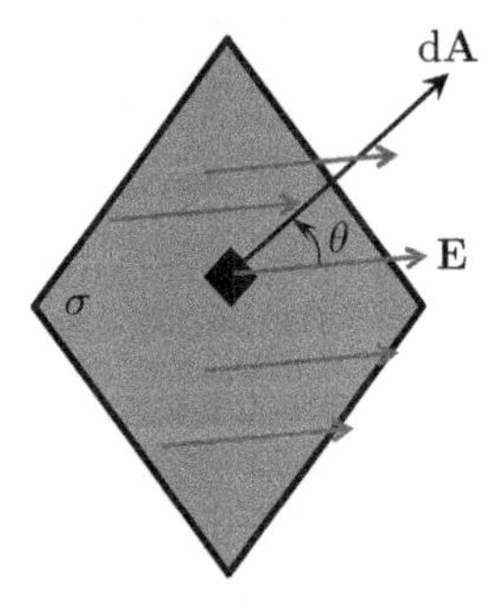


**Figure 8.6** Electric flux through a stationary surface.

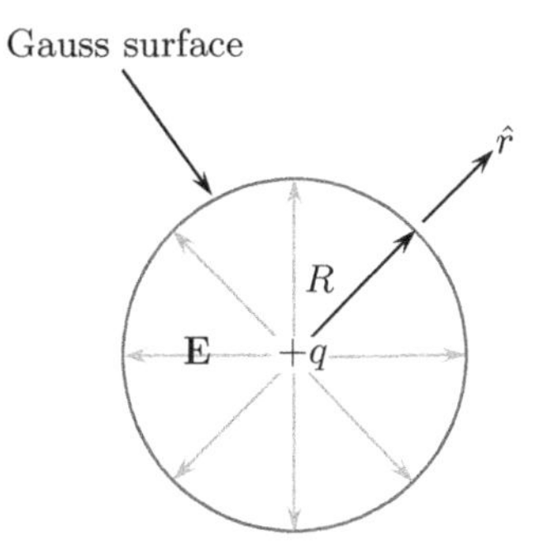


**Figure 8.7** Electric flux of a unit charge.

It is worth noting that, in the previous example, it would have not mattered if we chose the surface to be of any other shape, so long as it was closed and completely engulfed the charge. Furthermore, if we have multiple charges inside the surface, then the electric flux through the surface is the sum of the flux of all of the charges; that is

$$\oint \mathbf{E}.\mathrm{d}\mathbf{A} = \sum_{i=1}^{\mathcal{N}} \frac{Q_i}{\varepsilon} \tag{8.6}$$

where $\mathcal{N}$ is the total number of charges and $Q_i$ is the charge of each particle inside the surface. Equation (8.6) is due to Carl Friedrich Gauss, who originally stated it in 1813 [2]. It is now known as Gauss's law or Gauss's flux theorem.

**Example 8.5 Electric Field of a Charged Sphere**

Consider a sphere of radius $R$ that carries a uniformly-distributed charge $+Q$ on its surface, as shown in Figure 8.8. Use Gauss's law to find the electric field distribution inside and outside the sphere.

To solve the problem, we must first note that, due to the uniform distribution of the charge, the electric field lines are radial, spreading from the surface of the sphere inwards or outwards.

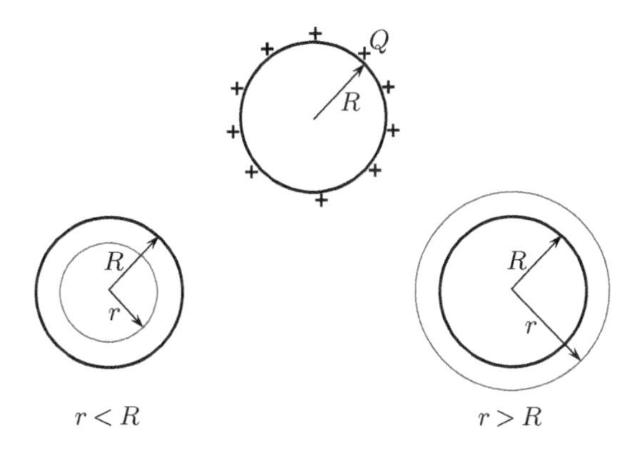


**Figure 8.8** Electric field in the vicinity of a charged sphere.

As such, if we choose Gauss's surface to be a concentric sphere, the integral on the left-hand side of Equation (8.6) simplifies to $EA$. To this end, we find the electric field inside the charged sphere by using a concentric sphere of radius $r < R$ to represent the Gauss surface. It follows from Equation (8.6) that

$$4\pi r^2 E = \sum_{i=1}^{\mathcal{N}} \frac{Q_i}{\varepsilon}.$$

It follows that, since there are no charges inside the chosen Gauss surface, the electric field vanishes inside the sphere.

To find the electric field outside the sphere, we use a concentric sphere of radius $r > R$ to represent the Gaussian surface. It follows that

$$4\pi r^2 E = \sum_{i=1}^{\mathcal{N}} \frac{Q_i}{\varepsilon} = \frac{Q}{\varepsilon}.$$

Hence, the electric field outside the surface is given by $E = \frac{Q}{4\pi\varepsilon r^2}$. This yields

$$E = \begin{cases} 0, & r < R \\ \frac{Q}{4\pi\varepsilon r^2}, & r \geq R \end{cases}$$

 **Flipped Classroom Exercise 8.1**

Find the electric field at a distance $y$ from an infinite sheet carrying a charge density $\sigma = Q/A$. Here, $Q$ represents the total charge on the sheet and $A$ is its area.

To solve this problem, take the following steps:

1. Define the Gaussian surface as a cylinder the sides of which are perpendicular to the charged sheet. Let the height of the cylinder above and below the surface be the same and equal to $y$.
2. Calculate the charge inside the cylinder.
3. Use Gauss's law to find the electric field, noting that the electric field is perpendicular to the surface and hence there are no electric field lines passing through the sides of the cylinder.
4. Show that the electric field is $E = \frac{\sigma}{2\varepsilon}$.

## 8.1.4 *Electrostatic Potential Energy*

To explain the concept of the electrostatic potential energy, we consider two positive charges $+q$, as shown in Figure 8.9. We place one of the charges at some point in space, and try to bring the other charge from very far away ($\infty$) and place it at a distance $R$ from the first charge. To achieve this task, we obviously need to do work against the repulsive force exerted by the stationary charge on the moving charge. The necessary work is given by

$$\mathcal{W} = \int_{\infty}^{R} \mathbf{F}.d\mathbf{r}. \tag{8.7}$$

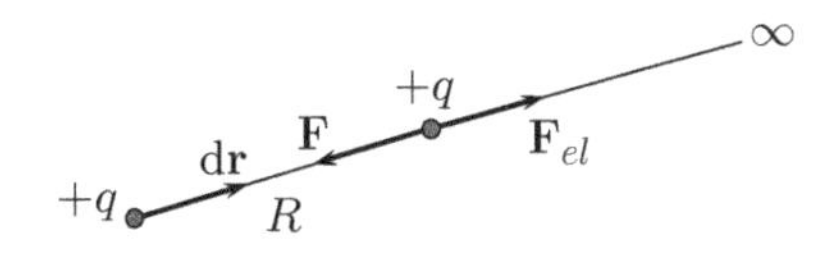


**Figure 8.9** Electrostatic potential energy.

Since the force exerted, **F**, is opposite to the repulsive electric force $\mathbf{F}_{12}$, then we have $\mathbf{F} = -\mathbf{F}_{12}$ and Equation (8.7) can be rewritten as

$$\mathcal{W} = \int_{R}^{\infty} \mathbf{F}_{12}.\mathbf{dr}. \tag{8.8}$$

Upon substituting for $\mathbf{F}_{12}$ using Coulomb's law, we obtain

$$\mathcal{W} = \frac{1}{4\pi\varepsilon}\frac{q_1 q_2}{R}. \tag{8.9}$$

If $q_1$ and $q_2$ are of the same sign, the work done is positive work; otherwise it is negative. The electrostatic potential energy stored in the charge is equal to the work done to bring the charge to its position; in other words, $\mathcal{W} = \mathcal{U}$. Note that, similar to any conservative field, the potential energy remains the same regardless of the path taken; that is, it is only a function of the start and end position of the charge.

For a collection of $\mathcal{N}$ charges, the total potential energy is equal to the sum of the potential energy needed to bring each charge to a given point in the vicinity of the other charges; that is

$$\mathcal{U} = \frac{1}{2}\sum_{i=1}^{\mathcal{N}} q_i \sum_{j=1}^{\mathcal{N}(i\neq j)} \frac{1}{4\pi\varepsilon}\frac{q_i q_j}{r_{ij}}, \tag{8.10}$$

where $r_{ij}$ is the distance between the charges $q_i$ and $q_j$. The factor of $\frac{1}{2}$ in Equation (8.10) demonstrates that the potential energy is shared mutually between the two charges.

### 8.1.5 Electric Potential (Voltage)

The electric potential is defined as the electrostatic potential energy per unit charge. The relationship between the electrostatic potential energy and electric potential is similar to the relationship between the electrostatic force and the electric field, in that the latter is equal to the former divided by a test charge $q$. In other words, the electric potential $\mathcal{V}_P$ at a point $P$ in space is given by

$$\boxed{\mathcal{V}_P = \int_{R}^{\infty} \frac{\mathbf{F}_{12}}{q}.\mathbf{dr} = \int_{R}^{\infty} \mathbf{E}_P.\mathbf{dr}} \tag{8.11}$$

where $q$ is a positive test charge. The electric potential is commonly referred to as the voltage and is measured in joules per coulomb (J/C) or volts.

It follows from Equation (8.3) that, for a positive charge $Q$, the electric potential $\mathcal{V}_P$ at a point $P$ located at a distance $R$ from the charge is given by

$$\mathcal{V}_P = \frac{Q}{4\pi\varepsilon R}. \tag{8.12}$$

Now, consider two points in space, $A$ and $B$, which are separated by a distance $R$. The voltage associated with these points can be expressed as

$$\mathcal{V}_A = \int_A^{\infty} \mathbf{E}.\mathrm{d}\mathbf{r}, \qquad \mathcal{V}_B = \int_B^{\infty} \mathbf{E}.\mathrm{d}\mathbf{r}, \tag{8.13}$$

where $\mathbf{E}$ is the electric field in the space containing $A$ and $B$. It follows that,

$$\boxed{\mathcal{V}_A - \mathcal{V}_B = -\int_B^{A} \mathbf{E}.\mathrm{d}\mathbf{r}} \tag{8.14}$$

Equation (8.14) is valid independent of the path taken to go between points $A$ and $B$. For a closed loop – that is, going from point $A$ to point $A$ – the integral vanishes, yielding

$$\boxed{\oint \mathbf{E}.\mathrm{d}\mathbf{r} = 0} \tag{8.15}$$

which implies that the voltage drop across a closed loop is always equal to zero (*Kirchoff's voltage law*).

Equation (8.14) can also be expressed in a differential form by letting $\mathcal{V}_A - \mathcal{V}_B = \mathrm{d}\mathcal{V}$ and assuming that the electric field is in the direction of $r$ only. This yields

$$\mathrm{d}\mathcal{V} = -\int_B^{A} E\mathrm{d}r \qquad \text{or} \qquad E = -\frac{\mathrm{d}\mathcal{V}}{\mathrm{d}r} \tag{8.16}$$

Equation (8.16) can be generalized to any set of coordinates and any electric field by replacing the derivative of the voltage with respect to $r$ with the gradient of the voltage with respect to the coordinates used to describe the electric field; that is

$$\boxed{\mathbf{E} = -\nabla\mathcal{V}} \tag{8.17}$$

where $\nabla\mathcal{V}$ represents the gradient of the electric potential. This result yields the more common unit of the electric field of volts per meter (V/m).

Note that, similar to a gravitational field, where particles go from high to low potentials, a positively-charged particle moves spontaneously from high electric potential (a high voltage) to a low electric potential (a low voltage). A negatively-charged particle, on the other hand, moves spontaneously from a low to a high voltage.

**Example 8.6 Charged Sphere**

Calculate the voltage at a point located at a distance $r$ from the center of a positively-charged sphere of radius $R$. Find the voltage when $R > r$ and $R < r$.

In Example 8.5, we obtained the electric field for a positively-charged sphere as

$$E = \begin{cases} 0, & r < R \\ \frac{Q}{4\pi\varepsilon r^2}, & r \geq R \end{cases}$$

Using the equation relating the voltage to the electric field, $\mathcal{V} = \int_r^{\infty} \mathbf{E}.d\mathbf{r}$, we obtain:

*For* $r < R$:

$$\mathcal{V}(r) = \int_r^R (0)\mathrm{d}r + \int_R^\infty \frac{Q}{4\pi\varepsilon r^2}\mathrm{d}r = \frac{Q}{4\pi\varepsilon R},$$

*For* $r \geq R$:

$$\mathcal{V}(r) = \int_r^\infty \frac{Q}{4\pi\varepsilon r^2}\mathrm{d}r = \frac{Q}{4\pi\varepsilon r},$$

Therefore, the voltage is given by

$$\mathcal{V} = \begin{cases} \frac{Q}{4\pi\varepsilon R}, & r < R \\ \frac{Q}{4\pi\varepsilon r}, & r \geq R. \end{cases}$$

### *8.1.6 Capacitance*

The capacitance is defined as the ability of a body to store charge. The capacitance of a given body is a function of its geometry and electric permittivity, a material property that reflects the ability of the material to store charge. Mathematically, the capacitance is defined as

$$C = \frac{q}{\mathcal{V}}, \tag{8.18}$$

where $q$ is the charge stored in the body and $\mathcal{V}$ is the electric potential necessary to store the charge. In other words, if a larger electric potential is needed to store a unit charge in a given body, the body has lower capacitance. The unit of capacitance is C/V, commonly referred to as a farad.

**Example 8.7 Capacitance of Two Parallel Plates**

Consider the parallel plate capacitor shown in Figure 8.10. It consists of one positively charged plate and one negatively charged plate with charge density $\sigma_0$. The plates are separated by a distance $d \ll w$, where $w$ is the length of the plates. Find (a) the electric field between the plates, and (b) the electric capacitance.

In Flipped Classroom Exercise 8.1, we found that the electric field of an infinitely long plate is given by $E = \frac{\sigma}{2\varepsilon}$. When putting two plates of different charge polarity close together, the electric field becomes the vectorial sum of the electric field resulting from each plate. As shown in Figure 8.11, the electric field of each plate adds between the two plates and cancels

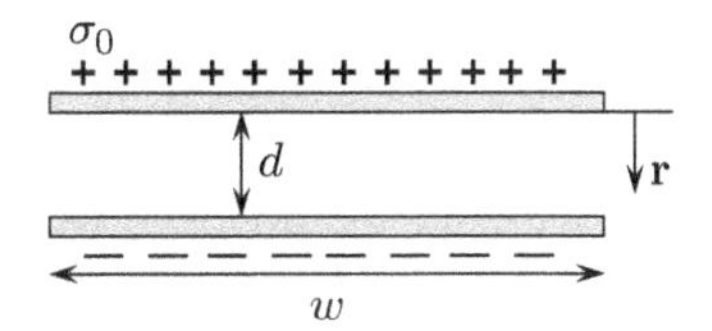


**Figure 8.10** A parallel plate capacitor.

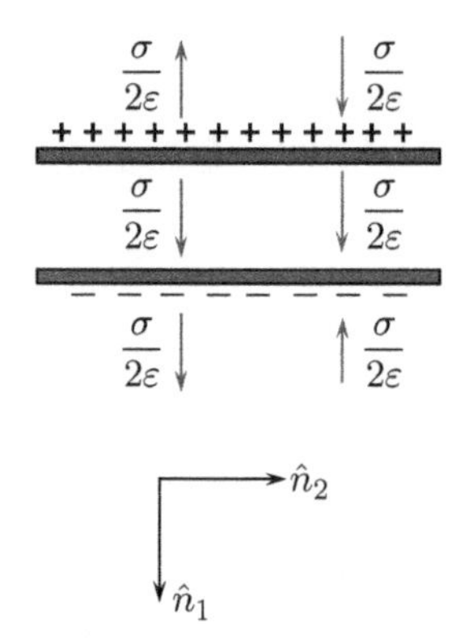


**Figure 8.11** The electric field of a parallel plate capacitor.

outside the two plates; that is

$$\mathbf{E}_{\text{outside}} = 0, \qquad \mathbf{E}_{\text{inside}} = \frac{\sigma}{\varepsilon}\hat{n}_1,$$

The voltage at a distance $r$ between the two plates measured from the top plate is given by $\mathcal{V} = \int \mathbf{E}.\mathrm{d}\mathbf{r}$. This yields

$$\mathcal{V} = \frac{\sigma}{\varepsilon}r.$$

Hence, the total voltage difference between the two plates is given by

$$\mathcal{V} = \frac{\sigma}{\varepsilon}d.$$

To find the capacitance, we use

$$C = \frac{q}{\mathcal{V}} = \frac{\varepsilon A}{d}.$$

**Example 8.8 Capacitance of Two Coaxial Cylinders**

A capacitor is composed of an inner cylinder of radius $a$ enclosed by an outer cylinder of radius $b$, as depicted in Figure 8.12. Assuming that the cylinders have a charge per unit length $\sigma$, and length $L$, find the capacitance of the system.

To find the capacitance, $C = q/\mathcal{V}$, we first need to find the electric potential $\mathcal{V}(r)$ as a function of the radius. To this end, we first find the electric field by defining a Gaussian surface as a coaxial cylinder of radius $a < r < b$ and length $L$. Using Gauss's flux theorem, we can write

$$\oint \mathbf{E}.\mathrm{d}\mathbf{A} = \sum_{i=1}^{\mathcal{N}} \frac{Q_i}{\varepsilon}.$$

Note that the electric field lines are perpendicular to the cylindrical surface, as shown in Figure 8.12. Therefore, $\mathbf{E}$ and $\mathrm{d}\mathbf{A}$ are parallel. It follows that the previous equation simplifies to

$$E(2\pi rL) = -\frac{\sigma L}{\varepsilon}$$

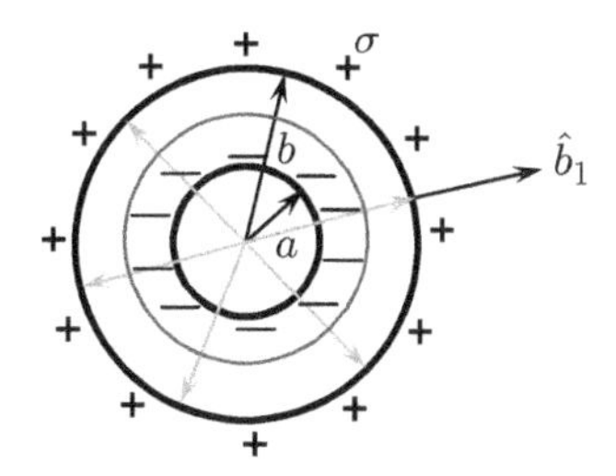


**Figure 8.12** A cylindrical capacitor.

or

$$\mathbf{E} = -\frac{\sigma}{2\pi\varepsilon r}\hat{b}_1 \qquad a < r < b.$$

Using the previous equation, the voltage between the two cylinders can be calculated as

$$\mathcal{V}(r) - \mathcal{V}(a) = -\int_a^r \mathbf{E}.\mathrm{d}\mathbf{r} = \int_a^r \frac{\sigma}{2\pi\varepsilon r}\hat{b}_1.dr\hat{b}_1,$$

where $\mathcal{V}(a)$ is the voltage on the surface of the inner cylinder. This yields

$$\mathcal{V}(r) - \mathcal{V}(a) = \frac{\sigma}{2\pi\varepsilon}\ln\frac{r}{a}.$$

The voltage difference between the outer and inner cylinders is therefore

$$\mathcal{V}(b) - \mathcal{V}(a) = \frac{\sigma}{2\pi\varepsilon}\ln\frac{b}{a},$$

and the capacitance is

$$C = \frac{q}{\mathcal{V}(b) - \mathcal{V}(a)} = \frac{2\pi\varepsilon}{\ln\frac{b}{a}}.$$

### *8.1.7 Motion in an Electric Field*

The knowledge gained so far in this chapter will now be used to model the motion of charged particles and rigid bodies in an electric field. To this end, we present next a series of examples involving the interaction of an electric field with a dynamical system.

**Example 8.9 Motion of Charged Simple Pendulum in Uniform Electric Field**

Obtain the equation of motion of a positively-charged pendulum moving in a uniform electric field for the two cases shown in Figure 8.13.

To find the equation of motion, we implement Newton's second law. To this end, we define the $B$-frame such that ${}^N\boldsymbol{\omega}^B = \dot{\theta}\hat{n}_3$, and use it to obtain the acceleration of the charged particle, which can be written as ${}^N\mathbf{a}^{m/O} = -l\dot{\theta}^2\hat{b}_1 + l\ddot{\theta}\hat{b}_2$.

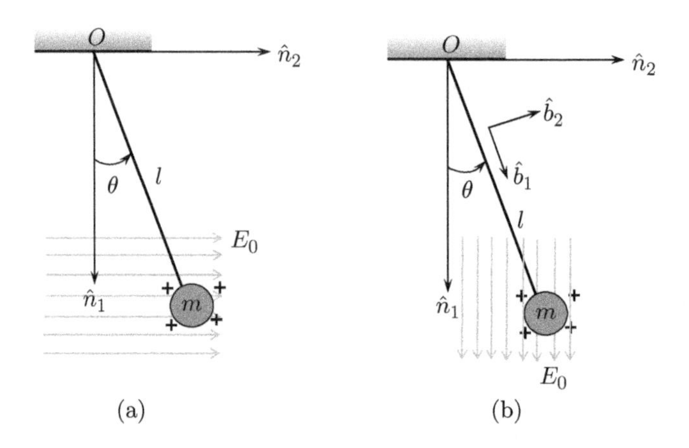


**Figure 8.13** Simple pendulum oscillating in a uniform field.

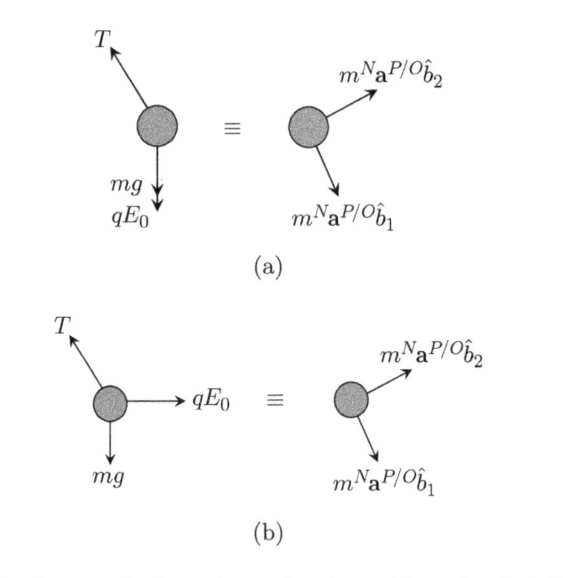


**Figure 8.14** Free-body diagram of a charged pendulum in a uniform electric field: (a) pointing in the horizontal direction; (b) pointing in the vertical direction.

For case (a) involving a horizontal field, we sum forces in the $\hat{b}_2$ direction, as shown in Figure 8.14a, and obtain the following equation of motion:

$$\ddot{\theta} + \frac{g}{l}\sin\theta = \frac{qE_0}{ml}\cos\theta.$$

For the vertical field, we sum forces in the $\hat{b}_2$ direction, as shown in Figure 8.14b, and find

$$\ddot{\theta} + \left(\frac{g}{l} + \frac{qE_0}{ml}\right)\sin\theta = 0.$$

**Example 8.10 Motion of a Proton in a Uniform Electric Field**

A proton is released from rest at a point $A$ in space where the voltage is $\mathcal{V}_A = 100$ V. Find its velocity as it approaches point $B$, of zero voltage.

To solve this problem we use energy balance. To this end, we write

$$\mathcal{U}_A + \mathcal{T}_A = \mathcal{U}_B + \mathcal{T}_B,$$

where $\mathcal{U}_A = q\mathcal{V}_A$, $\mathcal{T}_A = 0, \mathcal{U}_B = 0$, and $\mathcal{T}_B = \frac{1}{2}mv_B^2$. Using the charge and mass of the proton, $q = 1.6 \times 10^{-19}$ C and $m = 1.7 \times 10^{-27}$ kg, we find $v_B = 1.37 \times 10^5 \mathrm{m/s}$.

**Example 8.11 Motion of an Electric Dipole in a Uniform Electric Field**

Find the equation of motion of an electric dipole placed in a uniform electric field, $\mathbf{E} = E_0$, as shown in Figure 8.15. Each of the particles forming the dipole has a charge $q$ and mass $m$.

Due to the electric field, the charges are subject to the following equal but opposite forces:

$$\mathbf{F}^+ = qE_0\hat{n}_1, \qquad \mathbf{F}^- = -qE_0\hat{n}_1.$$

To find the equation of motion, we sum moments about the center of mass $O$ and obtain

$$\mathbf{M}_O = I_O\ddot{\theta},$$

$$-qdE_0 \sin\theta\hat{n}_3 = -m\frac{d^2}{2}\ddot{\theta}\hat{n}_3,$$

$$\ddot{\theta} - 2\frac{qE_0}{md}\sin\theta = 0.$$

**Example 8.12 Motion of a Charge in the Electric Field of a Dipole**

A particle of mass $m$ and charge $+q$ moves in the electric field of the dipole of charge $Q$ placed on the $\hat{n}_3$-axis, as shown in Figure 8.16. Find the equation governing its motion.

To find the equation of motion of the particle, we first define the rotating frames: the $C$-frame such that ${}^N\omega^C = \dot{\phi}\hat{n}_3$, and the $B$-frame such that ${}^C\omega^B = -\dot{\theta}\hat{c}_2$. We also use the three generalized coordinates $(\theta, \phi, r)$ to describe the dynamics. Using these definitions, we find the acceleration of the particle in space as follows:

$$\begin{aligned}{}^N\mathbf{a}^{P/O} &= (r\ddot{\theta} + 2\dot{r}\dot{\theta} - r\dot{\phi}^2 \sin\theta\cos\theta)\hat{b}_1 \\ &+ (2r\dot{\theta}\dot{\phi}\sin\theta\cos\theta + r\ddot{\phi}\sin\theta + 2\dot{r}\dot{\phi}\sin\theta)\hat{b}_2 \\ &- (\ddot{r} - r\dot{\theta}^2 - r\dot{\phi}^2\sin^2\theta)\hat{b}_3.\end{aligned}$$

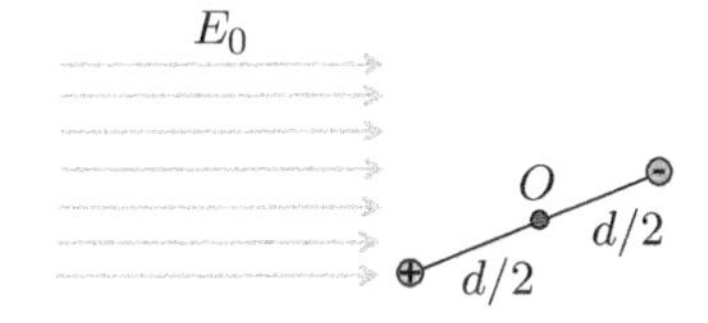


**Figure 8.15** Motion of an electric dipole in a uniform electric field.

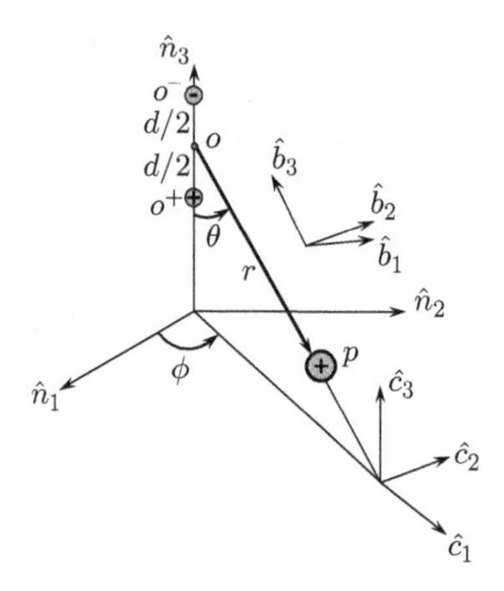


**Figure 8.16** Motion of a charge in the electric field of a dipole.

There are two forces acting on the particle: its weight $-mg\hat{n}_3$ and the electric force $Q\mathbf{E}$ due to the dipole. Here $\mathbf{E}$ is the electric field of the dipole at point $p$ and can be obtained using

$$\mathbf{E}^+ = \frac{Q}{4\pi\varepsilon}\frac{\left(\frac{d}{2}\hat{n}_3 - r\hat{b}_3\right)}{\|\mathbf{o}^+\mathbf{p}\|^3},$$

$$\mathbf{E}^- = -\frac{Q}{4\pi\varepsilon}\frac{\left(\frac{-d}{2}\hat{n}_3 - r\hat{b}_3\right)}{\|\mathbf{o}^-\mathbf{p}\|^3}.$$

Assuming that $\frac{d}{2} \ll r$, we can make the approximation that $\|\mathbf{o}^+\mathbf{p}\| \approx \|\mathbf{o}^-\mathbf{p}\| \approx r$. Thus, we can write

$$\mathbf{E}^+ = \frac{Q}{4\pi\varepsilon}\frac{\left(\frac{d}{2}\hat{n}_3 - r\hat{b}_3\right)}{r^3}, \qquad \mathbf{E}^- = -\frac{Q}{4\pi\varepsilon}\frac{\left(\frac{-d}{2}\hat{n}_3 - r\hat{b}_3\right)}{r^3}.$$

$$\mathbf{E} = \mathbf{E}^+ + \mathbf{E}^- = \frac{Q}{4\pi\varepsilon}\frac{d\hat{n}_3}{r^3} = \frac{Q}{4\pi\varepsilon}\frac{d\sin\theta\hat{b}_1 + d\cos\theta\hat{b}_3}{r^3}.$$

The electric force exerted by the electric field on the charged particle can therefore be written as

$$\mathbf{F}_{el} = q\mathbf{E} = \frac{Qq}{4\pi\varepsilon}\frac{d\sin\theta\hat{b}_1 + d\cos\theta_1\hat{b}_3}{r^3}.$$

Next, we apply Newton's second law along the three generalized coordinates used to describe the motion of the charge, obtaining

$$r\ddot{\theta} + 2\dot{r}\dot{\theta} - r\dot{\phi}^2\sin\theta\cos\theta + g\sin\theta = \frac{Qq}{4\pi\varepsilon m}\frac{d\sin\theta}{r^3},$$

$$2r\dot{\theta}\dot{\phi}\sin\theta\cos\theta + r\ddot{\phi}\sin\theta + 2\dot{r}\dot{\phi}\sin\theta = 0,$$

$$\ddot{r} - r\dot{\theta}^2 - r\dot{\phi}^2\sin^2\theta = -\frac{Qq}{4\pi\varepsilon m}\frac{d\cos\theta}{r^3} + g\cos\theta.$$

Note that the $\phi$ is a cyclic coordinate, and so can be eliminated from the dynamics by using the initial conditions.

### Example 8.13 Motion of an Electron Inside a Cylindrical Capacitance

The space between a pair of coaxial cylindrical conductors is evacuated. The radius of the inner cylinder is $a$, and the radius of the outer cylinder is $b$, as shown in Figure 8.17. The outer cylinder, called the anode, is given a positive potential $\mathcal{V}_o$ relative to the inner cylinder. We study the dynamics of electrons, with mass $m$ and charge $-e$, that are released from the surface of the inner cylinder:

(a) An electron is set free with negligible velocity at the surface of the inner cylinder. Determine its speed $v_b$ when it hits the anode.
(b) If the electron is released with any arbitrary velocity, find the equation governing its motion.

**(a)** ***With negligible velocity*** An electron will move spontaneously from lower towards higher voltages. Its kinetic energy, and thereby velocity at the anode, can be obtained using an energy balance:

$$\mathcal{T}_a + \mathcal{U}_a = \mathcal{T}_b + \mathcal{U}_b,$$

The kinetic energy is zero at $a$ and takes its maximum value of $\mathcal{T}_b = 1/2mv_b^2$ at $b$. The potential energy can be determined by choosing the outer cylinder as a datum, in which case, the potential energy at any point $a \leq r \leq b$ is equal to $\mathcal{U}(r) = -e(\mathcal{V}(r) - \mathcal{V}_o)$. Therefore, the potential energy at $a$ is equal to $\mathcal{U}_a = -e(\mathcal{V}(a) - \mathcal{V}_o) = e\mathcal{V}_o$, and is equal to zero at $b$. This yields

$$v_b = \sqrt{\frac{2e\mathcal{V}_o}{m}}.$$

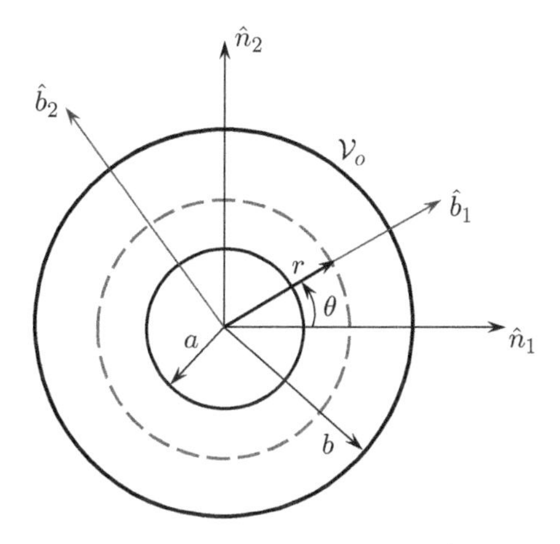


**Figure 8.17** Motion of an electron inside a cylindrical capacitor.

**(b)** ***With arbitrary velocity*** To find the equation of motion, we need to find the potential energy of the electron at any point $r$ between the two cylinders. The potential energy is given by

$$\mathcal{U}(r) = -e(\mathcal{V}(r) - \mathcal{V}_o).$$

where $\mathcal{V}(r) = -\int_a^r \mathbf{E}.\mathrm{d}\mathbf{r}$. The electric field between the two cylinders was obtained using Gauss's law in Example 8.8, and was shown to be $\mathbf{E} = -\frac{\sigma}{2\pi\varepsilon r}\hat{b}_1$. However, since in this example we do not know the charge distribution, $\sigma$, on the cylinders, we assume that the electric field takes the form

$$\mathbf{E} = -\frac{C_1}{r}\hat{b}_1,$$

where $C_1$ is a constant that depends on the charge distribution. Integrating the previous equation yields

$$\mathcal{V}(r) - \mathcal{V}(a) = -\int_a^r \mathbf{E}.\mathrm{d}\mathbf{r} = -C_1 \ln r + C_1 \ln a = -C_1 \ln \frac{r}{a}.$$

The constant $C_1$ is obtained using the boundary conditions $\mathcal{V}(b) = \mathcal{V}_o$. It follows that

$$\mathcal{V}(r) = \mathcal{V}_o \frac{\ln(r/a)}{\ln(b/a)},$$

and

$$\mathcal{U}(r) = -e\mathcal{V}_o\left(\frac{\ln(r/a)}{\ln(b/a)} - \mathcal{V}_o\right).$$

To find the equation of motion, we apply Lagrange's equations, knowing that the kinetic energy of the particle is given by

$$\mathcal{T} = \frac{1}{2}m\dot{r}^2 + \frac{1}{2}mr^2\dot{\theta}^2.$$

Using $\mathcal{L} = \mathcal{T} - \mathcal{U}$, we can write

$$\frac{\mathrm{d}}{\mathrm{d}t}\left(\frac{\partial \mathcal{L}}{\partial \dot{r}}\right) - \frac{\partial \mathcal{L}}{\partial r} = 0,$$

$$\ddot{r} + \frac{e\mathcal{V}_o}{ma\ln(a/b)}\frac{1}{r} - r\dot{\theta}^2 = 0,$$

$$\frac{\mathrm{d}}{\mathrm{d}t}\left(\frac{\partial \mathcal{L}}{\partial \dot{\theta}}\right) - \frac{\partial \mathcal{L}}{\partial \theta} = 0,$$

$$mr^2\dot{\theta} = \text{constant}.$$

Note that $\theta$ is a cyclic coordinate.

**Flipped Classroom Exercise 8.2**

Find the equation of motion of the electric tripole placed in a uniform electric field $\mathbf{E} = E_0$. Each of the tripole particles has a charge $q$ and mass $m$.

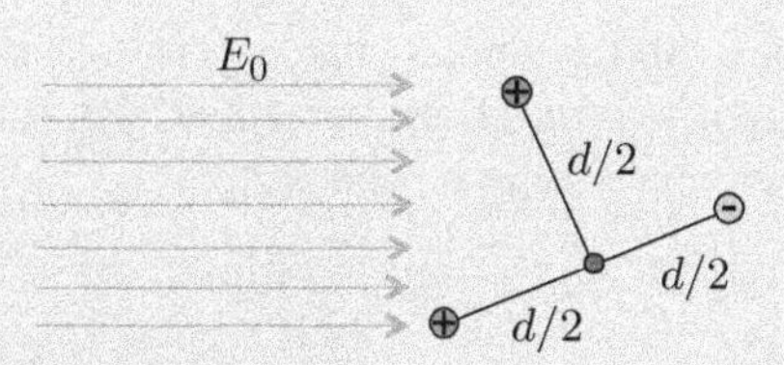


To solve this problem, take the following steps:

1. Define the degrees of freedom needed to fully describe the dynamics of the system.
2. Apply Newton's second law to capture the dynamics of the translational motion of the tripole.
3. Find the center of mass of the system.
4. Sum moments about the center of mass of the particles.
5. Show that the equations of motion can be written as

$$\ddot{x} = \frac{qE_0}{3m}, \qquad \ddot{\theta} = \frac{3qE_0}{4md}\left(\sin\theta + \frac{1}{3}\cos\theta\right),$$

where $x$ is the position of the center of mass in the horizontal direction, and $\theta$ is the angle that the center of mass makes with the horizontal.

**Example 8.14 Capacitive Micro-electromechanical Switch**

A micro-electromechanical switch is a very small device (of sub-millimeter dimensions) used as a very high-frequency (GHz) mechanical on-off switch in an electric circuit. The device consists of two parallel plates separated initially by a distance $d$, as shown in Figure 8.18. The lower plate is stationary and is commonly referred to as the electrode. The upper plate is suspended through an elastic element, a spring, and can be actuated by applying an electric potential (voltage) between the electrode and the moving plate. Assume that the moving plate is rigid and has mass $m$, and that the spring has stiffness $k$, find the equation of motion of the moving plate.

We will solve this problem using Lagrange's equations. The generalized coordinate describing the motion of the plate is $y$ and the velocity of its center of gravity with respect to a fixed point is $\dot{y}$. The kinetic energy is therefore $\mathcal{T} = \frac{1}{2}m\dot{y}^2$, while the potential energy is that due to the spring, $\mathcal{U}_s = \frac{1}{2}ky^2$, and the electrostatic potential energy $\mathcal{U}_e$.

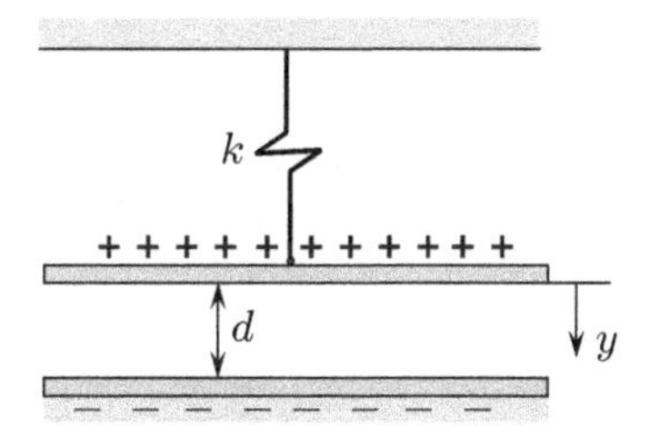


**Figure 8.18** A capacitive micro-electromechanical switch.

To obtain the electrostatic potential energy stored between the plates due to the electrostatic field, we note that, in general, for a point charge $q$, the electrostatic potential energy stored when moving the charge across an electric potential $\mathcal{V}$ is simply $\mathcal{U}_e = q\mathcal{V}$. Hence, for the plate, we can take a differential charge $\mathrm{d}q$ and write

$$\mathrm{d}\mathcal{U}_e = \mathcal{V}\mathrm{d}q.$$

However, $q = C\mathcal{V}$, which leads to

$$\mathrm{d}\mathcal{U}_e = \frac{q}{C}\mathrm{d}q.$$

The total potential energy is therefore

$$\mathcal{U}_e = \int_0^Q \frac{q}{C}\mathrm{d}q = \frac{1}{2}\frac{Q^2}{C} = \frac{1}{2}C\mathcal{V}^2,$$

where $C = \frac{\varepsilon A}{d-y}$. Here, $A$ is the area of the plate and $d - y$ is the distance between the two parallel plates. So the potential energy is

$$\mathcal{U}_e = \frac{1}{2}\mathcal{V}^2\frac{\varepsilon A}{d-y}.$$

Next, we use Lagrange's equation for conservative fields and write

$$\frac{\mathrm{d}}{\mathrm{d}t}\left(\frac{\partial \mathcal{L}}{\partial \dot{y}}\right) - \frac{\partial \mathcal{L}}{\partial y} = 0,$$

where $\mathcal{L} = \mathcal{T} - \mathcal{U}_s - \mathcal{U}_e$. This yields

$$m\ddot{y} + ky = \frac{1}{2}\mathcal{V}^2\frac{\varepsilon A}{(d-y)^2}.$$

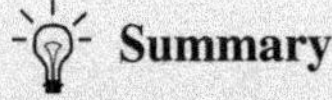

**Summary**

In the first part of this chapter, we introduced the following concepts:

- *The electrostatic force*: The electrostatic force, $\mathbf{F}_{12}$ between two charges, $q_1$ and $q_2$, of different polarities is described by Coulomb equation:

$$\mathbf{F}_{12} = \frac{1}{4\pi\varepsilon}\frac{q_1 q_2}{\|\mathbf{r}\|^2}\hat{r} = \frac{1}{4\pi\varepsilon}\frac{q_1 q_2}{\|\mathbf{r}\|^3}\mathbf{r},$$

  where $\mathbf{r}$ and $\|\mathbf{r}\|$ are, respectively, the position vector and the distance from $q_1$ to $q_2$, $\hat{r}$ is the unit vector in the direction of $\mathbf{r}$, and $\varepsilon$ is the electric permittivity of the space enclosing the charged particles.
- *The electric field:* The electric field at a point in space due to a positive charge $Q$ is the electric force exerted by $Q$ on a positive test charge $q$, divided by the test charge

$q$; that is,

$$\mathbf{E} = \frac{\mathbf{F}}{q} = \frac{1}{4\pi\varepsilon}\frac{Q}{\|\mathbf{r}\|^2}\hat{r}. \tag{8.19}$$

- *Gauss's flux theorem*: The electric field, **E**, due to $\mathcal{N}$ charges each of charge $Q_i$ can be obtained via

$$\oint \mathbf{E}.\mathrm{d}\mathbf{A} = \sum_{i=1}^{\mathcal{N}} \frac{Q_i}{\varepsilon},$$

where d**A** is the vector normal to any arbitrary surface enclosing $Q_i$.
- *Electrostatic potential energy:* The electrostatic potential energy stored between two charges of similar polarity separated by a distance $R$ is given by

$$\mathcal{U} = \int_R^\infty \mathbf{F}_{12}.\mathrm{d}\mathbf{r},$$

where $\mathbf{F}_{12}$ is the Coulomb repulsive force between the charges.
- *Electrostatic potential (voltage):* The electrostatic potential (voltage) at a given point $P$ in space is defined as the electrostatic potential energy per unit charge $q$.

$$\mathcal{V}_P = \frac{\mathcal{U}}{q} = \int_R^\infty \frac{\mathbf{F}_{12}}{q}.\mathrm{d}\mathbf{r} = \int_R^\infty \mathbf{E}.\mathrm{d}\mathbf{r},$$

The relationship between the electric field and the voltage can also be represented in the more general differential form:

$$\mathbf{E} = -\nabla \mathcal{V},$$

where $\nabla$ represents the gradient.

## 8.2 Electromagnetism

### 8.2.1 Electromagnetic Force

When discussing electrostatic forces, we showed that the force $\mathbf{F}(\mathbf{r}, t)$ on a stationary charge $q$ in an electric field, $\mathbf{E}(\mathbf{r}, t)$, is equal to $\mathbf{F} = q\mathbf{E}(\mathbf{r}, t)$. In the presence of an external magnetic field, $\mathbf{B}(\mathbf{r}, t)$, this relation can be generalized further to

$$\mathbf{F}(\mathbf{r}, t) = q\mathbf{E}(\mathbf{r}, t) + q^N\mathbf{v}^{P/O} \times \mathbf{B}(\mathbf{r}, t) \tag{8.20}$$

where ${}^N\mathbf{v}^{P/O}$ is the velocity of the particle $P$ with respect to an inertial point $O$. Equation (8.20) is known as the Lorentz law after the Dutch physicist Hendrik Lorentz, who derived it in this general form in 1895 [3]. The first term on the right-hand side of Equation (8.20) accounts for the influence of the electric filed, while the second term accounts for the influence of the magnetic field.

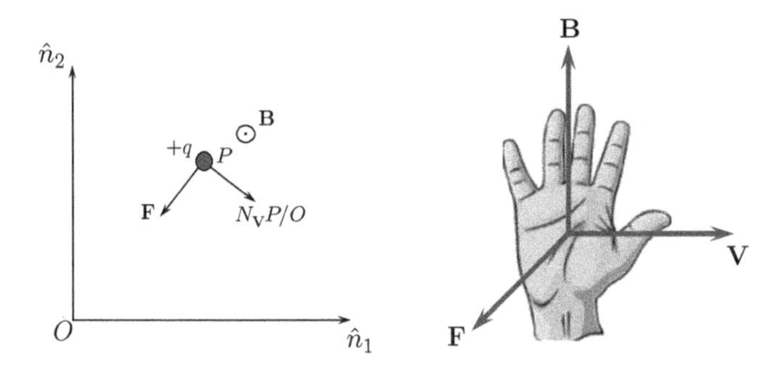


**Figure 8.19** Direction of the electromagnetic force acting on a charged particle.

Unlike the electrostatic force, which can affect a stationary particle, thereby changing its velocity, the magnetic force exerted by the magnetic field can only act on a moving particle. Furthermore, the magnetic force is perpendicular to the velocity of the particle, and therefore cannot exert any work. As a result, the magnetic force can only change the particle's direction; it cannot change its kinetic energy.

To find the direction of the electromagnetic force, consider the motion of a particle of charge $+q$ in a uniform magnetic field **B** pointing out of the page, as shown in Figure 8.19. The electromagnetic force **F** is perpendicular to the velocity of the particle and the magnetic field, and is in the direction shown in the figure. This direction can be determined using the right-hand rule for positively-charged particles. Specifically, as shown in Figure 8.19, if you point your thumb in the direction of the velocity vector and the rest of your fingers in the direction of the magnetic field, then the force points in a direction perpendicular to your palm. For a negatively-charged particle, you just need to invert the direction obtained for a positively-charged one.

### Example 8.15 Planar Motion of Charged Particle in a Uniform Magnetic Field

Consider the planar motion of a positively-charged particle $+q$ of mass $m$ in a uniform magnetic field $B\hat{n}_3$, as depicted in Figure 8.20. The particle is initially located at $\mathbf{r} = x_0\hat{n}_1 + y_0\hat{n}_2$, and is given an initial velocity $\mathbf{v} = v_{x0}\hat{n}_1 + v_{y0}\hat{n}_2$. Find the path of the particle as a function of time and show that it is a circle of radius $R = \frac{mv}{qB}$, where $v = \sqrt{v_{x0}^2 + v_{y0}^2}$. Show that the center of the circle is located at $\mathbf{r}_c = \left(x_0 - \frac{mv_{y0}}{qB}\right)\hat{n}_1 + \left(y_0 + \frac{mv_{x0}}{qB}\right)\hat{n}_2$.

To find the equation of motion of the particle, we apply Newton's second law. To this end, we express the position, velocity, and acceleration in the Cartesian frame as follows:

$$\mathbf{OP} = x\hat{n}_1 + y\hat{n}_2, \qquad {}^N\mathbf{v}^{P/O} = \dot{x}\hat{n}_1 + \dot{y}\hat{n}_2, \qquad {}^N\mathbf{a}^{P/O} = \ddot{x}\hat{n}_1 + \ddot{y}\hat{n}_2.$$

When neglecting gravity, the only force acting on the body is the magnetic component of the Lorentz force $\mathbf{F} = q(\dot{x}\hat{n}_1 + \dot{y}\hat{n}_2) \times B\hat{n}_3$. Summing forces in the $\hat{n}_1$- and $\hat{n}_2$-directions yields

$$m\ddot{x} = -qB\dot{y}, \qquad m\ddot{y} = qB\dot{x}.$$

These two linearly-coupled differential equations can be solved together to obtain

$$x(t) = -A_1\sin(\Omega t) + A_2\cos(\Omega t) + A_3,$$
$$y(t) = A_1\cos(\Omega t) + A_2\sin(\Omega t) + A_4,$$

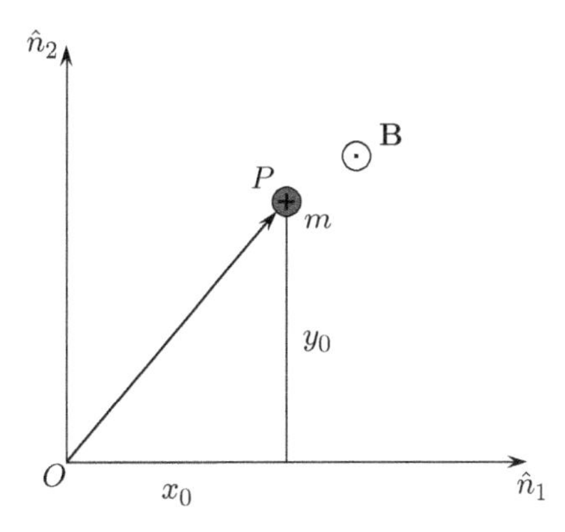


**Figure 8.20** Right-hand rule to determine the force acting on a positively-charged particle.

where $\Omega = \frac{qB}{m}$ is the frequency of motion (also known as the *cyclotrone* frequency) and $A_1, A_2, A_3, A_4$ are constants that can be obtained from the initial conditions. Upon substituting the initial conditions, we get

$$x(t) = \frac{mv_{x0}}{qB}\sin(\Omega t) + \frac{mv_{y0}}{qB}\cos(\Omega t) + x_0 - \frac{mv_{y0}}{qB},$$

$$y(t) = -\frac{mv_{x0}}{qB}\cos(\Omega t) + \frac{mv_{y0}}{qB}\sin(\Omega t) + y_0 + \frac{mv_{x0}}{qB}.$$

Squaring and adding these two equations yields

$$\left(x - x_0 + \frac{mv_{y0}}{qB}\right)^2 + \left(y - y_0 - \frac{mv_{x0}}{qB}\right)^2 = \frac{m^2}{q^2B^2}(v_{x0}^2 + v_{y0}^2).$$

This is the equation of a circle of radius $R = \frac{m\sqrt{v_{x0}^2+v_{y0}^2}}{qB}$ which is centered at $\left(x_0 - \frac{mv_{y0}}{qB},\right.$ $\left. y_0 + \frac{mv_{x0}}{qB}\right)$. Note that the radius of the circle increases with the magnitude of the linear momentum of the particle $mv$ and decrease as $qB$ increases. The radius $R$ is commonly referred to as the *Larmor radius*.

**Example 8.16 Three-dimensional Motion of a Charged Particle in a Uniform Magnetic Field**

Consider the general motion of a positively charged particle $+q$ of mass $m$ in a uniform magnetic field $B\hat{n}_3$. The particle is initially located at $\mathbf{r}_0 = x_0\hat{n}_1 + y_0\hat{n}_2 + z_0\hat{n}_3$, and is given an initial velocity $\mathbf{v}_0 = v_{x0}\hat{n}_1 + v_{y0}\hat{n}_2 + v_{z0}\hat{n}_3$. Neglecting gravity, find the path of the particle as a function of time.

To find the equation of motion of the particle, we apply Newton's second law. To this end, we express the position, velocity, and acceleration in the Cartesian frame as follows:

$$\mathbf{OP} = x\hat{n}_1 + y\hat{n}_2 + z\hat{n}_3,$$

$$^N\mathbf{v}^{P/O} = \dot{x}\hat{n}_1 + \dot{y}\hat{n}_2 + \ddot{z}\hat{n}_3,$$

$$^N\mathbf{a}^{P/O} = \ddot{x}\hat{n}_1 + \ddot{y}\hat{n}_2 + \ddot{z}\hat{n}_3.$$

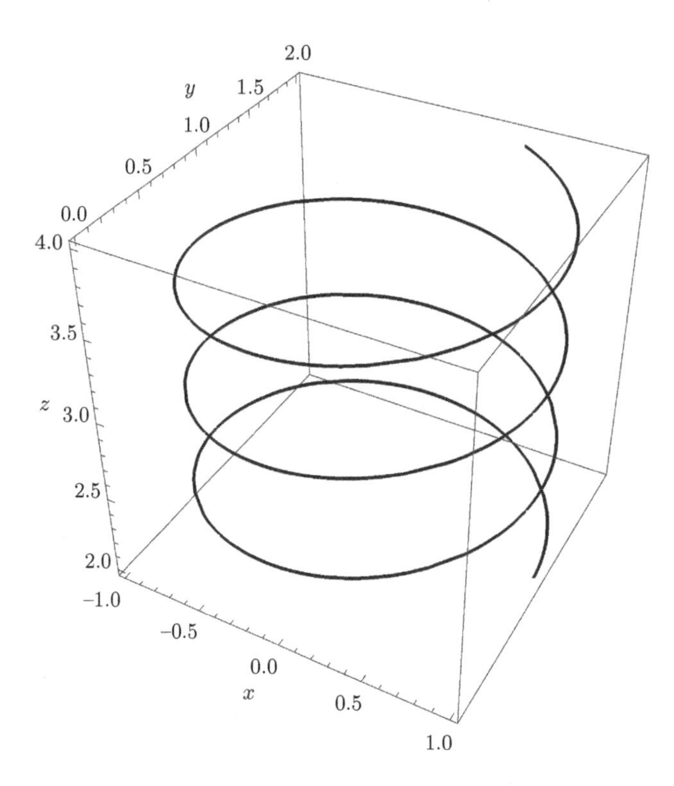


**Figure 8.21** Helical motion of a particle in a uniform magnetic field.

When neglecting gravity, the only force acting on the body is the Lorentz force $\mathbf{F} = q(\dot{x}\hat{n}_1 + \dot{y}\hat{n}_2 + \dot{y}\hat{n}_2) \times B\hat{n}_3$. Summing forces in the $\hat{n}_1$, $\hat{n}_2$, $\hat{n}_3$ directions yields

$$m\ddot{x} = -qB\dot{y}, \qquad m\ddot{y} = qB\dot{x}, \qquad m\ddot{z} = 0.$$

Note that the first two equations are the same as those obtained in the previous example, meaning that the motion of the particle is still a circle when projected into the $x$–$y$ plane. Nonetheless, the third equation, which describes the $z$-dynamics, can be solved to obtain $z(t) = v_{z0}t + z_0$. This implies that the position of the particle in the $\hat{n}_3$-direction changes linearly with time, leading to a helical motion of the particle, as shown in Figure 8.21.

 **Flipped Classroom Exercise 8.3**

Repeat the previous example to find the path of the particle when you include the gravitational forces in the negative $z$ direction. What will the shape of the path look like?

To solve this problem, you need to repeat the exact steps of Example 8.16.

### Example 8.17 Mass Spectrometer

A mass spectrometer is a device which can be used to separate the uranium isotopes $U_{235}$ and $U_{238}$ for enrichment purposes. As shown in Figure 8.22, the device operates by injecting a

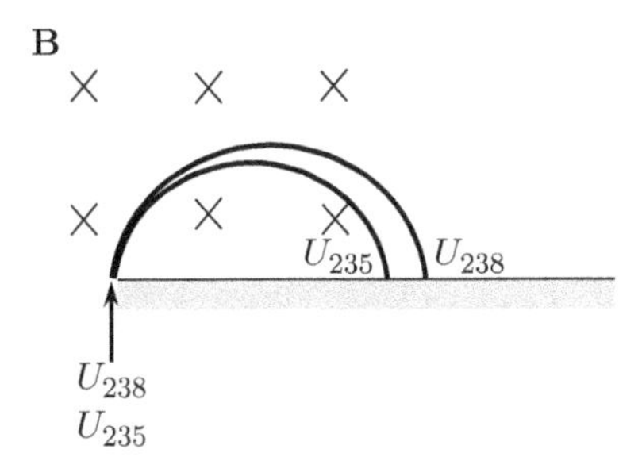


**Figure 8.22** Mass spectrometer used for separating uranium isotopes.

mixture of the two isotopes at a high velocity and applying a uniform magnetic field perpendicular to their motion. If the two isotopes enter the spectrometer with equal velocities, what is the ratio of the radii of their circular paths? Assume that the spectrometer is designed such that singly-ionized uranium ions pass first through a 500 m/s velocity selector, and then into a 2.85 T magnetic field. Find the diameter of the arc traced out by the $U_{235}$ and $U_{238}$.

We have already shown in Example 8.15, the radius of the circular motion of a charged particle in a magnetic field is equal to $R = \frac{mv}{qB}$. Therefore, under the same applied conditions, the ratio of the radii is equal to the ratio between the masses of the uranium atoms; that is, $R_{235}/R_{238} = m_{235}/m_{238} = 235/238 \approx 0.9874$. The diameter traced by the $U_{235}$ is $D_{235} = 2R_{235} = 2\frac{235\times1.66\times10^{27}\times500}{1.602\times10^{19}\times2.85} = 0.82$ mm, while the diameter traced by the $U_{235}$ is $D_{238} = D_{235}/0.9874 \approx 0.83$ mm

**Example 8.18 Planar Motion of a Charged Particle in Crossed Electric and Magnetic Fields**

Find the equations of motion of a positively-charged particle of mass $m$ in crossed electric and magnetic fields, as shown in Figure 8.23. Solve the resulting equation assuming zero initial conditions and plot the path of the particle.

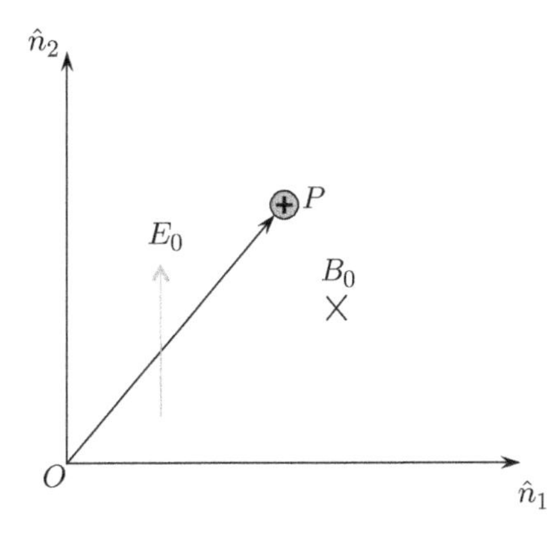


**Figure 8.23** Motion of a charged particle in crossed electric and magnetic fields.

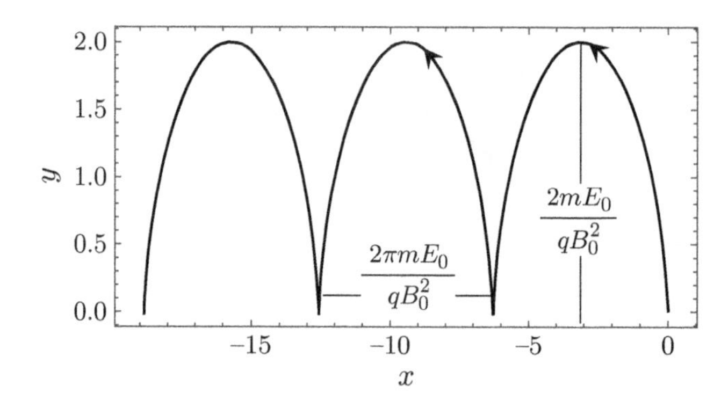


**Figure 8.24** Cycloidal motion of a charged particle in a cross field.

To solve this problem, we use Newton's second law of dynamics. To this end, we first express the velocity and acceleration of the particle in the Cartesian frame as

$$^{N}\mathbf{v}^{P/O} = \dot{x}\hat{n}_1 + \dot{y}\hat{n}_2, \qquad ^{N}\mathbf{a}^{P/O} = \ddot{x}\hat{n}_1 + \ddot{y}\hat{n}_2.$$

The only force acting on the particle is that due to the magnetic and electric fields, which can be written as

$$\mathbf{F} = qE_0\hat{n}_2 + q^{N}\mathbf{v}^{P/O} \times -B_0\hat{n}_3.$$

Applying Newton's second law along the two directions of motion, we obtain the following equations of motion:

$$m\ddot{x} + qB_0\dot{y} = 0,$$
$$m\ddot{y} - qB_0\dot{x} = qE_0.$$

These two linearly-coupled differential equations can be solved analytically to obtain:

$$x(t) = \frac{mE_0}{qB_0^2}\sin\left(\frac{qB_0}{m}t\right) - \frac{E_0}{B_0}t,$$
$$y(t) = \frac{mE_0}{qB_0^2}\left(1 - \cos\left(\frac{qB_0}{m}t\right)\right).$$

The motion described by these equations is a cycloid, as shown in Figure 8.24.

 **Flipped Classroom Exercise 8.4**

Consider the motion of an elastic pendulum consisting of a positively charged particle of mass $m$ and a spring of stiffness $k$ in a uniform magnetic field **B**, as shown in the figure. Assume that the unstretched length of the spring is $r_0$, find the equation of motion of the pendulum.

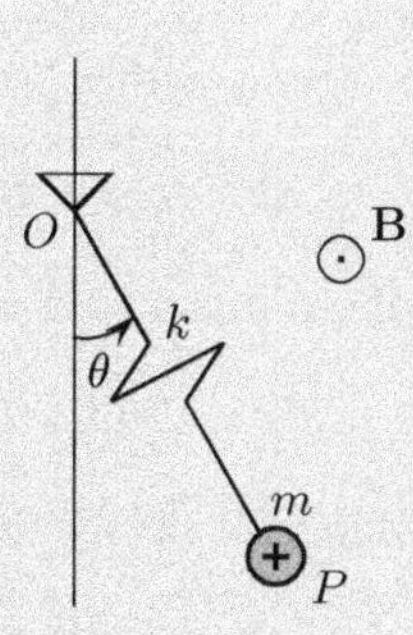


To find the equation of motion, take the following steps:

1. Find the velocity and acceleration of the particle ${}^{N}\mathbf{v}^{P/O}$ and ${}^{N}\mathbf{a}^{P/O}$.
2. Find the force exerted by the magnetic field on the particle.
3. Use Newton's second law to show that the equations of motion take the form

$$((r_0 + r)\ddot{\theta} + 2\dot{r}\dot{\theta}) + g\sin\theta + \frac{qB}{m}\dot{r} = 0,$$

$$(\ddot{r} - (r_0 + r)\dot{\theta}^2) + \frac{k}{m}r - g\cos\theta - \frac{(r_0 + r)}{m}q\dot{\theta} = 0.$$

## 8.2.2 *Forces on a Current-carrying Conductor*

We have already shown that a charge moving in a magnetic field will be subject to a force described by Equation (8.20). Since a current is the time rate change of charge, a current-carrying conductor placed in a magnetic field will also be subject to a force. To obtain an expression for this force, we consider an arbitrarily-shaped conductor carrying a current $I$ placed in a magnetic field $\mathbf{B}$, as shown in the Figure 8.25.

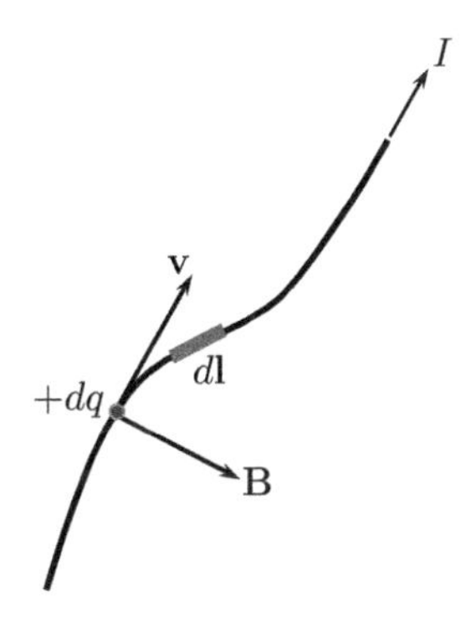


**Figure 8.25** A schematic of a current-carrying conductor placed in a magnetic field **B**.

Consider a differential charge $+dq$ in the conductor moving with a velocity $\mathbf{v}$. According to Lorentz's law, the differential force acting on the charge can be written as

$$\mathrm{d}\mathbf{F} = dq\mathbf{v} \times \mathbf{B}. \tag{8.21}$$

Since the current passing through the wire is $I = \mathrm{d}q/\mathrm{d}t$, we can write

$$\mathrm{d}\mathbf{F} = I\mathbf{v}\mathrm{d}t \times \mathbf{B}, \tag{8.22}$$

where $\mathbf{v}\mathrm{d}t = \mathrm{d}\mathbf{l}$. Here, $\mathrm{d}\mathbf{l}$ is the differential element in the direction of the wire. This yields

$$\mathrm{d}\mathbf{F} = I\mathrm{d}\mathbf{l} \times \mathbf{B}. \tag{8.23}$$

Integrating over the length of the conductor yields

$$\boxed{\mathbf{F} = \int_0^l I\mathrm{d}\mathbf{l} \times \mathbf{B}} \tag{8.24}$$

For a straight wire of length $l$, placed in uniform magnetic field $B$, the previous equation reduces to

$$\mathbf{F} = IlB\sin\theta. \tag{8.25}$$

### Example 8.19 Current Meter

One configuration of a current meter, also known as an ammeter, is shown in Figure 8.26. A current $I$ passes through a current loop placed in a uniform magnetic field $\mathbf{B}$. The magnetic field exerts a force on segments $AB$ and $DC$ of the loop. Using the right-hand rule, it can be shown that the forces are equal in magnitude but opposite in direction, resulting in a torque around the $\hat{n}_3$-axis as shown in Figure 8.26. This causes the loop to rotate around the $\hat{n}_3$-axis, until it is balanced by the restoring force in the torsional spring shown in the figure. This results in a steady-state rotation angle. The rotation angle is then calibrated to indicate the current passing through the loop.

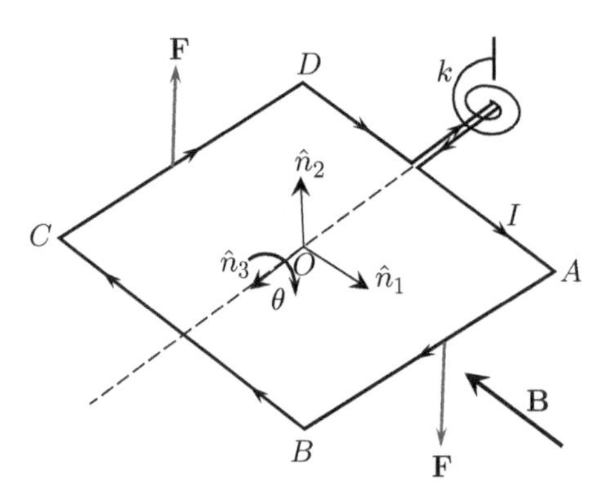


**Figure 8.26** Operation concept of a current meter (ammeter).

Consider a loop made of four segments, each of length $l$ and mass $m$. Find the equation of motion of the wire loop when a current $I$ passes through it. What is the steady-state rotation angle assuming there is a torsional viscous damper of damping coefficient $c$ acting on the suspension point?

To find the equation of motion, we sum moments about an axis passing through point $O$ and write

$$M_{3O} = I_{33O}\ddot{\theta}.$$

The moments acting on the wire are due to the electromagnetic force $\mathbf{M}_e = Bl^2 I\hat{n}_3$ and that due to the restoring force in the torsional spring $\mathbf{M}_s = -k\theta\hat{n}_3$ and viscous damping $\mathbf{M}_d = -c\dot{\theta}\hat{n}_3$. The moment of inertia about point $O$ is given by$I_{33O} = \frac{2}{3}ml^2$. This yields

$$\ddot{\theta} + \frac{3}{2}\frac{c}{ml^2}\dot{\theta} + \frac{3}{2}\frac{k}{ml^2}\theta = \frac{3B}{2m}I.$$

To find the steady-state rotation angle, we set $\ddot{\theta} = \dot{\theta} = 0$. This yields

$$\theta_{ss} = \frac{Bl^2}{k}I.$$

As such, by knowing $B$, $l$, and $k$ one can use the previous equation to obtain the current associated with any given steady-state rotation angle $\theta_{ss}$.

## 8.2.3 Electromagnetic Coupling

In 1820, the Danish scientist Hans Christian Ørsted discovered the first connection between electricity and magnetism when he noticed that a magnetized needle changes its direction when placed in the vicinity of a conductor carrying a current [4]. He concluded that this behavior must be due to a magnetic field generated by the motion of the current in the conductor.

The orientation of magnetized needles in the vicinity of a wire carrying a current is shown in Figure 8.27. The magnetic field forms a closed loop around the wire, with the wire at its center. The direction of the magnetic field is in the counter-clockwise direction when the current flows out of the page. This can be determined by directing your thumb in the direction of the current and curling the rest of the fingers. The direction of rotation is the direction of the induced magnetic field.

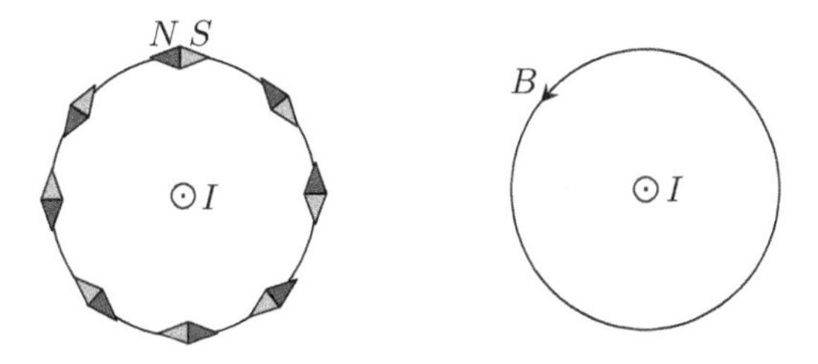


**Figure 8.27** Magnetic field lines form a closed loop around a current-carrying conductor.

In 1820, French physicists Jean-Baptiste Biot and Félix Savart related the current $I$ passing through a conductor to the induced magnetic field $\mathbf{B}$. They showed that the differential magnetic field $\mathrm{d}\mathbf{B}$ at a point $P$ induced by the motion of a current $I$ through a differential wire of length $\mathrm{d}\mathbf{l}$ is given by:

$$\mathrm{d}\mathbf{B} = \frac{\mu}{4\pi}\frac{I\mathrm{d}\mathbf{l} \times \mathbf{r}}{\|\mathbf{r}\|^3} = \frac{\mu}{4\pi}\frac{I\mathrm{d}\mathbf{l} \times \hat{r}}{\|\mathbf{r}\|^2} \tag{8.26}$$

where $\mu$ is the magnetic permeability of the medium containing the conductor, $\mathbf{r}$ is the position vector from $\mathrm{d}l$ to $P$, $\|\mathbf{r}\|$ is the length of $\mathbf{r}$, and $\hat{r} = \mathbf{r}/\|\mathbf{r}\|$ is the unit vector in the direction of $\mathbf{r}$.

When implementing the Biot–Savart law on an infinitely-long straight wire, as shown in Figure 8.28, the total magnetic field induced at point $P$ can be obtained as

$$B = \frac{\mu I}{4\pi}\int_{-\infty}^{\infty}\frac{\mathrm{d}l\|\mathbf{r}\|\sin\theta}{\|\mathbf{r}\|^3} = \frac{\mu I}{4\pi}\int_{-\infty}^{\infty}\frac{\mathrm{d}lR}{(l^2+R^2)^{3/2}} = \frac{\mu I}{2\pi R}. \tag{8.27}$$

Therefore, the magnitude of the magnetic field at a distant $R$ from an infinitely long wire is $B = \frac{\mu I}{2\pi R}$.

### Example 8.20 Magnetic Field at the Center of a Circular Conductor

Find the magnetic field at the center of a circular wire of radius $R$ carrying a current of magnitude $I$, as depicted in Figure 8.29.

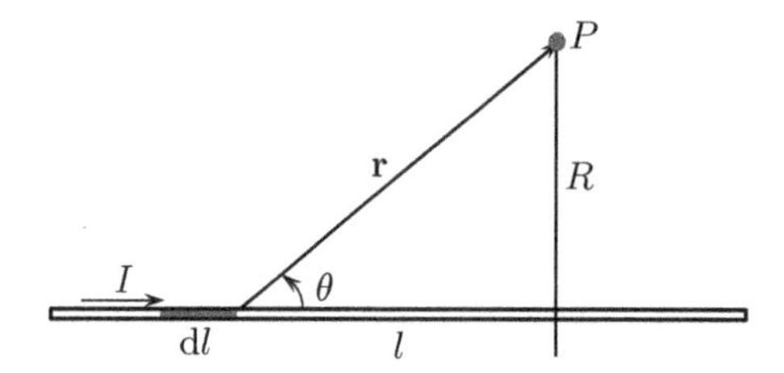


**Figure 8.28** Magnetic field induced at point $P$ by a current $I$ flowing through a conductor of length $l$.

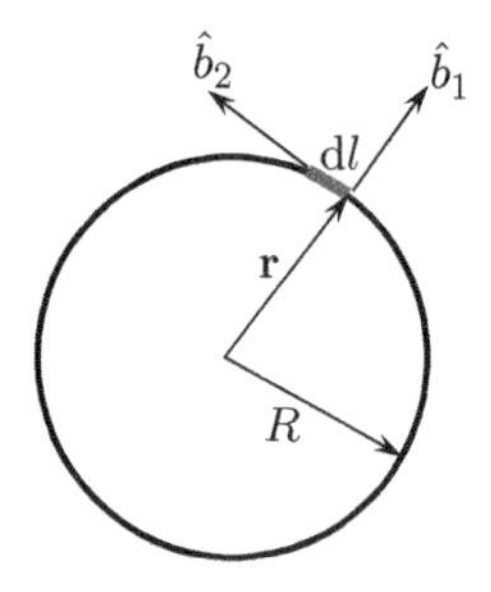


**Figure 8.29** Magnetic field at the center of a circular conductor.

Using the Biot-Savart law, we write

$$\mathbf{B} = \frac{\mu I}{4\pi} \int \frac{\mathbf{dl} \times \mathbf{r}}{\|\mathbf{r}\|^3}.$$

However, $\mathbf{dl} \times \mathbf{r} = dl\ R\hat{b}_2$ because $\mathbf{dl}$ is always normal to $\mathbf{r}$. Since $dl = R\mathrm{d}\theta$, it follows that

$$\mathbf{B} = \frac{\mu I}{4\pi} \int_0^{2\pi} \frac{R^2}{R^3} \mathrm{d}\theta = \frac{\mu I}{2R} \hat{b}_2.$$

**Flipped Classroom Exercise 8.5**

Find the magnetic field at point $P$ located at a distance $a$ from a finite wire of length $l$ carrying a current of magnitude $I$, as shown in the figure.

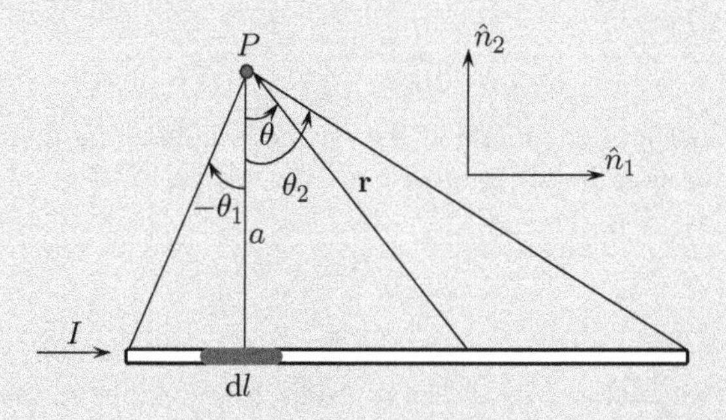


To solve this problem, we need to use the Biot–Savart law to find $d\mathbf{B}$ at point $a$. To this end, do the following:

1. Draw a vector $\mathbf{r}$ from any point on the wire to point $P$. Define the angle between the vector $\mathbf{r}$ and the straight vertical line $a$ as $\theta$.
2. Show that $\mathbf{dl} \times \mathbf{r} = \mathrm{d}l \cos\theta \hat{n}_3$.
3. Show that $\mathrm{d}l = a \sec^2\theta \mathrm{d}\theta$.
4. Substitute back into the Biot–Savart law, and show that

$$\mathbf{B} = \frac{\mu I}{4\pi a} \int \cos\theta \mathrm{d}\theta \hat{n}_3.$$

5. Depending on the location of point $P$, you can define the limits of the integral as $-\theta_1$, $\theta_2$, as shown in the figure. Show that $\mathbf{B}$ reduces to

$$\mathbf{B} = \frac{\mu I}{4\pi a}(\sin\theta_2 + \sin\theta_1)\hat{n}_3.$$

### 8.2.4 Ampere's Law

In the previous section, we showed, using the Biot–Savart law, that the magnetic field induced at a point $P$ due to the motion of a current $I$, in an infinite wire is given by $B = \frac{\mu I}{2\pi R}$, where $R$

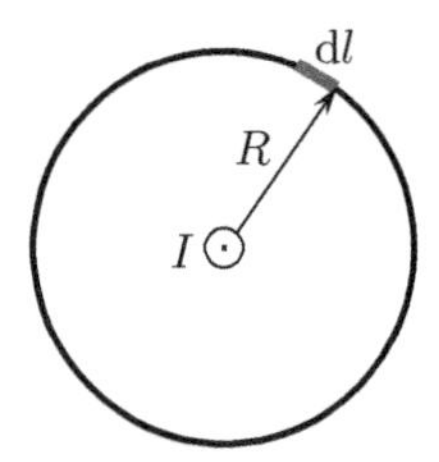


**Figure 8.30** Magnetic field along a closed circle of radius $R$.

is the vertical distance between the wire and point $P$. This result implies that the magnetic field is invariant on a circle of radius $R$ around the wire. Based on this understanding (Figure 8.30), we can write

$$\oint \mathbf{B}.\mathrm{d}\mathbf{l} = 2\pi RB, \tag{8.28}$$

where $B$ is the constant magnitude of the magnetic field along a circle of radius $R$. Using Equation (8.27) for an infinitely long wire, we can write $2\pi RB = \mu I$. This yields

$$\oint \mathbf{B}.\mathrm{d}\mathbf{l} = \mu I. \tag{8.29}$$

Equation (8.29) states that the integral of the magnetic field along a closed loop is proportional to the current enclosed by the loop. André-Marie Ampère was the first to realize that this is not only valid for a circular path but is equivalently valid for any closed loop enclosing a current $I$, provided that the electric field inside the loop is time invariant; that is, $\frac{\mathrm{d}\mathbf{E}}{\mathrm{d}t} = 0$. For instance, for the system shown in Figure 8.31, we can write

$$\oint \mathbf{B}.\mathrm{d}\mathbf{l} = \mu(I_1 - I_2 - I_3). \tag{8.30}$$

As such, in general, Ampere's law can be written as:

$$\boxed{\oint \mathbf{B}.\mathrm{d}\mathbf{l} = \mu I_{\text{enclosed}}} \tag{8.31}$$

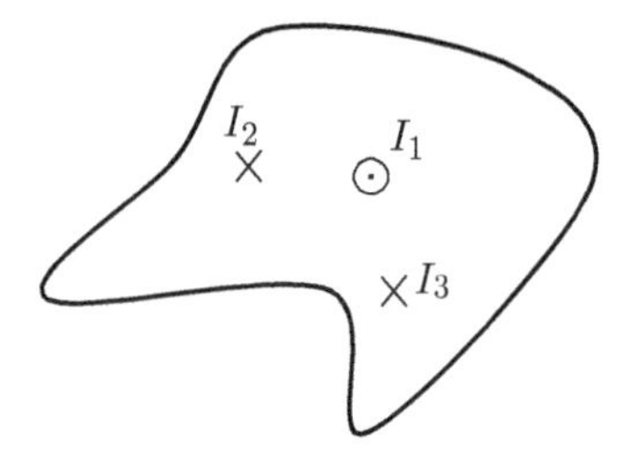


**Figure 8.31** Magnetic field along a closed loop enclosing a current.

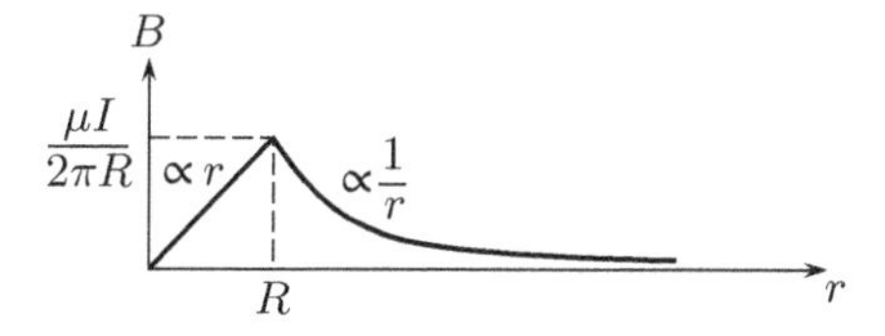


**Figure 8.32** Variation of the magnetic field induced by a cylindrical wire carrying a current $I$.

Note that Ampere's law share many similarities with Gauss's law for electric fields, as set out in Equation (8.6). In particular, when comparing Equation (8.31) to Equation (8.6), we can clearly see that the magnetic field in Ampere's law replaces the electric field in Gauss's law, and that the enclosed current replaces the enclosed charge in Gauss's law.

### Example 8.21 Magnetic Field inside a Wire

Consider an infinite wire of a cylindrical shape, of radius $R$ and carrying a current $I$. Find the magnetic field $B$ at radius $r < R$ measured from the center of the wire.

Using Ampere's law, we can write

$$\oint \mathbf{B}.\mathbf{dl} = \mu I_{\text{enclosed}},$$

where $I_{\text{enclosed}}$ is the current enclosed by the circle of radius $r$. Assume a constant current density in the wire can be written $I_{\text{enclosed}} = I(r^2/R^2)$. It follows that

$$2\pi r B = \mu I \frac{r^2}{R^2},$$

or

$$B = \frac{\mu I}{2\pi} \frac{r}{R^2}.$$

Figure 8.32 depicts variation of the magnetic field inside and outside a cylindrical wire of radius $R$. Note that the magnetic field increases linearly with $r$ inside the wire and is inversely proportional to $r$ outside the wire.

### Example 8.22 Magnetic Field inside a Solenoid

Calculate the magnetic field inside a solenoid of length $L$, radius $R$, and number of turns $N$, carrying a current $I$, as shown in Figure 8.33. Assume that $R \lll L$.

To solve this problem, we need to realize that the magnetic field inside a solenoid is nearly constant when the loops are very close together and $R \lll L$. It is also true that the magnetic field outside the solenoid is very small and can be neglected.

Taking a cross section of the solenoid, as shown in Figure 8.33, then choosing the closed loop $abcd$, we can use Ampere's law to write

$$\oint \mathbf{B}.\mathbf{dl} = \mu I_{\text{enclosed}},$$

$$\int_a^b \mathbf{B}.\mathbf{dl} + \int_b^c \mathbf{B}.\mathbf{dl} + \int_c^d \mathbf{B}.\mathbf{dl} + \int_d^a \mathbf{B}.\mathbf{dl} = \mu I_{\text{enclosed}},$$

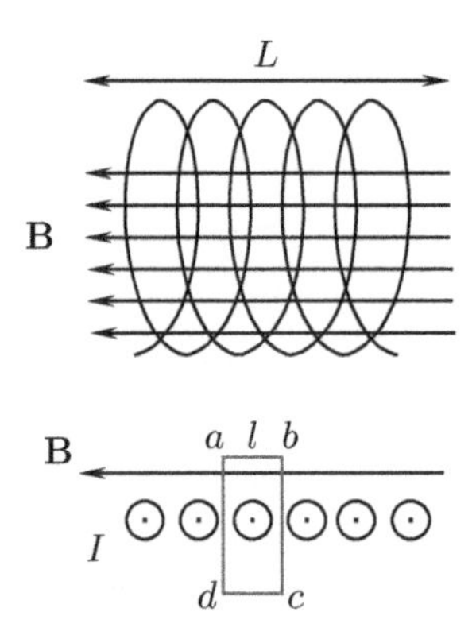


**Figure 8.33** Magnetic field inside a solenoid of length $L$ and carrying a current $I$.

Note that the second and fourth integrals are zero because **B** is perpendicular to $d\mathbf{l}$ along $bc$ and $da$. Furthermore, the third integral is zero because the magnetic field is negligible outside the solenoid. As such, we are left with

$$\int_a^b \mathbf{B}.\mathbf{dl} = \mu I_{\text{enclosed}},$$

$$B = \frac{\mu NI}{L}.$$

**Example 8.23 Conductor Moving in the Magnetic Field of a Stationary Conductor**

Consider the motion of a conductor, of mass $m$ and length $l$, in the magnetic field of another stationary conductor, of infinite length. The conductors are placed in the horizontal plane at a distance $r$ from each other. One of the conductors carries current $I_1$ while the other carries current $I_2$, as shown in Figure 8.34. The moving wire is connected to a spring of stiffness $k$. Find the equation of motion of the moving conductor.

To find the equation of motion, we first obtain the force exerted by the stationary conductor on the moving conductor. According to Lorentz's law, the force is given by

$$\mathbf{F} = I_2\mathbf{l}_2 \times \mathbf{B},$$

where $I_2$ is the magnitude of the current in the moving wire, $\mathbf{l}_2 = -l\hat{n}_2$, and **B** is the magnetic field generated by the stationary wire. As per the Biot–Savart law, the magnetic field in the plane shown at a distance $r$ from the stationary wire is given by

$$\mathbf{B} = \frac{\mu I_1}{2\pi r}\hat{n}_3,$$

As such, the force can be written as:

$$\mathbf{F} = -I_2 l\hat{n}_2 \times \frac{\mu I_1}{2\pi r}\hat{n}_3 = \frac{-\mu l I_1 I_2}{2\pi r}\hat{n}_1.$$

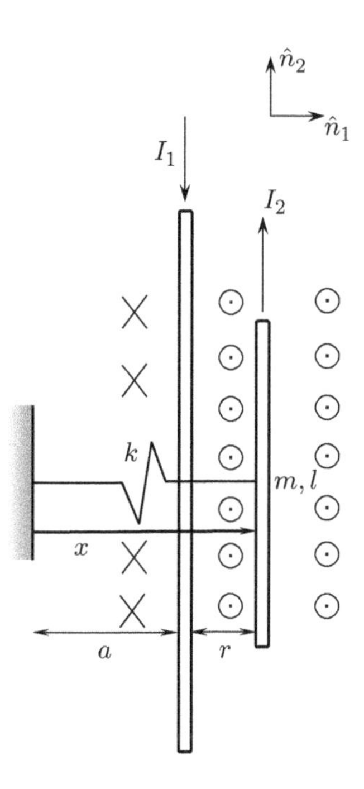


**Figure 8.34** Motion of a finite conductor in the magnetic field of an infinite conductor.

Next, we apply Newton's second law on the moving wire and obtain

$$m\ddot{x} + kx + \frac{\mu l I_1 I_2}{2\pi(x-a)} = 0,$$

where $x$ is the position of the moving wire as measured with respect to an inertial point.

### Flipped Classroom Exercise 8.6

Find the equation of motion of a particle of mass $m$ and charge $q$ oscillating in the magnetic field of an infinite wire carrying a current $I$. The particle is connected through an elastic link of stiffness $k$ and unstretched length $l_0$ to point $O$. Assume that motion occurs in the horizontal plane.

To solve this problem, take the following steps:

1. Define the rotating $B$-frame, such that ${}^N\boldsymbol{\omega}^B = \dot{\theta}\hat{n}_3$.
2. Use the Biot–Savart equation to show that the magnetic field acting on the particle is equal to $\mathbf{B} = -\frac{\mu I}{4\pi(h-(l_0+r)\sin\theta)}\hat{n}_3$.
3. Use the Lorentz equation to determine the magnetic force acting on the particle. Note that the magnetic force has two components: in the $\hat{b}_1$ and $\hat{b}_2$ directions.

4. Find the acceleration of the particle ${}^{N}\mathbf{a}^{P/O}$.
5. Apply Newton's second law to find the equations of motion of the particle.

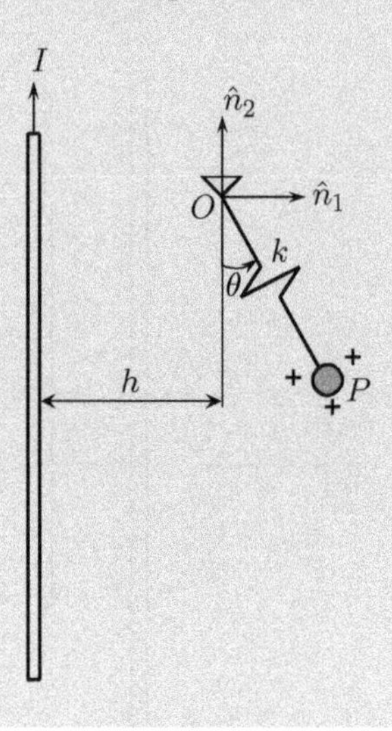


## 8.2.5 Faraday's Law of Induction

In 1831, 12 years after Ørsted illustrated that a current-carrying conductor produces a magnetic field, Michael Faraday investigated whether a magnetic field can produce a current in a coil. To this end, he experimented with a coil placed adjacent to a solenoid. He noticed that when the magnetic field was steady, no current flowed in the coil. However, whenever he opened or closed the switch in the circuit connected to the solenoid, a current would flow into the adjacent coil. He concluded that an electric field can be induced by a change in the magnetic field, a phenomenon known today as electromagnetic induction [5].

The direction of the resulting current in the wire loop is such that it produces a magnetic field that opposes the change in the external magnetic field. In other words, as shown in Figure 8.35, when the magnetic field increases, a current flows into the loop in a counter-clockwise direction such that it produces a magnetic field that opposes the increase in the external field. On the other hand, when the magnet moves away from the loop, the current flows in the clockwise direction because it must produce a magnetic field that compensates for the reduction in the external field. The process of finding the direction of the current is due to the Russian physicist Emil Lentz, and is known as Lentz's law [6].

Since the change of magnetic flux generates a current in a closed loop then there must be a potential difference across the loop to allow the current to flow. This potential difference, $\mathcal{E}$, is known as the back electromotive force (EMF) and is given by

$$\mathcal{E} = -\frac{d\Phi}{dt} \tag{8.32}$$

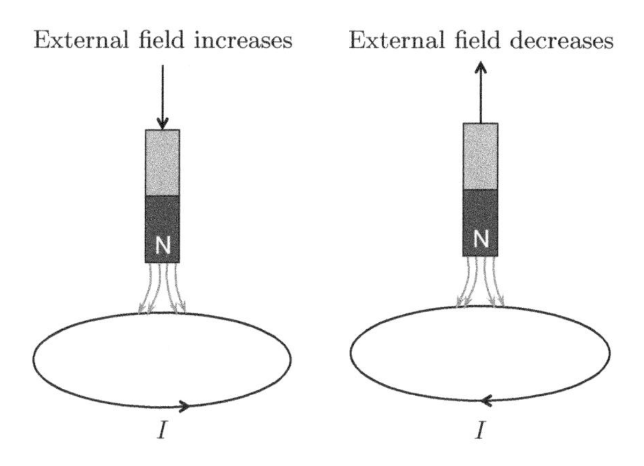


**Figure 8.35** Direction of the current induced in a wire loop as result of the change of magnetic field.

where $\Phi$ is known as the magnetic flux and is defined as

$$\Phi = \iint_{\sigma(t)} \mathbf{B}(\mathbf{r}, t).\mathrm{d}\mathbf{A}. \tag{8.33}$$

Here, $\mathbf{B}(\mathbf{r}, t)$ is the magnetic field, $\sigma(t)$ is the surface enclosed by the wire loop (in general it can be time varying), and $\mathrm{d}\mathbf{A}$ is a differential element of area on the surface $\sigma(t)$. Note that the dot product implies that only magnetic field lines parallel to the normal to the surface contribute to the magnetic flux. Therefore, the back EMF is not only proportional to the rate of change of the magnetic field but also to the area enclosed by the wire loop.

To simplify our understanding of Equation (8.33), consider the flat constant surface $\sigma$ formed by the boundaries of a thin wire, as shown in Figure 8.36. A uniform magnetic field $\mathbf{B}$ intersects the surface at a constant angle $\theta$ relative to the normal to the surface $\mathrm{d}\mathbf{A}$. Since $\sigma(t)$ is constant and $\mathbf{B}$ is uniform in space and time, the integral in Equation (8.33) simplifies to

$$\Phi = \mathbf{B}.\mathbf{A} = BA\cos\theta. \tag{8.34}$$

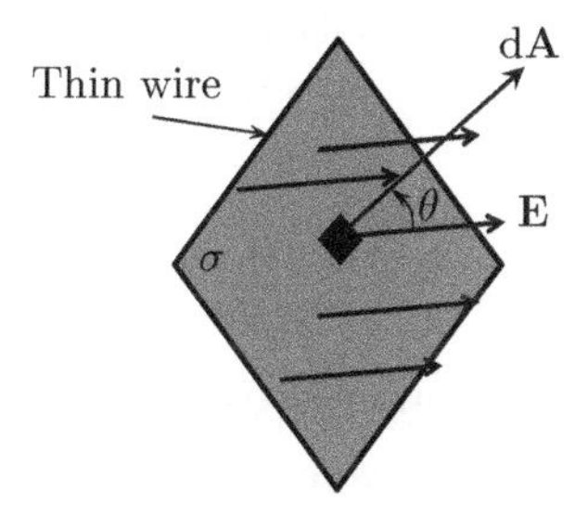


**Figure 8.36** Magnetic flux through a thin wire loop.

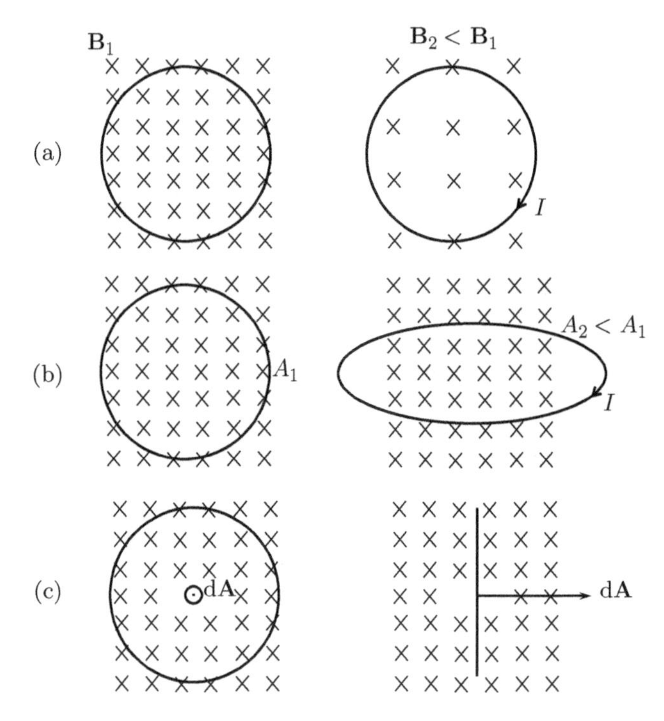


**Figure 8.37** Possible scenarios through which a current can be induced in a closed loop following changes in the magnetic flux.

For this simplified case, the back EMF can be written as

$$\mathcal{E} = -\frac{\mathrm{d}\Phi}{\mathrm{d}t} = -\left(\frac{\mathrm{d}B}{\mathrm{d}t}A\cos\theta + B\frac{\mathrm{d}A}{\mathrm{d}t}\cos\theta - BA\sin\theta\frac{\mathrm{d}\theta}{\mathrm{d}t}\right). \tag{8.35}$$

It is evident from that a back EMF, and thereby a current, can flow in a closed loop when either

- the magnetic field changes, $\frac{\mathrm{d}B}{\mathrm{d}t} \neq 0$
- the area enclosed by the wire loop changes, $\frac{\mathrm{d}A}{\mathrm{d}t} \neq 0$
- the angle between the magnetic field and the normal to the area changes, $\frac{\mathrm{d}\theta}{\mathrm{d}t} \neq 0$
- or any combination of these.

Figure 8.37 demonstrates the three possible scenarios for current generation in a closed loop following changes in the magnetic flux.

### Example 8.24 Induction due to Change in Area

A square loop with sides of length $l$ is placed in a uniform magnetic field **B** that points into the page. The loop is pulled from its two edges and turned into a rhombus, as shown in Figure 8.38. Assuming that the total resistance of the loop is $R$, find the average induced current in the loop and its direction. Plot the variation of the current with $\psi$.

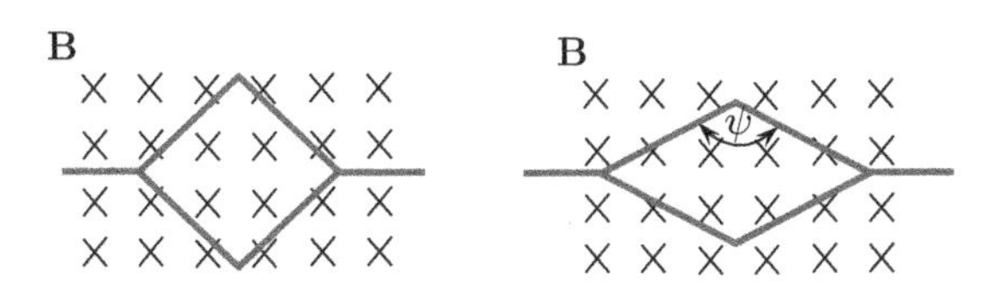


**Figure 8.38** Current induced in a variable-angle rhombus.

To find the current, we first find the magnetic flux through the loop

$$\Phi = \mathbf{B}.\mathbf{A} = BA\cos\theta.$$

Assuming $A$ is positive, pointing out of the page, then $\theta = \pi$ and $A = l^2 \sin\psi$. To find the back EMF, we differentiate the previous equation once with respect to time and obtain

$$\frac{\mathrm{d}\Phi}{\mathrm{d}t} = Bl^2 \cos\psi\dot{\psi}.$$

As the rhombus is pulled, $\psi$ increases, which means that $\dot{\psi} > 0$. Furthermore, since $0 \leq \psi \leq \pi$, then $\cos\psi \leq 0$, which implies that $\frac{\mathrm{d}\Phi}{\mathrm{d}t} < 0$ or that $\Phi$ is decreasing. As such, the direction of the current should be such that it produces a magnetic field that compensates for the reduction in the external field, so the current will be clockwise. The magnitude of the current is

$$I = \frac{\mathcal{E}}{R} = \frac{Bl^2}{R}\cos\psi\dot{\psi}.$$

It is interesting to note that the current goes to zero when $\psi = \frac{\pi}{2}$.

### Example 8.25 Power-generating Conductor

The conductor of mass $m$ and length $l$ shown in Figure 8.39 is subject to an initial velocity $\dot{x}(0) = v_0$ to the left, as shown. Find:

(a) the direction of the induced current in the circuit;
(b) the magnitude of the induced voltage;
(c) the equation of motion of the conductor;
(d) the velocity $v(t)$.

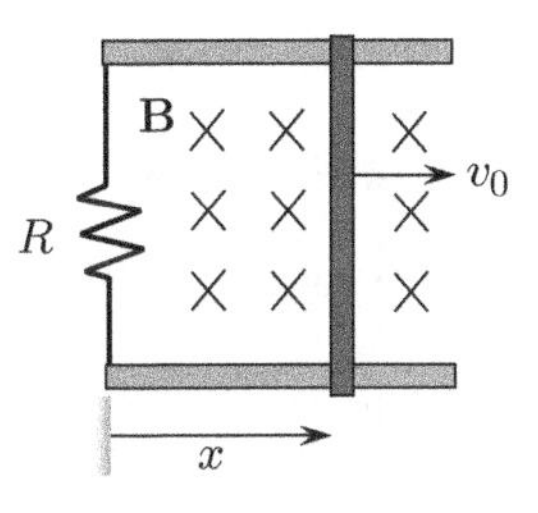


**Figure 8.39** Motion of a conductor in a magnetic field.

**(a)** ***Direction of the induced current*** We need to determine whether $\Phi$ is increasing or decreasing. To this end, we need to find the sign of $\frac{\mathrm{d}\Phi}{\mathrm{d}t}$. Since **B** and the angle between the normal to the area and **B** both remain unchanged, the change in area governs the change in the magnetic flux. Thus, we can write

$$\frac{\mathrm{d}\Phi}{\mathrm{d}t} = B\frac{\mathrm{d}A}{\mathrm{d}t}\cos\theta.$$

Assuming that the normal to the area points out of the page, then $\theta = \pi$ and $\cos\theta = -1$, which implies

$$\frac{\mathrm{d}\Phi}{\mathrm{d}t} = -B\frac{\mathrm{d}A}{\mathrm{d}t}.$$

Since the area is increasing, $\frac{\mathrm{d}A}{\mathrm{d}t} > 0$, when the wire moves in the direction of $x$, then $\frac{\mathrm{d}\Phi}{\mathrm{d}t} < 0$. This implies that $\Phi$ is decreasing. Therefore, the direction of the current is such that it compensates for the reduced flux. In other words, the current has to be in the counter-clockwise direction.

**(b)** ***Magnitude of the induced voltage*** The back EMF is given by

$$\mathcal{E} = -\frac{\mathrm{d}\Phi}{\mathrm{d}t} = B\frac{\mathrm{d}A}{\mathrm{d}t} = Bl\frac{\mathrm{d}x}{\mathrm{d}t}.$$

**(c)** ***Equation of motion*** We apply Newton's second law on the conductor and write

$$m\ddot{x} = -F,$$

where $F = BlI$ is the force on the conductor due to the applied field. For the circuit dynamics, we apply Kirchoff's law and obtain

$$-\mathcal{E} + IR = 0, \qquad I = \frac{Bl}{R}\frac{\mathrm{d}x}{\mathrm{d}t}.$$

Substituting the expression for $I$ in Newton's equation, yields

$$\ddot{x} = -\frac{B^2l^2}{mR}\dot{x}.$$

**(d)** ***Velocity*** Letting $v = \dot{x}$, we obtain

$$\dot{v} = -\frac{B^2l^2}{mR}v.$$

This is a first-order linear differential equation whose solution can be written as

$$v(t) = v_0\exp^{-\tau t},$$

where $\tau = \frac{B^2l^2}{mR}$. Thus, upon receiving the initial condition, the wire comes to rest exponentially under the influence of the magnetic force.

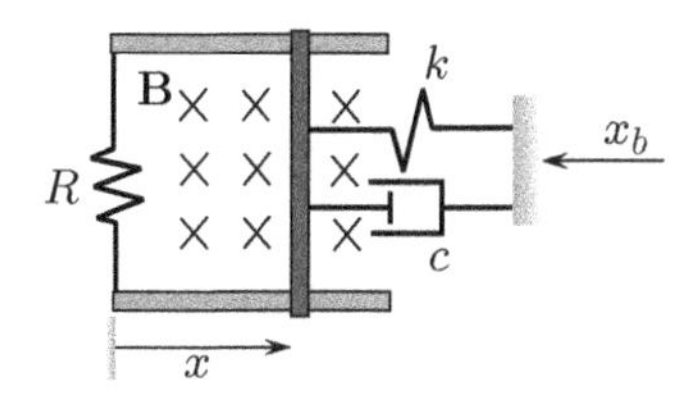


**Figure 8.40** A vibration-based electromagnetic energy harvester.

**Example 8.26 Vibration Energy Harvester**

Figure 8.40 is a schematic of an electromagnetic energy harvester, a tool to collect energy from external vibrations. It consists of a conductor of length $l$ and mass $m$ sliding on conducting rails. When the conductor is subject to an external base excitation $x_b$, it starts oscillating on top of the rails, resulting in a change in the loop area. Assuming that the total resistance of the conductor and rail is $R$ and that the inductance of the loop is negligible, derive a model for the dynamics of the harvester.

Assume that the normal to the area $A$ is pointing out of the page and that the wire deflection is positive to the right. Based on these assumptions, the direction of the current can be determined to be in the counter-clockwise direction. As such, the force acting on the conductor due to the magnetic field is to the left. Next, apply Newton's second law on the conductor and Kirchhoff's voltage law across the closed circuit loop to obtain

$$m\ddot{x} + c(\dot{x} + \dot{x}_b) + k(x + x_b) = -BlI, \qquad I = \frac{Bl}{R}\dot{x}.$$

Letting $x_r = x + x_b$, we obtain

$$m\ddot{x}_r + (c + \alpha)\dot{x}_r + kx_r = m\ddot{x}_b, \qquad I = \frac{Bl}{R}(\dot{x}_r - \dot{x}_b),$$

where $\alpha = \frac{B^2l^2}{R}$. Knowing the nature of the base excitation $x_b(t)$, the above equations can be solved for $x_r$ and $I$. The power harvested and dissipated in the load, $R$, can be obtained via $P = I^2R$.

 **Flipped Classroom Exercise 8.7**

Find the equation of motion of a wire of mass $m$, sliding on an inclined surface, as shown in the figure. The wire, which is connected to a spring of stiffness $k$, slides on two conducting rails connected to a resistive load $R$.

To solve this problem, take the following steps:

1. Find the magnetic flux through the wire loop formed by the sliding wire.
2. Determine the direction and magnitude of the current in the loop.

3. Determine the direction and magnitude of the electromagnetic force acting on the wire.
4. Apply Newton's second law to the wire and Kirchoff's voltage law across the closed circuit loop to determine the equations of motion of the wire.

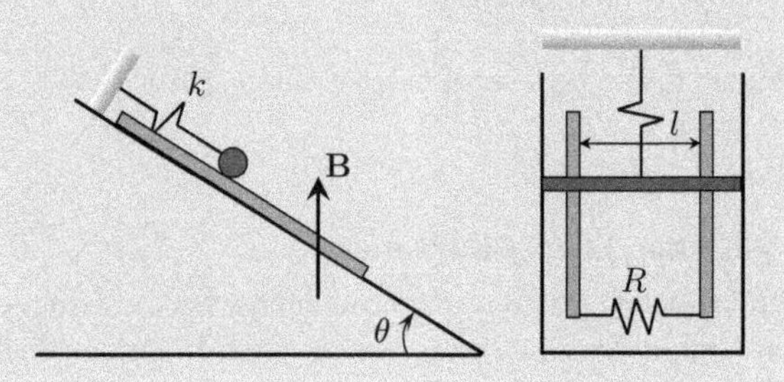


## 8.3 Lagrangian Formulation for Electrical Elements

In order to use Lagrange's approach to model the dynamics of electro-mechanical systems, it is essential to learn how to define the energy associated with electrical elements. The three electrical components that constantly appear in electrical systems are the resistor, capacitor, and inductor. If we choose the charge $q$ as our generalized coordinate, then one can describe Lagrange's equation for non-conservative electrical systems as follows:

$$\frac{\mathrm{d}}{\mathrm{d}t}\left(\frac{\partial \mathcal{L}}{\partial \dot{q}}\right) - \frac{\partial \mathcal{L}}{\partial q} + \frac{\partial \mathcal{D}}{\partial \dot{q}} = 0. \tag{8.36}$$

In what follows, we obtain the definition of the Lagrangian $\mathcal{L}$ and the dissipative function $\mathcal{D}$ for the three most common electrical elements.

### *8.3.1 Capacitor*

Much like a spring in a mechanical system, a capacitor is a device that stores electric potential energy as a charge in an electric circuit. The total charge stored in a capacitor is $q = C\mathcal{V}$, where $C$ is its capacitance and $\mathcal{V}$ is the voltage drop across it.

The energy stored in the capacitor can be obtained by using the energy stored in a differential charge and integrating it over the total charge; that is,

$$\mathrm{d}\mathcal{U}_e = \mathcal{V}\mathrm{d}q = \frac{q}{C}\mathrm{d}q. \tag{8.37}$$

Integrating the previous equation over the total charge, we obtain

$$\mathcal{U}_e = \frac{1}{2}\frac{q^2}{C}. \tag{8.38}$$

### 8.3.2 *Inductor*

The inductor is analogous to the mass in a mechanical system. The inductor serves to oppose any changes in the circuit current by transforming electric energy into a magnetic field or vice versa. Specifically, when the current passing through the inductor changes, it produces a voltage (a back EMF) across its terminals that opposes the change in current that created it.

The voltage drop $\mathcal{V}$, across the inductor, can be related to the current $I$ passing through it using

$$\mathcal{V} = L\frac{\mathrm{d}I}{\mathrm{d}t}, \tag{8.39}$$

where $L$ is the inductance. Since the inductor opposes any changes in the circuit current, a voltage source connected in series with the inductor must do work against the inductor to increase its terminal voltage. The work is given by

$$\mathcal{W} = \int \mathcal{P}\mathrm{d}t = \int \mathcal{V}I\mathrm{d}t = \int LI\frac{\mathrm{d}I}{\mathrm{d}t}\mathrm{d}t = \int LI\mathrm{d}I = \frac{1}{2}LI^2. \tag{8.40}$$

The work done by the source $\mathcal{W}$ is equal to the energy stored in the inductor $\mathcal{T}_e$, which in turn can be represented as a kinetic energy expression using Lagrange's equations. This is because the effect of the inductor on the circuit dynamics is analogous to that of the inertia in a mechanical system. It follows that the electric energy stored in the inductor is

$$\mathcal{T}_e = \frac{1}{2}LI^2 = \frac{1}{2}L\dot{q}^2. \tag{8.41}$$

### 8.3.3 *Resistor*

Much like a viscous damper in mechanical systems, a resistor is an energy dissipative element. It dissipates electrical energy as heat, causing a voltage drop across it. According to Ohm's law, the voltage drop $\mathcal{V}$ across a resistive element $R$ can be related to the current $I$ passing through it via

$$\mathcal{V} = IR. \tag{8.42}$$

To include the energy dissipated by the resistor in Lagrange's equation, we define the power dissipation function

$$\mathcal{D}_e = \frac{1}{2}R\dot{q}^2 \tag{8.43}$$

and treat the resistor in the same way we treat the viscous damper in mechanical systems.

**Example 8.27 Circuit Dynamics using Lagrange's Equation**

Use Lagrange's approach to find the equation governing the dynamics of the current in the series $RLC$ circuit shown in Figure 8.41.

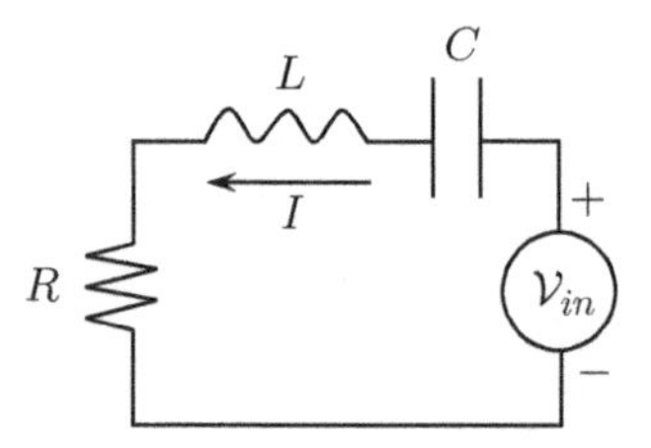


**Figure 8.41** Dynamics of the current $I$ in an $RLC$ circuit.

To find the equation of motion, we state the energy for each circuit element in terms of the charge $q$.

- The energy stored in the inductor is given by $\mathcal{T}_e = \frac{1}{2}L\dot{q}^2$
- The electrical potential energy stored in the capacitor is $\mathcal{U}_e = \frac{1}{2}\frac{q^2}{C}$
- The power dissipated by the resistor is $\mathcal{D}_e = \frac{1}{2}R\dot{q}^2$
- The energy generated by the source is $\mathcal{U}_s = -q\mathcal{V}_{in}(t)$.

Note the negative sign used to describe the energy generated by the source. This is because the source is doing work to generate the current. The Lagrangian is therefore, $\mathcal{L} = \mathcal{T}_e - \mathcal{U}_e - \mathcal{U}_s$. Using Lagrange's equation, we obtain

$$\frac{\mathrm{d}}{\mathrm{d}t}\left(\frac{\partial \mathcal{L}}{\partial \dot{q}}\right) - \frac{\partial \mathcal{L}}{\partial q} + \frac{\partial \mathcal{D}}{\partial \dot{q}} = 0,$$

$$L\ddot{q} + R\dot{q} + \frac{q}{C} = \mathcal{V}_{in}(t).$$

Expressing the previous equation in terms of $I$, we obtain

$$L\ddot{I} + R\dot{I} + \frac{1}{C}I = \dot{\mathcal{V}}_{in}(t).$$

### Example 8.28 Capacitive Microphone

A capacitive microphone consists of a moving charged plate placed at a distance $d$ from and parallel to a fixed plate (electrode) forming a capacitor, as shown in Figure 8.42. When the moving plate is subjected to acoustic waves, it vibrates and thereby changes the effective capacitance of the system. This process generates a current in the circuit and this can be used to amplify the sound applied. Find the equations of motion describing the deflection $x$ of the plate and the charge $q$ in the circuit using Lagrange's equation.

To find the equation of motion, we define our generalized coordinates as the deflection $x$ and the charge $q$. We then state the energy associated with each element in the system. The kinetic energy and kinetic energy-like terms are, respectively, due to the inertia of the moving plate and the electric energy stored in the inductor; that is

$$\mathcal{T} = \frac{1}{2}m\dot{x}^2 + \frac{1}{2}L\dot{q}^2.$$

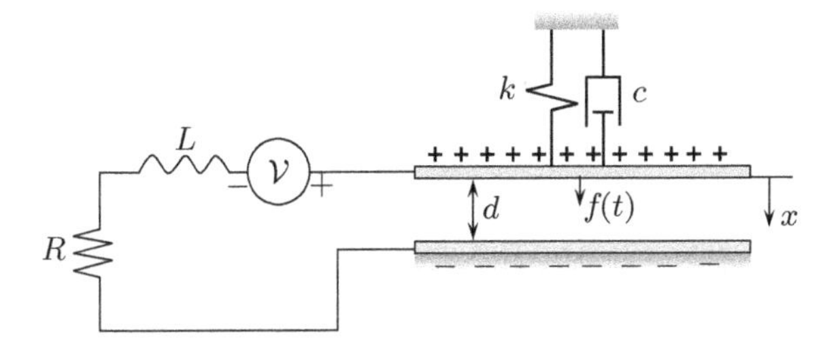


**Figure 8.42** Schematic of a capacitive microphone.

The total potential energy is due to the mechanical energy stored in the spring, the electric energy stored in the capacitor, and the energy generated by the voltage source; that is

$$\mathcal{U} = \frac{1}{2}kx^2 + \frac{1}{2C}q^2 - q\mathcal{V}(t),$$

where $C = \frac{\varepsilon A}{d-x}$. Here, $A$ is the area of the plates and $\varepsilon$ is the electric permittivity of the space between the capacitor plates.

The total dissipated power is due to mechanical and electrical losses, and is given by

$$\mathcal{D} = \frac{1}{2}c\dot{x}^2 + \frac{1}{2}R\dot{q}^2.$$

It follows that the Lagrangian can be written as

$$\mathcal{L} = \frac{1}{2}L\dot{q}^2 + \frac{1}{2}m\dot{x}^2 - \frac{1}{2}kx^2 - \frac{d-x}{2\varepsilon A}q^2 + q\mathcal{V}(t).$$

Using Lagrange's equation, we obtain the following equation for the mechanical sub-system:

$$\frac{\mathrm{d}}{\mathrm{d}t}\left(\frac{\partial \mathcal{L}}{\partial \dot{x}}\right) - \frac{\partial \mathcal{L}}{\partial x} + \frac{\partial \mathcal{D}}{\partial \dot{x}} = f(t)\hat{n}_2 \cdot \frac{d\dot{x}}{d\dot{x}}\hat{n}_2,$$

$$m\ddot{x} + c\dot{x} + kx - \frac{q^2}{2\varepsilon A} = f(t),$$

where the term on the right-hand side of the equation represents the generalized force due to the force $f(t)$. For the electrical sub-system, we can write

$$\frac{\mathrm{d}}{\mathrm{d}t}\left(\frac{\partial \mathcal{L}}{\partial \dot{q}}\right) - \frac{\partial \mathcal{L}}{\partial q} + \frac{\partial \mathcal{D}}{\partial \dot{q}} = 0,$$

$$L\ddot{q} + R\dot{q} + \frac{q(d-x)}{\varepsilon A} = \mathcal{V}(t).$$

**Summary**

In what follows, we summarize the main concepts of electromagnetism.

- *The Lorentz force*: The general force $\mathbf{F}$ acting on a charged particle $q$ moving under the influence of electric and magnetic fields can be written as

$$\mathbf{F}(\mathbf{r}, t) = q\mathbf{E}(\mathbf{r}, t) + q^N\mathbf{v}^{P/O} \times \mathbf{B}(\mathbf{r}, t),$$

where $\mathbf{r}$ is the position vector of the particle in space, $\mathbf{E}(\mathbf{r}, t)$ is the electric field, $\mathbf{B}(\mathbf{r}, t)$ is the magnetic field, and ${}^{N}\mathbf{v}^{P/O}$ is the velocity of the particle $P$ with respect to an inertial point $O$.

- *Force on a current-carrying conductor:* The force $\mathbf{F}$ acting on a current-carrying conductor placed in a magnetic field $\mathbf{B}$ is given by

$$\mathbf{F} = \int_0^l I\mathrm{d}\mathbf{l} \times \mathbf{B},$$

where $l$ is the length of the conductor, $\mathrm{d}\mathbf{l}$ is a differential vector in the direction of $l$, and $I$ is the magnitude of the current in the conductor. For a straight wire placed in a uniform magnetic field, $B$, the previous equation reduces to

$$\mathbf{F} = BlI \sin\theta.$$

- *Biot–Savart law:* The Biot–Savart law states that the differential magnetic field $\mathrm{d}\mathbf{B}$ at point $P$ induced by the motion of a current $I$ through a differential wire $\mathrm{d}\mathbf{l}$ is given by

$$\mathrm{d}\mathbf{B} = \frac{\mu}{4\pi}\frac{I\mathrm{d}\mathbf{l} \times \mathbf{r}}{\|\mathbf{r}\|^3} = \mathrm{d}\frac{\mu}{4\pi}\frac{I\mathrm{d}\mathbf{l} \times \hat{r}}{\|\mathbf{r}\|^2}.$$

- *Ampere's law:* Ampere's law states that the circulation of the induced magnetic field $\mathbf{B}$ about a closed loop enclosing a current $I_{\mathrm{enclosed}}$ is proportional to the current enclosed by the loop; that is

$$\oint \mathbf{B}.\mathrm{d}\mathbf{l} = \mu I_{\mathrm{enclosed}},$$

where $\mu$ is the magnetic permeability of the space containing the conductors. Note that this equation is valid only when the electric field inside the loop is time invariant; that is $\frac{\mathrm{d}\mathbf{E}}{\mathrm{d}t} = 0$.

- *Faraday's law of induction:* Any change in the magnetic flux $\Phi$ through a surface enclosed by a wire loop generates a current in the loop. The generated current in the loop is due to an induced voltage difference $\mathcal{E}$ across two points on the loop. The voltage difference is known as the back EMF and is related to the change of flux via

$$\mathcal{E} = -\frac{\mathrm{d}\Phi}{\mathrm{d}t}.$$

- *Lagrange's equation for electrical systems:* By using the charge $q$ as a generalized coordinate, one can utilize Lagrange's equation to obtain the equations governing the dynamics of an electrical system. To this end, we define the following functions to describe the energy stored and power dissipated in electrical elements:
  - Capacitor: The capacitor is an energy storage element which has an electric potential energy function $\mathcal{U}_e = \frac{1}{2C}q^2$, where $C$ is the capacitance.
  - Inductor: Much like an inertial mass in a mechanical system, the inductor transforms electric energy into magnetic field and vice versa. The electric energy stored in the inductor is therefore kinetic-like energy and is given by $\mathcal{T}_e = \frac{1}{2}L\dot{q}^2$, where $L$ is the inductance.

– Resistor: The resistor is an energy dissipative element for which we can define a power dissipation function of the form $\mathcal{D}_e = \frac{1}{2}R\dot{q}^2$, where $R$ is the resistance.

## 8.4 Maxwell's Equations

Everything we have learned so far in this chapter can be summarized into the four, widely-celebrated Maxwell equations. In this section, we will derive these equations based on the knowledge already gained in this chapter.

### *8.4.1 Maxwell's First Equation*

Maxwell's first equation is based on Gauss's law, which states that the electric flux through a closed surface is proportional to the charge enclosed by the surface; that is

$$\oint \mathbf{E}.\mathrm{d}\mathbf{A} = \sum_{i=1}^{\mathcal{N}} \frac{Q_i}{\varepsilon}, \tag{8.44}$$

where $\mathbf{E}$ is the electric field, $\mathrm{d}\mathbf{A}$ is a unit vector normal to the flux area, $Q_i$ are the charges enclosed by the surface, $\mathcal{N}$ is the number of charges, and $\varepsilon$ is the permittivity of the space. Assuming that the charge enclosed $Q_i$ is distributed over a volume $V$, we can write

$$Q_i = \int_V \rho \mathrm{d}V, \tag{8.45}$$

where $\rho$ is the charge per unit volume. It follows that

$$\oint \mathbf{E}.\mathrm{d}\mathbf{A} = \frac{1}{\varepsilon}\int_V \rho \mathrm{d}V, \tag{8.46}$$

Using Gauss's (divergence) theorem, the term on the left-hand side of the equation can be rewritten as

$$\oint \mathbf{E}.\mathrm{d}\mathbf{A} = \int_V \nabla.\mathbf{E}\mathrm{d}V, \tag{8.47}$$

Substituting the previous equation back into Equation (8.46) yields Maxwell's first equation.

$$\boxed{\nabla.\mathbf{E} = \frac{\rho}{\varepsilon}} \tag{8.48}$$

### *8.4.2 Maxwell's Second Equation*

Maxwell's second equation states that there are no magnetic monopoles. In other words, magnets can only exist as a dipole. As such, the total magnetic flux through a closed surface is zero because all the magnetic fields lines leaving the north pole of the magnet will come back and enter the south pole. Thus, we can write

$$\oint \mathbf{B}.\mathrm{d}\mathbf{A} = 0. \tag{8.49}$$

Using Gauss's divergence theorem, the previous equation can be simplified to

$$\int \nabla.\mathbf{B}\mathrm{d}V = 0, \tag{8.50}$$

or

$$\boxed{\nabla.\mathbf{B} = 0} \tag{8.51}$$

This is known as Maxwell's second equation.

### *8.4.3 Maxwell's Third Equation*

Maxwell's third equation originates from Faraday's law of induction, which states that the voltage $\mathcal{E}$ induced in a closed loop is proportional to the rate of change of the magnetic flux $\Phi$ that the loop encloses; that is,

$$\mathcal{E} = -\frac{\mathrm{d}\Phi}{\mathrm{d}t} = -\frac{\mathrm{d}}{\mathrm{d}t}\int_{\sigma} \mathbf{B}.\mathrm{d}\mathbf{A}, \tag{8.52}$$

where $\mathbf{B}$ is the magnetic field and $\sigma$ is the area enclosed by the loop. Using the common knowledge that the voltage induced in the loop is given by

$$\mathcal{E} = \oint \mathbf{E}.\mathrm{d}\mathbf{r}, \tag{8.53}$$

where $\mathbf{dr}$ is a differential element in the circuit loop, and using the previous two equations, we can write

$$\oint \mathbf{E}.\mathbf{dr} = -\frac{\mathrm{d}}{\mathrm{d}t}\int_{\sigma} \mathbf{B}.\mathrm{d}\mathbf{A}. \tag{8.54}$$

Using Stokes theorem, the integral on the left-hand side can be manipulated such that

$$\oint (\nabla \times \mathbf{E}).\mathbf{dA} = -\frac{\mathrm{d}}{\mathrm{d}t}\int_{\sigma} \mathbf{B}.\mathrm{d}\mathbf{A}. \tag{8.55}$$

It follows that

$$\boxed{\nabla \times \mathbf{E} = -\frac{\mathrm{d}\mathbf{B}}{\mathrm{d}t}} \tag{8.56}$$

The previous equation is known as Maxwell's third equation.

### *8.4.4 Maxwell's Fourth Equation*

Maxwell's fourth equation originates from *Ampere's* law described in Section 8.2.4 which states that the circulation of the magnetic field along a closed loop is proportional to the electric current enclosed by the loop; i.e.

$$\oint \mathbf{B}.\mathrm{d}\mathbf{l} = \mu I_{\text{enclosed}}. \tag{8.57}$$

Defining the enclosed current in terms of the electric current density, **J**, per unit area enclosed by the loop, we can write

$$I_{\text{enclosed}} = \int_{\sigma} \mathbf{J}.\mathrm{d}\mathbf{A}. \tag{8.58}$$

Applying Stokes' theorem on the left-hand side of Ampere's law, we can write

$$\int_{\sigma} (\nabla \times \mathbf{B}).\mathrm{d}\mathbf{A} = \mu \int_{\sigma} \mathbf{J}.\mathrm{d}\mathbf{A}. \tag{8.59}$$

It follows that

$$\nabla \times \mathbf{B} = \mu \mathbf{J}. \tag{8.60}$$

This equation has a major issue, which only becomes apparent when taking its divergence to obtain

$$\begin{aligned} \nabla.(\nabla \times \mathbf{B}) &= \mu \nabla.\mathbf{J}, \\ \nabla.\mathbf{J} &= 0. \end{aligned} \tag{8.61}$$

When the divergence of **J** is zero, the electric current flowing into any region is always equal to the electric current flowing out of the region; in other words there is no divergence. This seems reasonable until we place a capacitor in a closed circuit loop. A capacitor in its simple form consists of two parallel plates separated by air. As such, in principle, a capacitor should not allow a current to pass through it. This means that the divergence of **J** across the capacitor plates should not be zero because whatever current is entering the positive side of the capacitor plate should not be able to pass through the air barrier. Nevertheless, we have learned in circuit theory that a time-varying current does indeed flow through the capacitor.

In 1865, Maxwell realized that that current can flow through the capacitor as a result of the change of the electric field inside the capacitor. Therefore, he amended the right-hand side of Equation (8.60) as:

$$\boxed{\nabla \times \mathbf{B} = \mu \left( \mathbf{J} + \varepsilon \frac{\mathrm{d}\mathbf{E}}{\mathrm{d}t} \right)} \tag{8.62}$$

where the added term reflects currents generated due to the change in the electric field.

In summary, Maxwell's fourth equation states that the circulation of the magnetic field about a closed loop is proportional to the electric current enclosed by it, plus currents induced due to the time rate of change of the electric field that the loop encloses.

## 8.5 Lagrangian Formulation of the Lorentz Force

The reason for introducing Maxwell's four equations was not only to summarize theories of electromagnetism in an elegant way, but also to permit the derivation of a Lagrangian formulation of the Lorentz force, which takes the general form

$$\mathbf{F}(\mathbf{r}, t) = q\mathbf{E}(\mathbf{r}, t) + q^{N}\mathbf{v}^{P/O} \times \mathbf{B}(\mathbf{r}, t). \tag{8.63}$$

In order to derive a potential energy expression to represent the Lorentz force, we should be able to write the force as the negative gradient of some scalar potential function $\mathcal{U}_{\text{em}}$; that is,

$$\mathbf{F}(\mathbf{r}, t) = -\nabla \mathcal{U}_{\text{em}}. \tag{8.64}$$

This can be easily done for the electrostatic component of the force, because the electric field is simply the gradient of the electric potential, which is a scalar quantity: $\mathbf{E} = -\nabla \mathcal{V}$. On the other hand, the magnetic component of the force cannot be easily described in terms of the gradient of a scalar potential because it depends on the velocity of the charge ${}^{N}\mathbf{v}^{P/O}$.

In what follows, we present an analysis that uses Maxwell's equations to derive a Lagrangian associated with the Lorentz force.

By virtue of Maxwell's second equation, $\nabla.\mathbf{B} = 0$, we can express the magnetic field as the curl of some vectorial potential function $\mathbf{G}$; that is,

$$\mathbf{B} = \nabla \times \mathbf{G}, \tag{8.65}$$

Using Maxwell's third equation, we can also write

$$\nabla \times \mathbf{E} = -\frac{\partial}{\partial t}(\nabla \times \mathbf{G}), \tag{8.66}$$

or

$$\nabla \times \left(\mathbf{E} + \frac{\partial \mathbf{G}}{\partial t}\right) = 0. \tag{8.67}$$

This implies that we can represent the term in brackets as the negative gradient of some scalar potential function $\Psi$ such that

$$-\nabla \Psi = \left(\mathbf{E} + \frac{\partial \mathbf{G}}{\partial t}\right). \tag{8.68}$$

Using this to eliminate $\mathbf{E}$ from the Lorentz force, we obtain

$$\mathbf{F}(\mathbf{r}, t) = q\left(-\nabla \Psi - \frac{\partial \mathbf{G}}{\partial t} + {}^{N}\mathbf{v}^{P/O} \times (\nabla \times \mathbf{G})\right). \tag{8.69}$$

Next, we use the identity $\mathbf{A} \times (\mathbf{B} \times \mathbf{C}) = \mathbf{B}(\mathbf{A}.\mathbf{C}) - \mathbf{C}(\mathbf{A}.\mathbf{B})$ to simplify the last term of the equation and obtain

$$\mathbf{F}(\mathbf{r}, t) = q\left(-\nabla \Psi - \frac{\partial \mathbf{G}}{\partial t} + \nabla({}^{N}\mathbf{v}^{P/O}.\mathbf{G}) - \mathbf{G}({}^{N}\mathbf{v}^{P/O}.\nabla)\right). \tag{8.70}$$

Note that

$$\frac{\mathrm{d}\mathbf{G}}{\mathrm{d}t} = \frac{\partial \mathbf{G}}{\partial t} + \mathbf{G}({}^{N}\mathbf{v}^{P/O}.\nabla), \tag{8.71}$$

which further implies that

$$\mathbf{F} = q\left(-\nabla(\Psi - {}^N\mathbf{v}^{P/O}.\mathbf{G}) - \frac{\mathrm{d}\mathbf{G}}{\mathrm{d}t}\right). \tag{8.72}$$

Assuming the system has no constraints, we can use the Cartesian coordinates $(x_1, x_2, x_3)$ to replace the generalized coordinates. This allows the previous equation to be written in the Cartesian form:

$$F_i = \frac{\mathrm{d}}{\mathrm{d}t}(-qG_i) - q\frac{\partial}{\partial x_i}(\Psi - {}^N\mathbf{v}^{P/O}.\mathbf{G}), \qquad i = 1, 2, 3. \tag{8.73}$$

Next, we define a velocity-dependent potential function $\mathcal{U}_{\mathrm{em}}$ of the form

$$\mathcal{U}_{\mathrm{em}} = q\Psi - q^N\mathbf{v}^{P/O}.\mathbf{G} \tag{8.74}$$

and substitute it into Lagrange's equation to obtain the following electromagnetic force expression:

$$F_i = \frac{\mathrm{d}}{\mathrm{d}t}\left(\frac{\partial \mathcal{U}_{\mathrm{em}}}{\partial \dot{x}_i}\right) - \frac{\partial \mathcal{U}_{\mathrm{em}}}{\partial x_i}. \tag{8.75}$$

Substituting Equation (8.74) into Equation (8.75) leads to Equation (8.73). This implies that Equation (8.74) is indeed the actual velocity-dependent scalar potential of the Lorentz force.

**Example 8.29 Motion of a Particle in a Magnetic Field**

Use Lagrange's equations to find the general equation of motion for a particle in a uniform magnetic field **B**, where the vector field **G** is given by $\mathbf{G} = -\frac{1}{2}\mathbf{r} \times \mathrm{B}$.

We first note that, in the absence of electrostatic field, the electrostatic component $\Psi$ of the electromagnetic potential vanishes. Furthermore, for the assumed vector field $\mathbf{G} = -\frac{1}{2}\mathbf{r} \times \mathrm{B}$, we can use the established definition of the Lagrangian of a particle in a magnetic field to write

$$\mathcal{L} = \mathcal{T} - \mathcal{U}_{\mathrm{em}} = \frac{1}{2}m\dot{\mathbf{r}}^2 + q^N\mathbf{v}^{P/O}.\mathbf{G} = \frac{1}{2}m\dot{\mathbf{r}}^2 - \frac{1}{2}q.(\mathbf{r} \times \mathbf{B}).$$

Applying Lagrange's equation along the three Cartesian coordinates $(x_1, x_2, x_3)$, we can write

$$\frac{\mathrm{d}}{\mathrm{d}t}\left(\frac{\partial \mathcal{L}}{\partial \dot{x}_i}\right) - \frac{\partial \mathcal{L}}{\partial x_i} = 0, \qquad i = 1, 2, 3,$$

where

$$\frac{\mathrm{d}}{\mathrm{d}t}\left(\frac{\partial \mathcal{L}}{\partial \dot{x}_i}\right) = m\ddot{x}_i - \frac{q}{2}({}^N\mathbf{v}^{P/O} \times \mathbf{B})_i, \qquad i = 1, 2, 3,$$

$$\frac{\partial \mathcal{L}}{\partial x_i} = \frac{q}{2}(\mathbf{v}^{P/O} \times \mathbf{B})_i, \qquad i = 1, 2, 3.$$

It follows that

$$m\ddot{x}_i = q(\mathbf{v}^{P/O} \times \mathbf{B})_i, \qquad i = 1, 2, 3,$$

or

$$m\ddot{\mathbf{r}} = q(\mathbf{v}^{P/O} \times \mathbf{B}).$$

This implies that the mass times the acceleration of the particle is equal to the magnetic component of the Lorentz force, a fact that would have been easily realized when using Newton's second law of dynamics.

**Example 8.30 Motion of a Conductor in a Magnetic Field**

A conductor of mass $m$ and length $l$, as shown Figure 8.43, is immersed in a constant uniform magnetic field $\mathbf{B} = -B_0\hat{n}_3$. The motion of the wire is constrained by a spring of stiffness $k$ and a linear viscous damper of damping coefficient $c$. Find the equation of motion of the wire and the current in the circuit using Lagrange's approach.

To use Lagrange's equation, we need first to define all the energy fields present in the system using the two generalized coordinates, namely $x$ (wire displacement) and $q$ (charge). The kinetic energy $\mathcal{T}$ is only due to the motion of the wire and is given by $\mathcal{T} = \frac{1}{2}m\dot{x}^2$. There are two dissipative energy fields: the first is due to the mechanical viscous damping $\mathcal{D}_{\mathrm{m}} = \frac{1}{2}c\dot{x}^2$ and the second is due to the electrical damping in the resistor $\mathcal{D}_{\mathrm{e}} = \frac{1}{2}R\dot{q}^2$. There are also two potential energy storing elements: the spring $\mathcal{U}_{\mathrm{m}} = \frac{1}{2}kx^2$ and the electromagnetic field $\mathcal{U}_{\mathrm{em}}$.

For a current-carrying conductor, the potential energy stored in the electromagnetic field can be written as

$$\mathcal{U}_{\mathrm{em}} = l\mathbf{I}.\mathbf{G}.$$

This follows directly from Equation (8.74). Since the magnetic field is pointing in the $\hat{n}_3$-direction, the correct choice of the vector potential is $\mathbf{G} = B_0x\hat{n}_2$. It follows that

$$\mathcal{U}_{\mathrm{em}} = l(I\hat{n}_2).(B_0x\hat{n}_2) = B_0lIx = Bl\dot{q}x.$$

Next, we express the Lagrangian as

$$\mathcal{L} = \mathcal{T} - \mathcal{U} = \frac{1}{2}m\dot{x}^2 - \frac{1}{2}kx^2 + Bl\dot{q}x$$

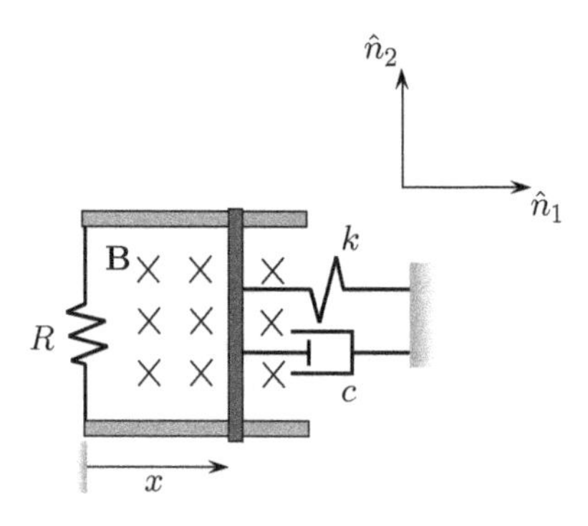


**Figure 8.43** Motion of current-carrying conductor in a magnetic field.

The total power dissipation is given by

$$\mathcal{D} = \frac{1}{2}c\dot{x}^2 + \frac{1}{2}R\dot{q}^2.$$

The electro-mechanical equations can be obtained using Lagrange's equations as

$$\frac{\mathrm{d}}{\mathrm{d}t}\left(\frac{\partial \mathcal{L}}{\partial \dot{x}}\right) - \frac{\partial \mathcal{L}}{\partial x} + \frac{\partial \mathcal{D}}{\partial \dot{x}} = 0,$$

$$m\ddot{x} + c\dot{x} + kx = -BlI,$$

$$\frac{\mathrm{d}}{\mathrm{d}t}\left(\frac{\partial \mathcal{L}}{\partial \dot{q}}\right) - \frac{\partial \mathcal{L}}{\partial q} + \frac{\partial \mathcal{D}}{\partial \dot{q}} = 0,$$

$$-Bl\dot{x} + R\dot{q} = 0, \qquad I = \frac{Bl\dot{x}}{R}.$$

## Exercises

8.1 A cyclotron is a device used to accelerate electrons by applying magnetic and electric fields. The device works by emitting an electron with initial velocity $v_0$ in an electric field $E_0$, as shown in Figure 8.44. The electron accelerates under the influence of the electric field up to a point where it enters a $D$-shaped region in which a uniform magnetic field $B_0\hat{n}_3$ is applied. The field serves to move the electron along a circular path down to the point where it re-enters the electric field. At this point, the electric field reverses direction and accelerates the particle again. The process continues until the electron reaches the desired speed. Find the frequency at which the electric field needs to switch direction to achieve the desired effect.

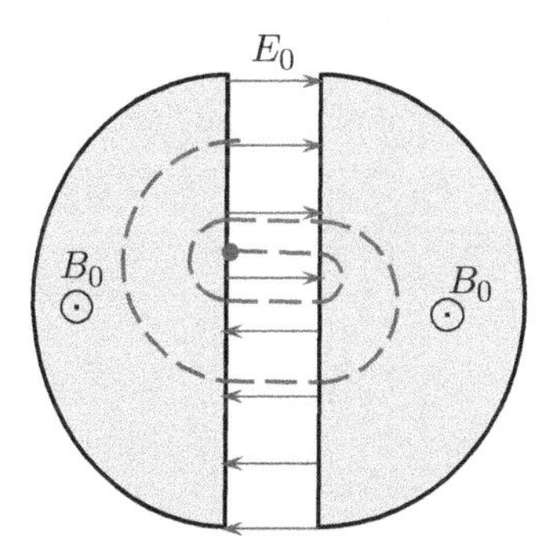


**Figure 8.44** Exercise 8.1.

8.2 We want to study the motion of an electron of mass $m$ and charge $q$ in the space between a pair of coaxial cylindrical conductors (cylindrical capacitance) as shown in Figure 8.45. The space between the cylinders is evacuated. The radius of the inner cylinder is $a$, and the radius of the outer cylinder is $b$. A static homogeneous magnetic field $B_0$ is applied parallel to the cylinder axis, directed out of the plane of the figure.

An electron is released from the surface of the inner cylinder with an initial velocity $v_0$ in the radial direction. For magnetic fields larger than a critical value, $B_{cr}$, the electron will not reach the outer cylinder. Show that $B_{cr} = \frac{2bmv_0}{q(b^2-a^2)}$.

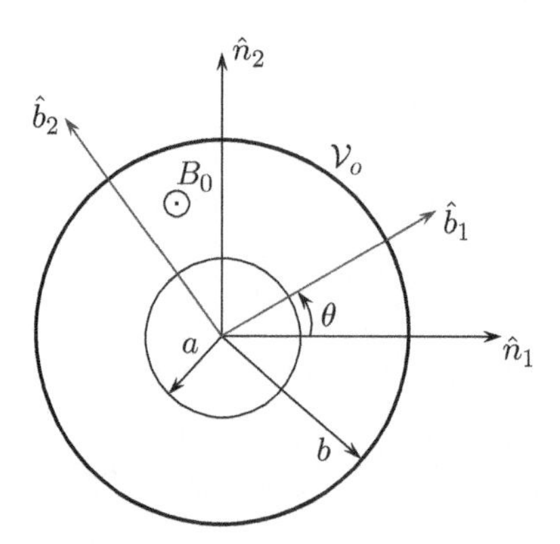


**Figure 8.45** Exercise 8.2.

8.3 For the setup discussed in the Exercise 8.2, we apply a voltage $\mathcal{V}_0$ at the outer cylinder in addition to the already existing uniform magnetic field $B_0\hat{n}_3$. The electron is released with negligible velocity in the radial direction. Show that the quantity $L - \frac{1}{2}qB_0^2r^2$ is conserved where $L$ is the angular momentum of the particle.

Again, for magnetic fields larger than a critical value $B_{cr}$, the electron will not reach the outer cylinder. Show that, in the presence of the voltage $\mathcal{V}_0$, the critical magnetic field above which the electron does not reach the outer cylinder is given by $B_{cr} = \frac{2b}{(b^2-a^2)}\sqrt{\frac{2m\mathcal{V}_0}{q}}$.

8.4 A particle of mass $m$ and charge $q$ is suspended on a string of length $l$, which is fixed at point $O$ (Figure 8.46). Under the point of suspension, there is an infinite plane grounded conductor. When $\theta = 0$, the distance between the particle and ground is $d$. Assuming that oscillation occurs in the horizontal place, find the equation of motion of the pendulum.

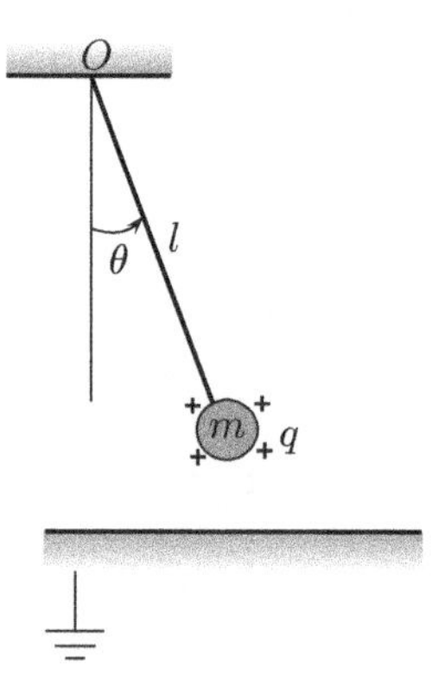


**Figure 8.46** Exercise 8.4.

8.5 A beam of protons enters a uniform magnetic field $B$ at a velocity $v_0$, making an angle $\theta$ with the field (Figure 8.47). Show that the beam will move in a helical path in the field with a radius $R_h = \frac{mv_0 \sin\theta}{qB}$ and a pitch $P_h = \frac{2\pi m v_0 \cos\theta}{qB}$.

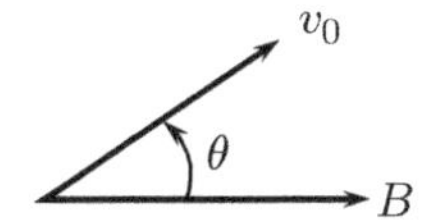


**Figure 8.47** Exercise 8.5.

8.6 A charged particle of mass $m$ and charge $q$ moves in a sector of magnetic field of uniform density $B$ and width $d$, as shown in Figure 8.48. Show that the angle $\theta$ at which the particle exits the magnetic field is given by $\theta = \arcsin \frac{qbd}{mv}$.

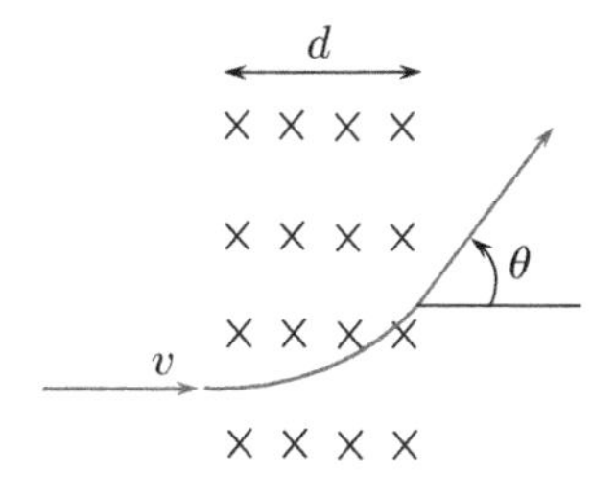


**Figure 8.48** Exercise 8.6.

8.7 A charged particle of mass $m$ and charge $q$ is accelerated by a potential voltage $\mathcal{V}_0$ and enters a magnetic field at an angle $\theta$, as shown in Figure 8.49. Find the time it will take the particle to exit the field.

**Figure 8.49** Exercise 8.7.

8.8 Two long parallel wires each of mass $m$ were initially placed adjacent to each other. As shown in Figure 8.50, the wires are suspended by two cables of length $l$. What happens when currents of equal magnitude and opposite direction flow into the wires? Find their equations of motion.

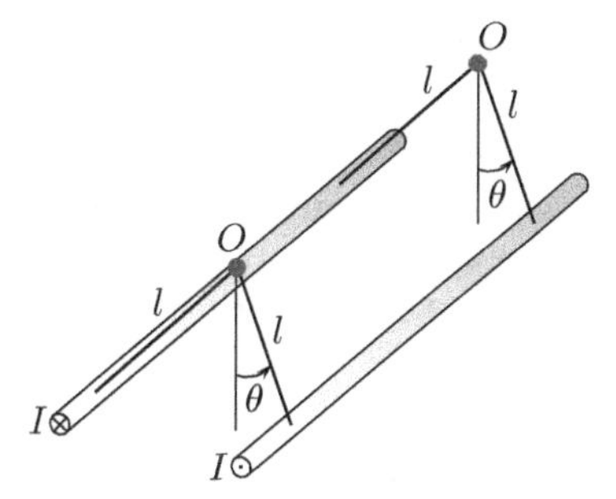


**Figure 8.50** Exercise 8.8.

8.9 Consider the motion of a conductor strip, of mass $m$, length $l$, and width $d$, in the magnetic field of another stationary conductor of infinite length, as shown in Figure 8.51. The conductors are placed in the horizontal plane at a distance $r$. One of the conductors carries current $I_1$ while the other carries current $I_2$. The moving strip is connected to a spring of stiffness $k$. Find the equation of motion of the strip.

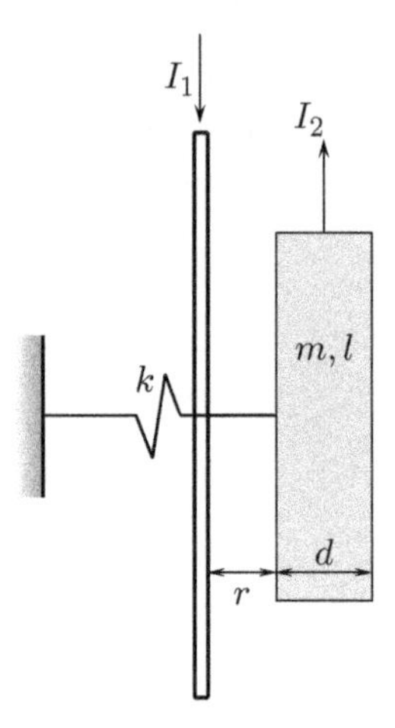


**Figure 8.51** Exercise 8.9.

8.10 A horizontal wire of mass $m$ and length $l$ is placed on top of two conductors that are connected to a circuit, as shown in Figure 8.52. The switch in the circuit is closed for a small time $\delta t$. Find the maximum height that the wire jumps.

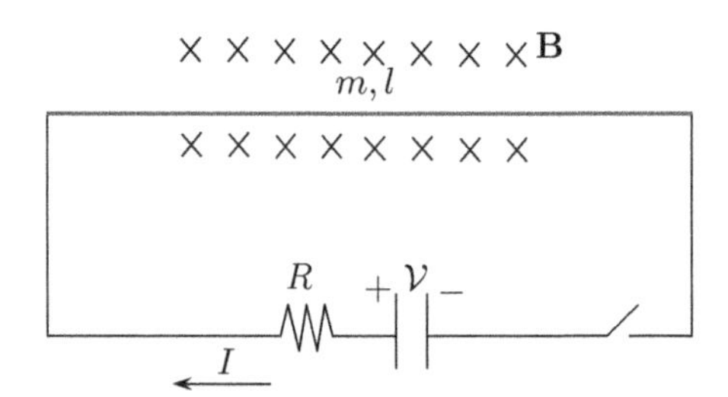


**Figure 8.52** Exercise 8.10.

8.11 A horizontal wire of mass $m$ and length $l$ is suspended by two massless rods, each of length $d$. The rods are hinged such that they are allowed to rotate around the axis shown in Figure 8.53. The whole setup is placed at a distance $r$ from a wire of infinite length. Find the equation of motion of the rod when it is given a small initial angular velocity about the axis of rotation.

8.12 A rectangular loop of wire of mass $m$, width $w$, length $l$, and resistance $R$ is moving in and out of a uniform magnetic field $-B_0\hat{n}_3$, as shown in Figure 8.54. Find the equation of motion of the loop and the equation governing the current in the circuit.

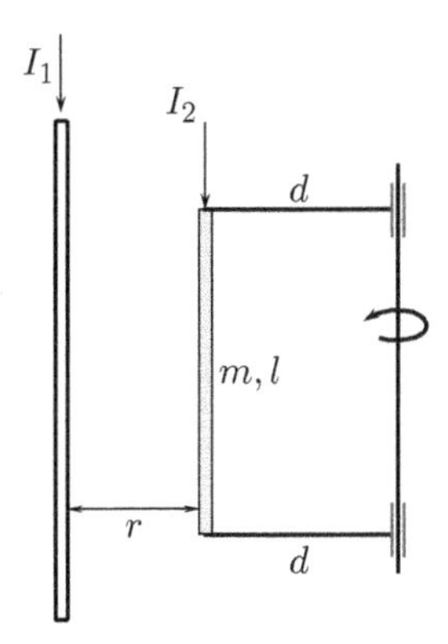


**Figure 8.53** Exercise 8.11.

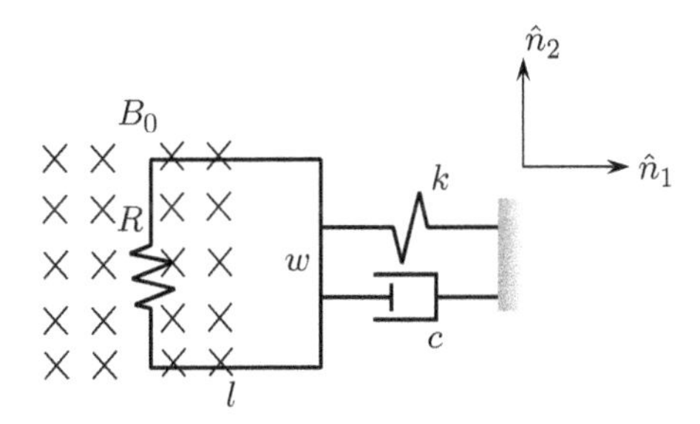


**Figure 8.54** Exercise 8.12.

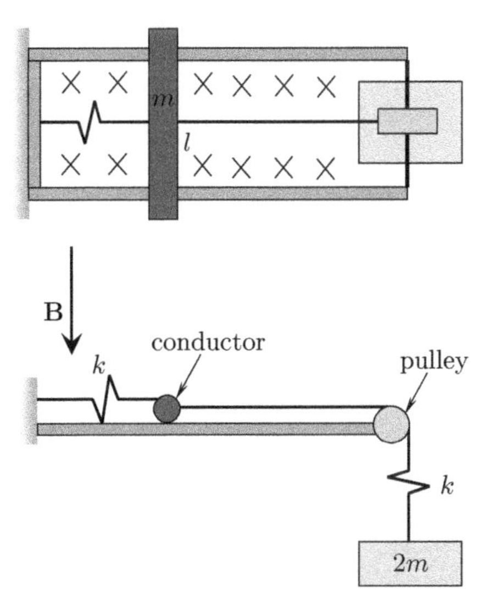


**Figure 8.55** Exercise 8.13.

8.13 A conducting bar of mass $m$ and resistance $R$ is pulled in the horizontal direction across two frictionless parallel rails a distance $l$ apart by an elastic string. The string passes over a frictionless pulley and is connected to a block of mass $2m$, as shown in Figure 8.55. A uniform magnetic field is applied vertically downward. Find the equation of motion of the bar and the current in the closed circuit loop.

8.14 Use Lagrange's equation to find the equation governing the voltage dynamics for the parallel $RLC$ circuit shown in Figure 8.56.

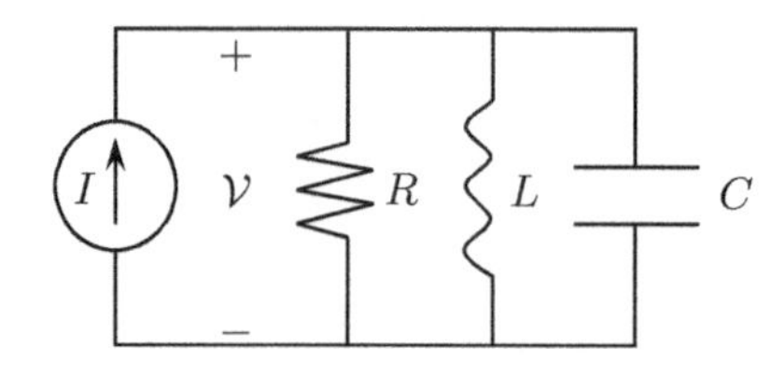


**Figure 8.56** Exercise 8.14.

## References

1. Coulomb, C.A. (1788) *Premier mémoire sur l'électricité et le magnétisme, Histoire de l'Académie Royale des Sciences*, 1788, 569–577.
2. Gauss, C.F. (1813) *Theoria Attractionis Corporum Sphaeroidicorum Ellipticorum Homogeneorum Methodo Nova Tractata*, Ch. 55, pp. 545–552.
3. Dahl P.F. (1997) *Flash of the Cathode Rays: A History of J.J. Thomson's Electron*, CRC Press.
4. Ørsted H.C. (1997) *Selected Scientific Works of Hans Christian Oersted*, trans. Jelved K., Jackson A.D., and Knudsen O., Princeton University Press.
5. Ulaby F. (2007) *Fundamentals of Applied Electromagnetics*, 5th edn. Pearson: Prentice Hall.
6. Lentz E. (1834) "Ueber die Bestimmung der Richtung der durch Elektodynamische Vertheilung Erregten Galvanischen Ströme", *Annalen der Physik und Chemie*, 107 (31), 483–494.

# 9

# Introduction to Analysis Tools

Chapters 2–8 focused on a single objective: finding the equations governing the motion of a system consisting of particles and rigid bodies, a process commonly known as *mathematical modeling*. Nonetheless, while models constitute the fundamental basis upon which everything else is built, the process of fully understanding and controlling the dynamic behavior of a system cannot be fully realized without analyzing its response, a process commonly known as *mathematical analysis*.

As you may have noticed in the previous chapters, models of particles and rigid bodies in motion are described by one or more non-linear ordinary differential equations. As such, analyzing the dynamic behavior of a system of particles and/or rigid bodies necessitates learning the tools commonly used to analyze non-linear differential equations. This subject will be the core of this chapter. The analysis of non-linear differential equations can be divided into local and global categories. In local analysis, we inspect the dynamics of the system in the vicinity of some important points where it is possible to implement analytical tools. In a global analysis, we try to draw a global picture of the dynamics using numerical techniques.

The reader should bear in mind that this chapter is only introductory in nature; the analysis of non-linear differential equations is a wide subject, and cannot be covered in a single chapter. For more details, interested readers should consult the book by Nayfeh and Balachandran [1] or that by Strogatz [2].

## 9.1 Basic Definitions

A system of ordinary differential equations has one or more dependent variables and a single independent variable. The dependent variables are called dependent because their evolution depends on some other variables, while the independent variables are those whose evolution does not depend on anything else. The process of solving a non-linear differential

*Dynamics of Particles and Rigid Bodies: A Self-Learning Approach*, First Edition. Mohammed F. Daqaq.

Companion website: www.wiley.com/go/daqaq/dynamics

equation entails finding how the dependent variable evolves with the independent variable. For example, in the equation of a simple pendulum

$$\ddot{\theta} + \frac{g}{l}\sin\theta = 0, \tag{9.1}$$

$\theta$ is the dependent variable, while time – which is contained within the derivative – is the independent variable. The process of solving Equation (9.1) entails finding the dependence of $\theta$ on time; in other words, $\theta(t)$. Generally, for models involving the dynamic behavior of particles and rigid bodies, time is always the only independent variable because Newton's second law governs the temporal variation of these systems, while the spatial variation is not of interest.

Ordinary differential equations can be generally classified into linear and non-linear. All the terms in a linear differential equation vary linearly with the dependent variable, while a non-linear differential equation contains terms that have non-linear dependence on the dependent variable. For instance, Equation (9.1) is non-linear because the term $\sin\theta$ is a non-linear function of the dependent variable $\theta$. Ordinary differential equations can also be classified according to their order, which represents the order of the highest derivative in the equation. Thus, Equation (9.1) is of the second order.

Non-linear ordinary differential equations can be classified into autonomous and non-autonomous. Autonomous equations are those in which time does not appear explicitly (appears only as part of the derivative of the dependent variables), whereas non-autonomous equations are the ones in which time appears explicitly. For example, Equation (9.1) is an autonomous differential equation because time only appears as part of the derivative.

The asymptotic behavior of a set of differential equations as $t \to \infty$ is known as the *steady-state* response. Such response could be bounded or unbounded, static or dynamic. Prior to reaching the steady-state, the behavior of a dynamic system is called *transient*.

Any set of higher-order (order greater than one) non-linear differential equations can be expressed in the general form

$$\dot{\mathbf{x}} = \mathbf{G}(\mathbf{x}), \tag{9.2}$$

where $\mathbf{x} \in \mathbb{R}^n$ is known as the state vector; $\mathbf{G} \in \mathbb{R}^n$ is known as the vector field. For example, Equation (9.1) can be transformed into the general form of Equation (9.2) by letting $x_1 = \theta$, $x_2 = \dot{\theta}$, which yields

$$\begin{aligned} \dot{x}_1 &= x_2, \\ \dot{x}_2 &= -\frac{g}{l}\sin x_1. \end{aligned} \tag{9.3}$$

Thus, we can write

$$\dot{\mathbf{x}} = \mathbf{G}(\mathbf{x}), \qquad \text{where} \qquad \mathbf{x} = \begin{bmatrix} x_1 \\ x_2 \end{bmatrix}, \qquad \mathbf{G}(\mathbf{x}) = \begin{bmatrix} x_2 \\ -\frac{g}{l}\sin x_1 \end{bmatrix}. \tag{9.4}$$

Even non-autonomous differential equations can be expressed in the form of Equation (9.2). For instance, consider the system

$$\ddot{x} + \dot{x} + x^3 = \cos(3t). \tag{9.5}$$

Let $x_1 = x$, $x_2 = \dot{x}$, and $x_3 = t$. This yields

$$\dot{\mathbf{x}} = \mathbf{G}(\mathbf{x}), \qquad \text{where} \qquad \mathbf{x} = \begin{bmatrix} x_1 \\ x_2 \\ x_3 \end{bmatrix} \qquad \mathbf{G}(\mathbf{x}) = \begin{bmatrix} x_2 \\ -x_2 - x_2^3 + \cos 3x_3 \\ 1 \end{bmatrix}. \tag{9.6}$$

## 9.2 Equilibrium Solutions of Dynamical Systems

An important aspect of the response of a dynamical system is its steady-state behavior. We are always interested to understand the characteristics of the response after a sufficiently long time. Does this system approach a static point or does it continue to move? Is the response bounded, or does it grow without bounds? Many such questions can only be answered if we learn how to find the equilibrium points of the system; in other words, the final static configuration of the system.

Since the equilibrium solutions represent the static behavior of the system, they refer to the points where $\dot{\mathbf{x}} = 0$, or where the vector field vanishes. These points can be obtained by solving

$$\mathbf{G}(\mathbf{x}_0) = 0, \tag{9.7}$$

where $\mathbf{x}_0$ represent the vector of equilibrium points. Equation (9.7) represents a set of $n$ algebraic equations that can be solved together for the vector $\mathbf{x}_0$. The equilibrium solutions can also be referred to as fixed points, stationary solutions, critical points, constant solutions, and steady-state solutions.

As a physical example, let us find the equilibrium points of a pendulum. To this end, we set $\mathbf{G}(\mathbf{x}_0) = 0$. This yields $x_2 = 0$, $\sin x_1 = 0$. It follows then that the equilibrium solutions are $(x_{10}, x_{20}) = ((i-1)\pi, 0)$, $i = 1, 2, \dots, \infty$. In terms of the original dynamics, these solutions are $(\theta_0, \dot{\theta}_0) = ((i-1)\pi, 0)$, $i = 1, 2$. Physically, any equilibrium point beyond $i = 2$ is just a repetition of one of the first two equilibrium points, as shown in Figure 9.1.

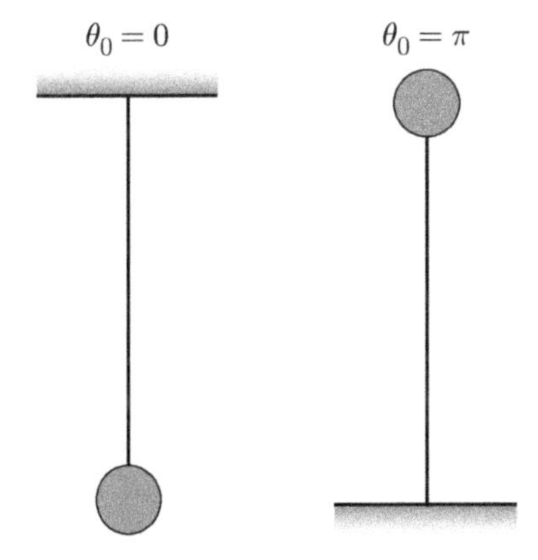


**Figure 9.1** Equilibrium configurations of a simple pendulum.

**Example 9.1 Equilibrium Points of a Rotating Rigid Body**

Find the equilibrium points of the rotating T-shaped rod modeled in Example 3.10.

The equation of motion is found to take the following form:

$$\ddot{\theta} + \left(\frac{16}{27}\frac{g}{l} - \frac{79}{81}\Omega^2 \cos\theta\right) \sin\theta = 0.$$

In the general form of Equation (9.2), the state vector and the vector field can be written as

$$\mathbf{x} = \begin{bmatrix} x_1 \\ x_2 \end{bmatrix} \qquad \mathbf{G}(\mathbf{x}) = \begin{bmatrix} x_2 \\ -\left(\frac{16}{27}\frac{g}{l} - \frac{79}{81}\Omega^2 \cos x_1\right) \sin x_1 \end{bmatrix}.$$

Setting the vector field to zero, we obtain the following equilibrium points:

$$\begin{aligned}(x_{10}, x_{20}) &= (\theta_0, \dot{\theta}_0) = ((i-1)\pi, 0), \qquad i = 1, 2, \ldots, \infty, \\ (x_{10}, x_{20}) &= (\theta_0, \dot{\theta}_0) = (\pm \arccos \frac{48}{79}\frac{g}{l\Omega^2}, 0), \qquad \Omega^2 \geq \frac{48}{79}\frac{g}{l}.\end{aligned}$$

This implies that there are four *possibly* realizable equilibrium solutions. Two of these are independent of the angular velocity of rotation and correspond to $\theta_0 = 0$ and $\theta_0 = \pi$, while the third and fourth correspond to an equilibrium angle whose magnitude increases with the angular velocity of rotation $\theta_0 = \pm \arccos \frac{48}{79}\frac{g}{l\Omega^2}$ and only exist when $\Omega^2 \geq \frac{48}{79}\frac{g}{l}$. Note that we used the expression "possibly" realizable because some equilibrium solutions cannot be physically realized, as will be discussed next.

## 9.3 Stability and Classification of Equilibrium Solutions

The equilibrium solutions of a given dynamical system represent the steady-state points at which the trajectories of a dynamical system *may* eventually settle after a sufficiently long time. Nonetheless, what determines whether the dynamic trajectories can actually end up at these points is the type of the equilibrium solution: mainly whether it is *stable* or *unstable*. A stable equilibrium point will attract any dynamic trajectories initiated close to it and hence can be called an *attractor*. In contrast, an unstable equilibrium point repels any dynamic trajectories, no matter how close they are to it. As such, these are often called *repellors*. For non-linear systems with multiple stable equilibrium points, the final destination of the dynamic trajectories is determined by the set of initial conditions that lead to one equilibrium point versus another. These set of initial conditions are commonly referred to as the *basin of attraction* of a given equilibrium point.

To determine whether an equilibrium point is stable or not, we agitate the system slightly around the equilibrium by introducing a small perturbation. We then investigate the dynamic of the perturbation itself after a sufficiently long time. If the perturbation dies out, then the equilibrium point is stable; otherwise it is unstable. In other words, to assess the stability of a given equilibrium solution, we let

$$\mathbf{x}(t) = \mathbf{x}_0 + \mathbf{y}(t), \tag{9.8}$$

where $\mathbf{x}_0$ is the equilibrium solution and $\mathbf{y}(t)$ is the introduced perturbation. Substituting Equation (9.8) into Equation (9.2), we obtain

$$\dot{\mathbf{y}}(t) = \mathbf{G}(\mathbf{x}_0 + \mathbf{y}(t)). \tag{9.9}$$

Assuming that the vector field $\mathbf{G}$ is continuous and twice differentiable, and since the perturbation is small by definition, we can expand Equation (9.9) in a Taylor series around $\mathbf{x}_0$. This yields

$$\dot{\mathbf{y}} = \mathbf{G}(\mathbf{x}_0) + \mathbf{D}_\mathbf{x}\mathbf{G}(\mathbf{x}_0)\mathbf{y}(t) + \mathcal{O}(\mathbf{y}^2), \tag{9.10}$$

where

$$\mathbf{D}_\mathbf{x}\mathbf{G}(\mathbf{x}_0) = \left.\begin{bmatrix} \dfrac{\partial G_1}{\partial x_1} & \dfrac{\partial G_1}{\partial x_2} & \cdots & \dfrac{\partial G_1}{\partial x_n} \\ \dfrac{\partial G_2}{\partial x_1} & \dfrac{\partial G_2}{\partial x_2} & \cdots & \dfrac{\partial G_2}{\partial x_n} \\ \vdots & \vdots & \vdots & \vdots \\ \dfrac{\partial G_n}{\partial x_1} & \dfrac{\partial G_n}{\partial x_2} & \cdots & \dfrac{\partial G_n}{\partial x_n} \end{bmatrix}\right|_{\mathbf{x}=\mathbf{x}_0}, \tag{9.11}$$

and $\mathcal{O}(\mathbf{y}^2)$ represents the order of $\mathbf{y}^2$. By virtue of Equation (9.7), the term $\mathbf{G}(\mathbf{x}_0)$ vanishes, so Equation (9.10) reduces to

$$\dot{\mathbf{y}} \approx \mathbf{D}_\mathbf{x}\mathbf{G}(\mathbf{x}_0)\mathbf{y} \approx \mathbf{A}\mathbf{y}, \tag{9.12}$$

where terms of order $\mathcal{O}(\mathbf{y}^2)$ are assumed to be sufficiently small to be neglected. Here, $\mathbf{A}$ is the matrix of partial derivatives evaluated at the equilibrium points, also known as the *Jacobian* matrix.

Our goal next is to determine if every component in the perturbation vector $\mathbf{y}(t)$ decays to zero with time. If this is the case, then the associated equilibrium point is stable. To this end, we first note that the matrix $\mathbf{A}$ has constant entries. As such, Equation (9.12) represents a set of linearly-coupled ordinary differential equations whose solution can be written in the general form:

$$\mathbf{y}(t) = e^{t\mathbf{A}}\mathbf{y}_0, \tag{9.13}$$

where $\mathbf{y}_0 \in \mathbb{R}^n$ is the vector of initial conditions at $t = 0$. Examining Equation (9.13), it becomes evident that the entities of matrix $\mathbf{A}$ play a critical role in characterizing whether the exponential function $e^{t\mathbf{A}}\mathbf{y}_0$ decays to zero as time evolves. However, this role is not yet clear. To clarify this role, let us introduce the transformation

$$\mathbf{y} = \mathbf{P}\mathbf{z}, \tag{9.14}$$

where $\mathbf{P} \in \mathbb{R}^{n\times n}$ is a non-singular matrix. Substituting Equation (9.14) into Equation (9.12) yields

$$\mathbf{P}\dot{\mathbf{z}} = \mathbf{A}\mathbf{P}\mathbf{z}, \tag{9.15}$$

or

$$\dot{\mathbf{z}} = \mathbf{P}^{-1}\mathbf{A}\mathbf{P}\mathbf{z}. \tag{9.16}$$

The matrix $\mathbf{P}^{-1}\mathbf{A}\mathbf{P}$ can be diagonalized when the *geometric multiplicity* and *algebraic multiplicity* coincide; in other words, there is a linearly independent eigenvector for each eigenvalue even if the eigenvalues are not distinct. If this condition is satisfied, one can choose $\mathbf{P}$ such that $\mathbf{P}^{-1}\mathbf{A}\mathbf{P}$ is a diagonal matrix with the diagonal terms being the eigenvalues of $\mathbf{A}$. The matrix $\mathbf{P}$ that can realize this diagonalization is that consisting of the right eigenvectors of $\mathbf{A}$ as the columns of $\mathbf{P}$. As such, one can write

$$\dot{\mathbf{z}} = \Lambda\mathbf{z}, \tag{9.17}$$

where

$$\Lambda = \begin{bmatrix} \lambda_1 & 0 & 0 & \dots & 0 \\ 0 & \lambda_2 & 0 & \dots & 0 \\ 0 & 0 & \lambda_3 & \dots & 0 \\ \vdots & \vdots & \vdots & \vdots & \vdots \\ 0 & 0 & 0 & \dots & \lambda_n \end{bmatrix}.$$

Here $\lambda_i$ are the eigenvalues of $\mathbf{A}$. Using Equations (9.13)–(9.15) and (9.18), one can write

$$\mathbf{y}(t) = \mathbf{P}e^{t\Lambda}\mathbf{P}^{-1}\mathbf{y}_0. \tag{9.18}$$

Since $\Lambda$ is diagonal, Equation (9.18) can be further simplified to

$$\mathbf{y}(t) = \mathbf{P}\begin{bmatrix} e^{\lambda_1 t} & 0 & 0 & \dots & 0 \\ 0 & e^{\lambda_2 t} & 0 & \dots & 0 \\ 0 & 0 & e^{\lambda_3 t} & \dots & 0 \\ 0 & 0 & 0 & \dots & e^{\lambda_n t} \end{bmatrix}\mathbf{P}^{-1}\mathbf{y}_0. \tag{9.19}$$

Examining Equation (9.19), it becomes clear that:

- $\mathbf{y}(t) \to 0$ as $t \to \infty$ if and only if all eigenvalues $\lambda_i$ of the matrix $\mathbf{A}$ have negative real parts;
- $\mathbf{y}(t) \to \infty$ as $t \to \infty$ if at least one eigenvalue has a positive real part.

As a consequence, an equilibrium solution is locally asymptotically stable if all the eigenvalues of the Jacobian matrix evaluated at the equilibrium points have negative real parts (located in the left-hand side of the complex plane). It is unstable if at least one eigenvalue has a positive real part (located on the right-hand side of the complex plane).

It is worth noting that, even if the eigenvalues of the Jacobian matrix are not distinct, there still exists a matrix $\mathbf{P}$ such that $\mathbf{P}^{-1}\mathbf{A}\mathbf{P} = \mathbf{J}$, where $\mathbf{J}$ is a *Jordan canonical* matrix with off-diagonal entries. This matrix can also be used to study the stability of the system.

The equilibrium solutions can be classified according to the location of the eigenvalues $\lambda_k = \zeta_k \pm \imath\omega_k$ in the complex plane:

***Hyperbolic*** An equilibrium solution is hyperbolic when all the eigenvalues of the Jacobian matrix have non-zero real parts. Based on the nature of the eigenvalues, a hyperbolic fixed point can be divided into three major types:

- *Sink*, if all $\zeta_k < 0$. A sink is also called a stable focus if at least one eigenvalue has $\omega_k \neq 0$, and is called a stable node if all $\omega_k = 0$.
- *Source*, if all $\zeta_k > 0$. A source is also called an unstable focus if at least one eigenvalue has $\omega_k \neq 0$, and is called an unstable node if all $\omega_k = 0$.
- *Saddle*, if some $\zeta_k > 0$ and some $\zeta_k < 0$.

***Non-hyperbolic*** An equilibrium solution is non-hyperbolic if at least one eigenvalue has a zero real part. A non-hyperbolic fixed point is unstable if one or more of the eigenvalues has a positive real part and is neutrally stable if the rest of the eigenvalues have negative real parts. If all the eigenvalues are purely imaginary a non-hyperbolic fixed point is called a *center*.

**Equilibrium Points**

Finding the equilibrium points of dynamical systems and analyzing their stability is one of the most common techniques to analyze the local behavior of a system. Despite being a local analysis, it is often possible to establish a good understanding of the global dynamics of a system by classifying its equilibrium points.

**Example 9.2 Classification of Equilibrium Solutions of a Simple Pendulum**

Classify the stability and type of the equilibrium solutions of the simple pendulum considered in Example 9.1. The equation of motion of the simple pendulum is given by

$$\ddot{\theta} + \frac{g}{l}\sin\theta = 0.$$

Rewriting Equation (9.2) in the general form:

$$\begin{cases} \dot{x}_1 &= x_2 \\ \dot{x}_2 &= -\frac{g}{l}\sin x_1 \end{cases}$$

we find the equilibrium solutions as $(x_{10}, x_{20}) = ((i-1)\pi, 0)$, $i = 1, 2, \dots, \infty$. To classify the stability and types of these solutions, we construct the Jacobian matrix as

$$\mathbf{A} = \begin{bmatrix} 0 & 1 \\ -\frac{g}{l}\cos x_{10} & 0 \end{bmatrix},$$

for which the eigenvalues can be obtained by solving the equation

$$|\mathbf{A} - \lambda_i \mathcal{I}| = 0,$$

Solving the previous equation yields the following eigenvalues:

$$\lambda_{1,2} = \pm\imath\sqrt{\frac{g}{l}\cos x_{10}},$$

where $\imath^2 = -1$. Substituting $x_{10} = (i-1)\pi$ into Equation (9.2) yields

$$\lambda_{1,2} = \pm\imath\sqrt{\frac{g}{l}\cos(i-1)\pi}, \qquad i = 1, 2, 3, \dots, \infty.$$

Note that when $i$ is odd, $\lambda_{1,2} = \pm\imath\sqrt{\frac{g}{l}}$, meaning the fixed point is a neutrally stable center. On the other hand, when $i$ is even, or when the pendulum is in the upright position, $\lambda_{1,2} = \pm\sqrt{\frac{g}{l}}$, meaning the fixed point is an unstable saddle.

The resulting classification makes sense since, if you give the pendulum a small perturbation about $\theta_0 = 0$, it will continue to oscillate forever (no dissipation is included in the model). On the other hand, if you give the pendulum a small perturbation about the upright position, the dynamic trajectories will quickly diverge away from that position.

**Example 9.3 Classification of the Equilibrium Solutions of a Rotating Rod**

Classify the equilibrium points of the rotating T-shaped rod modeled in Example 3.10. The equation of motion is given by

$$\ddot{\theta} + \left(\frac{16}{27}\frac{g}{l} - \frac{79}{81}\Omega^2 \cos\theta\right)\sin\theta = 0.$$

The state vector and the vector field can be written as

$$\mathbf{x} = \begin{bmatrix} x_1 \\ x_2 \end{bmatrix} \qquad \mathbf{G}(\mathbf{x}) = \begin{bmatrix} x_2 \\ -\left(\frac{16}{27}\frac{g}{l} - \frac{79}{81}\Omega^2 \cos x_1\right)\sin x_1 \end{bmatrix}.$$

Setting the vector field to zero, we obtain the following equilibrium points:

$$\begin{aligned}
(x_{10}, x_{20}) &= (\theta_0, \dot{\theta}_0) = ((i-1)\pi, 0), i = 1, 2, \ldots, \infty \\
(x_{10}, x_{20}) &= (\theta_0, \dot{\theta}_0) = (\pm \arccos \frac{48}{79}\frac{g}{l\Omega^2}, 0) \qquad \Omega^2 \geq \frac{48}{79}\frac{g}{l}.
\end{aligned}$$

To analyze the stability of the resulting equilibrium points, we find the Jacobian matrix

$$\mathbf{A} = \begin{bmatrix} 0 & 1 \\ \frac{1}{81}\left(79\Omega^2 \cos 2x_{10} - \frac{48g}{l}\cos x_{10}\right) & 0 \end{bmatrix},$$

for which the eigenvalues can be obtained by solving the equation

$$|\mathbf{A} - \lambda_i \mathcal{I}| = 0.$$

This yields the following eigenvalues:

$$\lambda_{1,2} = \pm\frac{\sqrt{-79l\Omega^2 - 48g\cos x_{10} + 158l\Omega^2\cos^2 x_{10}}}{9\sqrt{l}},$$

Therefore:

- for $x_{10} = (i-1)\pi$ and $i$ odd,

$$\lambda_{1,2} = \pm\frac{\sqrt{79l\Omega^2 - 48g}}{9\sqrt{l}}.$$

  which implies that when $\Omega^2 > \frac{48g}{79l}$ and $i$ is odd, $x_{10} = (i-1)\pi$ is a saddle; otherwise, it is a center.
- for $x_{10} = (i-1)\pi$, and $i$ even,

$$\lambda_{1,2} = \pm\frac{\sqrt{79l\Omega^2 + 48g}}{9\sqrt{l}},$$

  so for $x_{10} = (i-1)\pi$, and $i$ even, which represents the upright position of the rod, it is always an unstable saddle.

- for $x_{10} = \arccos\left(\frac{48}{79}\frac{g}{l\Omega^2}\right)$, $\Omega^2 > \frac{48}{79}\frac{g}{l}$,

$$\lambda_{1,2} = \pm \frac{\sqrt{-79l\Omega^2 + \frac{2304g^2}{79l\Omega^2}}}{9\sqrt{l}}$$

It follows that, when $\Omega^2 \geq \frac{48}{79}\frac{g}{l}$ which is the condition necessary for this equilibrium position to exist, $\lambda_{1,2}$ are purely imaginary numbers. Therefore, $x_{10} = \arccos\left(\frac{48}{79}\frac{g}{l\Omega^2}\right)$ is a center.

**Example 9.4 Classification of Equilibrium Solutions of a Simple Pendulum with Damping**

Let us reconsider the simple pendulum but with an additional torsional viscous damper at the pivot. The addition of the damper changes the equation of motion to the form

$$\ddot{\theta} + \frac{c}{ml^2}\dot{\theta} + \frac{g}{l}\sin\theta = 0,$$

where $c$ is the damping coefficient. Rewriting the previous equation in the general form:

$$\begin{cases} \dot{x}_1 &= x_2 \\ \dot{x}_2 &= -\frac{c}{ml^2}x_2 - \frac{g}{l}\sin x_1 \\ x_1 &= \theta \end{cases}$$

we find the equilibrium solutions as $(x_{10}, x_{20}) = ((i-1)\pi, 0), i = 1, 2, \ldots, \infty$. Note that the equilibrium solutions do not change with the addition of the damper.

To classify the stability and types of these solutions, we construct the Jacobian matrix:

$$\mathbf{A} = \begin{bmatrix} 0 & 1 \\ -\frac{g}{l}\cos x_{10} & -\frac{c}{ml^2} \end{bmatrix}.$$

The eigenvalues can be obtained by solving the equation

$$|\mathbf{A} - \lambda_i \mathcal{I}| = 0.$$

Solving the previous equation for $\lambda_i$ yields the following eigenvalues:

$$\lambda_{1,2} = \pm\frac{1}{2}\left(-\frac{c}{ml^2} \pm \sqrt{\frac{c^2}{ml^2} + \frac{4g\cos x_{10}}{l}}\right).$$

Hence, when $i$ is odd and $\frac{c^2}{ml^2} > \frac{4g}{l}$, $\lambda_{12}$ are both real and negative, which implies that the fixed point is a stable node. On the other hand, when $i$ is odd and $\frac{c^2}{ml^2} < \frac{4g}{l}$, $\lambda_{12}$ are complex conjugates with negative real parts, so the fixed point is a stable focus. When $i$ is even, $\lambda_{1,2}$ are real with opposite signs, so the fixed point is an unstable saddle.

**Example 9.5 Particle Sliding on A Smooth Surface**

Consider the motion of two particles connected through a cable of length $l$, as shown in Figure 9.2. The first particle of mass $m$ is allowed to slide in the radial and tangential directions on the smooth surface, while the second particle of mass $\alpha m$ moves in the vertical direction under the influence of gravity. At time $t = 0$ and $r = r_i$, the first particle was subjected to an

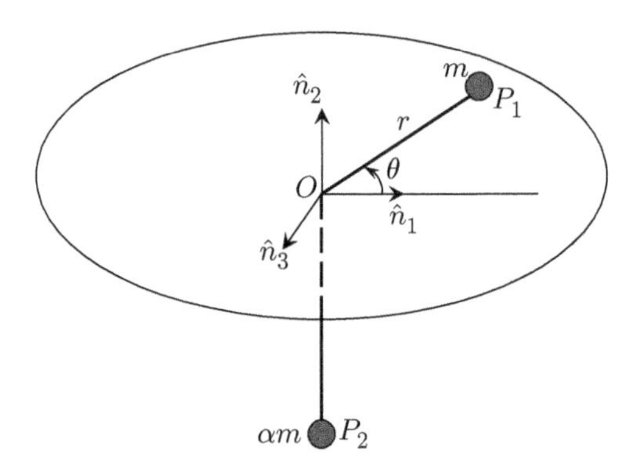


**Figure 9.2** Two particles connected through a rigid cable of length $l$.

initial angular velocity $\omega_0 \hat{n}_2$. Find the steady-state position of the first particle - in other words, the value of $r$ – after a very long time. Is this position a stable position?

To solve this problem, we first obtain the equations of motion of the system using Newton's second law. To this end, we define a rotating $B$-frame such that ${}^N\boldsymbol{\omega}^B = \dot{\theta}\hat{b}_2$, and write the acceleration of both particles with respect to point $O$ as

$$
\begin{aligned}
{}^N\mathbf{a}^{P_1/O} &= (\ddot{r} - r\dot{\theta}^2)\hat{b}_1 + (r\ddot{\theta} - 2\dot{r}\dot{\theta})\hat{b}_3, \\
{}^N\mathbf{a}^{P_2/O} &= \ddot{r}\hat{b}_2.
\end{aligned}
$$

Applying Newton's second law to both particles, we can write

$$
\begin{aligned}
\mathbf{F}\hat{b}_1 &= m{}^N\mathbf{a}^{P_1/O}\hat{b}_1, & T &= -m\ddot{r} + mr\dot{\theta}^2, \\
\mathbf{F}\hat{b}_3 &= m{}^N\mathbf{a}^{P_1/O}\hat{b}_3, & r\ddot{\theta} - 2\dot{r}\dot{\theta} &= 0, \\
\mathbf{F}\hat{b}_2 &= \alpha m{}^N\mathbf{a}^{P_2/O}\hat{b}_2, & T - \alpha mg &= \alpha m\ddot{r}.
\end{aligned}
$$

This yields the following equations of motion:

$$
\begin{aligned}
(1+\alpha)\ddot{r} - r\dot{\theta}^2 + \alpha g = 0, \\
r\ddot{\theta} - 2\dot{r}\dot{\theta} = 0.
\end{aligned}
$$

Note that the second of these can be integrated to obtain $\dot{\theta} = c_\theta r^2$, where $c_\theta$ is the constant of integration. This can be obtained by using the initial condition $\dot{\theta}(r_i) = \omega_0$, and yields

$$
\dot{\theta} = \omega_0 \left(\frac{r}{r_i}\right)^2.
$$

Using this equation, we can eliminate the angle $\theta$ from the dynamics, which, in turn, implies that $\theta$ is a cyclic coordinate. The dynamics of the system can then be reduced to

$$
\ddot{r} - \frac{\omega_0^2}{r_i^4(1+\alpha^2)} r^3 + \alpha g = 0.
$$

To obtain the equilibrium points, we put the previous equation in the general form:

$$\begin{aligned}\dot{x}_1 &= x_2,\\ \dot{x}_2 &= \frac{\omega_0^2}{r_i^4(1+\alpha^2)}x_1^3 - \alpha g.\end{aligned}$$

This yields the equilibrium solution $(x_{10}, x_{20}) = \left(\left(\frac{\alpha(1+\alpha^2)gr_i^4}{\omega_0^2}\right)^{1/3}, 0\right)$. The stability of this steady-state solution can be assessed by constructing the Jacobian matrix:

$$\mathbf{A} = \begin{bmatrix} 0 & 1 \\ \frac{3x_{10}^2\omega_0^2}{(1+\alpha^2)r_i^4} & 0 \end{bmatrix},$$

for which the eigenvalues can be obtained by solving the equation

$$|\mathbf{A} - \lambda_i \mathcal{I}| = 0.$$

The resulting eigenvalues are

$$\lambda_{1,2} = \pm\sqrt{\frac{3x_{10}^2\omega_0^2}{(1+\alpha^2)r_i^4}}.$$

Since the term under the square root is always positive, the equilibrium position $r_0 = x_{10}$ is an unstable saddle.

**Flipped Classroom Exercise 9.1**

For the system shown in the figure, find the equilibria and classify them. Use $l = 9.81$ m, $k = 1$ N/m, $m = 1$ kg, and $c = 1/9.81$ kg/sec. Assume that the moving block to which the spring and damper are connected is massless, and that all interacting surfaces are smooth.

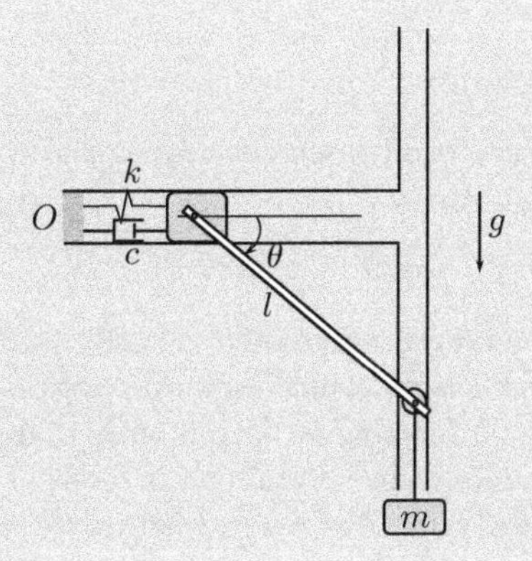

To solve this problem, take the following steps:

1. Use $\theta$ as the generalized coordinate, then find the equation of motion of the system. Show that it takes the form:

$$\ddot{\theta}\cos^2\theta - \dot{\theta}^2\cos\theta\sin\theta + \frac{cl}{m}\sin\theta\dot{\theta} + \frac{k}{m}(1-\cos\theta)\sin\theta - \frac{g}{l}\cos\theta = 0.$$

2. Put the equation in the general form $\dot{\mathbf{x}} = \mathbf{G}(\mathbf{x})$ by letting $x_1 = \theta$, and $x_2 = \dot{\theta}$.
3. Find the equilibrium points by letting $\dot{x}_2 = 0$, $\dot{x}_1 = 0$. Show that the equilibrium points are: $(x_{10}, x_{20}) = (-0.47, 0), (x_{10}, x_{20}) = (0.88, 0)$. Note that the answers are presented in radians.
4. Construct the Jacobian matrix and find the eigenvalues.
5. Show that the first equilibirum point is an unstable focus, while the second equilibrium point is a stable focus.

## 9.4 Phase-plane Representation of the Dynamics

One insightful way of representing and analyzing the dynamics of non-linear systems is by drawing the dynamic trajectories in the phase plane. In a phase plane, we define a region in the space of $(x, \dot{x})$, then draw the trajectories of the dynamics starting from a large set of initial conditions. This results in a unique graphical representation of the dynamics that can be used to draw conclusion about the global behavior of the system.

To learn more about the phase-plane representation, we divide the analysis into two parts. First, we deal with conservative systems for which the phase space can be constructed graphically using analytical tools. Subsequently, we study non-conservative systems, for which numerical techniques are required. Since, the phase-plane representation of systems containing more than one degree of freedom is not very insightful, the discussions will only be limited to systems involving a single degree of freedom.

### *9.4.1 Conservative Systems*

We learned in Chapter 5 that, for a conservative dynamical system, the sum of the kinetic and potential energy is constant; that is

$$\mathcal{T}(\dot{q}, q) + \mathcal{U}(q) = \mathcal{H}, \tag{9.20}$$

where $q$ is some generalized coordinate used to describe the dynamics, $\mathcal{T}(\dot{q}, q)$ is the kinetic energy, $\mathcal{U}(q)$ is the potential energy, which for most systems can be assumed to be a function of $q$ only, and $\mathcal{H}$ is a constant denoting the Hamiltonian of the system.

In general, Equation (9.20) can be solved for $\dot{q}$ in terms of $q$ for different values of $\mathcal{H}$. For the simple and common case when $\mathcal{T} = \frac{1}{2}\alpha\dot{q}^2$, where $\alpha > 0$ is a constant, one can solve for $\dot{q}$ and obtain

$$\dot{q} = \pm\frac{2}{\alpha}\sqrt{\mathcal{H} - \mathcal{U}(q)}. \tag{9.21}$$

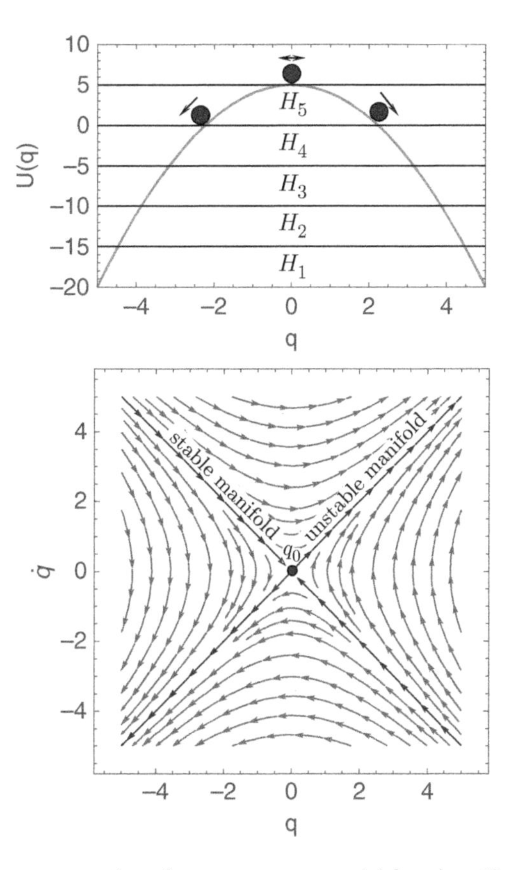


**Figure 9.3** Phase-plane representation of a concave up potential function. (*See color plate section for the color representation of this figure.*)

Thus, for a given energy level $\mathcal{H}$ and a known potential energy function, $\mathcal{U}(q)$, Equation (9.21) can be solved for $\dot{q}$ and used to plot the dynamic trajectories in the $(\dot{q}, q)$ plane for different initial energy levels (initial conditions). Such a plot is known as the phase plane.

To illustrate this procedure, we construct the phase-plane representation for the generic concave potential energy function shown in Figure 9.3. The phase plane of the system dynamics in the $(\dot{q}, q)$ plane can be generated by first selecting a low energy level, $\mathcal{H}_1$, and plotting $\dot{q}$ versus $q$ using Equation (9.21). The plus and minus sign in the equation lead to two trajectories for each value of $\mathcal{H}$. The process is then repeated for higher energy levels $\mathcal{H}_2$, $\mathcal{H}_3$, …, $\mathcal{H}_n$, resulting in the phase plane shown in Figure 9.3. For a quick sketch of the phase-plane representation, one can roughly approximate the shape of the trajectories of $\dot{q}$ from the graph without obtaining an exact value.

To find the direction of the trajectories in the phase space, we differentiate Equation (9.21) with respect to $q$ and obtain

$$\alpha\dot{q}\frac{\mathrm{d}\dot{q}}{\mathrm{d}q} = -\frac{d\mathcal{U}}{dq}, \qquad \alpha > 0. \tag{9.22}$$

This equation can be used to deduce the following:

1. When $\dot{q} > 0$ and $\frac{d\mathcal{U}}{\partial q} > 0$, $\frac{d\dot{q}}{dq} < 0$, meaning $\dot{q}$ is decreasing with $q$ and hence the trajectories point to the left.
2. When $\dot{q} > 0$ and $\frac{d\mathcal{U}}{\partial q} < 0$, then $\frac{d\dot{q}}{dq} > 0$, meaning $\dot{q}$ is increasing with $q$ and hence the trajectories point to the right.
3. When $\dot{q} < 0$ and $\frac{d\mathcal{U}}{\partial q} > 0$, then $\frac{d\dot{q}}{dq} > 0$, meaning $\dot{q}$ is decreasing with $q$ and hence the trajectories point to the right.
4. When $\dot{q} < 0$ and $\frac{d\mathcal{U}}{\partial q} < 0$, then $\frac{d\dot{q}}{dq} < 0$, meaning $\dot{q}$ is decreasing with $q$ and hence the trajectories point to the left.

The resulting phase plane is shown in Figure 9.3, where it can be seen that there is an equilibrium point at $(q_0, \dot{q}_0) = (0, 0)$. The presence of an equilibrium point at $(q_0, \dot{q}_0) = (0, 0)$ can also be deduced by inspecting the potential energy function and noting that it takes its maximum value at $q_0 = 0$. An extremum (minimum or maxmium) in the potential energy function is an equilibrium point of the system because, at the equilibrium point, $\frac{d\mathcal{U}}{dq} = -\mathbf{G}(q_0) = 0$. The potential energy function can also be used to conclude that the equilibrium point is unstable because any perturbation to the dynamic trajectories represented here by the motion of a ball on the potential energy surface will diverge away from the equilibrium point.

In the phase plane, any trajectories passing close to $q_0$ are repelled away from it. As such, the fixed point, $(\dot{q}_0, q_0) = (0, 0)$, is an unstable repellor. Further inspection of Figure 9.3 reveals that the motion trajectories near $(0, 0)$ form a saddle shape, meaning the equilibrium point is a saddle. It is also worth mentioning that there are two trajectories (manifolds) passing through the saddle. The one that enters the saddle is called a *stable manifold* while the one which exits the saddle is called an *unstable manifold*.

As a second example, we consider constructing the phase-space representation of the twin-well potential function shown in Figure 9.4. The potential function has three extrema: one local maximum and two global minima. These extrema correspond to the three equilibrium points of the system. The local maximum corresponds to an unstable fixed point because any perturbation to a ball placed on it will cause the ball to slide down to one of the wells on either side of the maximum. The local minima are marginally stable centers because if we give a ball placed at the bottom of the well a small perturbation it will continue to oscillate forever with a magnitude proportional to the initial energy level.

The phase-plane representation of the dynamics highlights the three fixed points; the one in the middle is an unstable saddle, as clearly evidenced by the shape of the trajectories passing through it. The ones on either side of the saddle are marginally stable centers, as evidenced by the shape of the trajectories around them.

The dynamic trajectories, highlighted in red, which leave the saddle point and come back to it are called *homo-clinic* orbits. In Figure 9.4, there are two *homo-clinic* orbits on either side of the saddle. The union of these orbits is called a *separatrix* because it separates initial conditions leading to qualitatively different dynamic responses. Inside the separatrix, all initial conditions lead to motions inside one well, also known as intra-well motion. Outside the separatrix, all initial conditions lead to motions between the two wells, also known as inter-well motions.

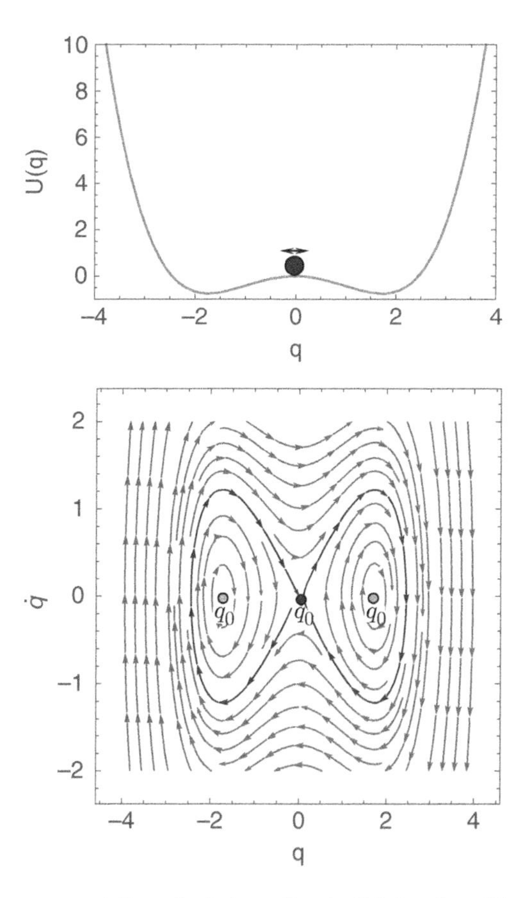


**Figure 9.4** Phase-plane representation of a twin-well potential function. (*See color plate section for the color representation of this figure.*)

### Example 9.6 Phase-plane Representation of Simple Pendulum Motion

Draw the phase plane representation of the motion of a simple pendulum. Use $\frac{g}{l} = 1$.

For the simple pendulum, the potential energy function is $U(\theta) = \frac{g}{l}\cos\theta$. This can be used to plot the phase-plane representation of the dynamics, as shown in Figure 9.5. As you can see, there are an infinite number of equilibrium points. Those that are at $0, 2\pi, 4\pi, \ldots$ are marginally stable centers, while those that are $\pi, 3\pi, 5\pi, \ldots$ are unstable saddles. The trajectories highlighted in red, which leave one saddle and enter another saddle, are called *hetero-clinic orbits*. The union of these hetero-clinic orbits is also called a separatrix because it separates regions of initial conditions that lead to qualitatively different dynamic behaviors. Here, initial conditions inside the separatrix lead to oscillations around $\theta_0 = 0$, while initial conditions outside the separatrix lead to full rotational motion of the pendulum about its pivot.

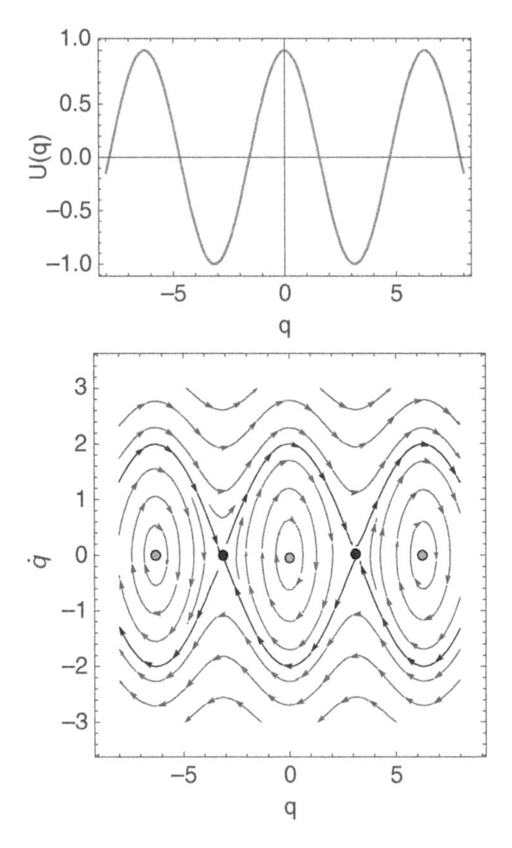


**Figure 9.5** Phase-plane representation of simple pendulum motion. (*See color plate section for the color representation of this figure.*)

### Example 9.7 Conductor Moving in the Magnetic Field of a Stationary Conductor

Consider the motion of a conductor of mass $m$, and length $l$ in the magnetic field of another stationary conductor of infinite length. The conductors are placed in the horizontal plane at a distance $r$, as shown in Figure 9.6. One of the conductors carries current $I_1$ while the other carries current $I_2$. The moving wire is connected to a spring of stiffness $k$. Find the equilibrium solutions of the system, classify them and draw the phase plane for different values of the design parameter.

In Example 8.23, we found the equation of motion as

$$m\ddot{x} + kx + \frac{\mu l I_1 I_2}{2\pi(x-a)} = 0, \qquad x \neq a.$$

To simplify the subsequent analysis, we non-dimensionalize the previous equation by letting $x = x/a$ and $t = \omega_n \tau$, where $\omega_n = \sqrt{\frac{k}{m}}$. This yields

$$\ddot{x} + x + \frac{\mu l I_1 I_2}{2\pi a^2 k}\frac{1}{x-1} = 0, \qquad x \neq 1$$

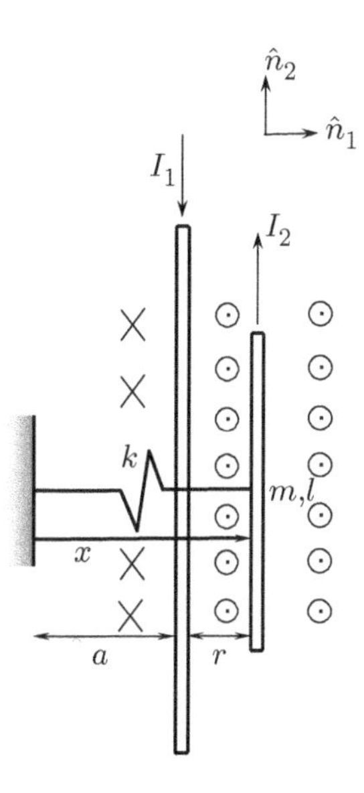


**Figure 9.6** Motion of a finite conductor in the magnetic field of an infinite conductor.

where the overdots are derivatives with respect to the non-dimensional time $\tau$. Letting $\Delta = \frac{\mu l I_1 I_2}{2\pi a^2 k}$, we obtain the following non-dimensional equation:

$$\ddot{x} + x - \frac{\Delta}{1-x} = 0, \qquad x \neq a.$$

The equilibrium solutions of the system can be found as

$$(x_{10}, x_{20}) = \left(\frac{1}{2}(1 \pm \sqrt{1-4\Delta}), 0\right), \qquad \Delta \leq \frac{1}{4}.$$

To classify the equilibrium solutions, we find the Jacobian matrix and calculate the eigenvalues as

$$\lambda_{1,2} = \pm\sqrt{\frac{\Delta}{\left(\frac{1}{2} \pm \sqrt{\frac{1}{4} - \Delta}\right)^2} - 1}.$$

By inspecting the eigenvalues, we notice that the stability of the fixed points is dependent on the parameter $\Delta$. As such, we divide the domain of $\Delta$ into the following subdomains:

- $\Delta < 0$: In this case, the currents $I_1$ and $I_2$ are in the same direction. There are two fixed points – one greater than zero and the other less than zero – meaning that the moving wire could settle at one of two points located on either side of the stationary wire. The eigenvalues $\lambda_{1,2}$ associated with both fixed points are purely imaginary complex conjugates, so the equilibrium points are centers. In other words, if we give the wire a small initial condition about one of these equilibrium points, the wire will continue to oscillate around it.

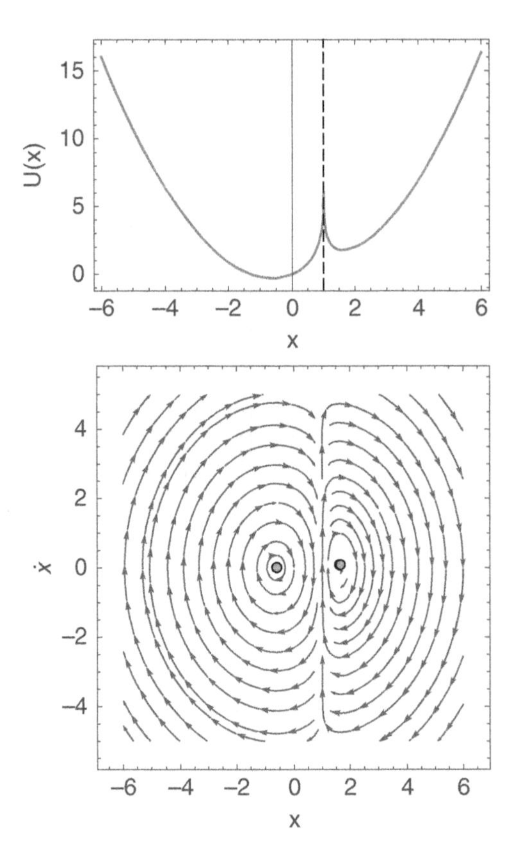


**Figure 9.7** Potential energy function and phase plane representation of the moving wire for $\Delta = -1$.

To obtain the phase plane, the potential energy function of the system is first obtained by integrating the restoring force, which yields

$$\mathcal{U}(x) = \frac{1}{2}x^2 + \Delta \ln|1 - x|.$$

Using the potential energy function, the phase-plane representation is illustrated in Figure 9.7 for the case $\Delta = -1$.

- $\Delta = 0$: In this case, there is only one fixed point: $x_{10} = 0$. This is expected because when $\Delta = 0$ at least one of the wires does not carry any current, thereby eliminating the magnetic force altogether. Therefore, the only force is that associated with the restoring spring. For $x_{10} = 0$, the eigenvalues $\lambda_{1,2}$ are equal to $\pm\imath$ meaning $x_{10} = 0$ is a center, as shown in Figure 9.8.
- $0 < \Delta < \frac{1}{4}$: In this case, the currents flow in the same direction and there are two positive equilibrium solutions. When $x_{10} = \frac{1}{2} + \sqrt{\frac{1}{4} - \Delta}$, $\lambda_{1,2}$ are complex conjugates with zero real parts, and so the fixed point is a center. On the other hand, when $x_{10} = \frac{1}{2} - \sqrt{\frac{1}{4} - \Delta}$, $\lambda_{1,2}$ are real with opposite signs, meaning the equilibrium point is a saddle. The potential

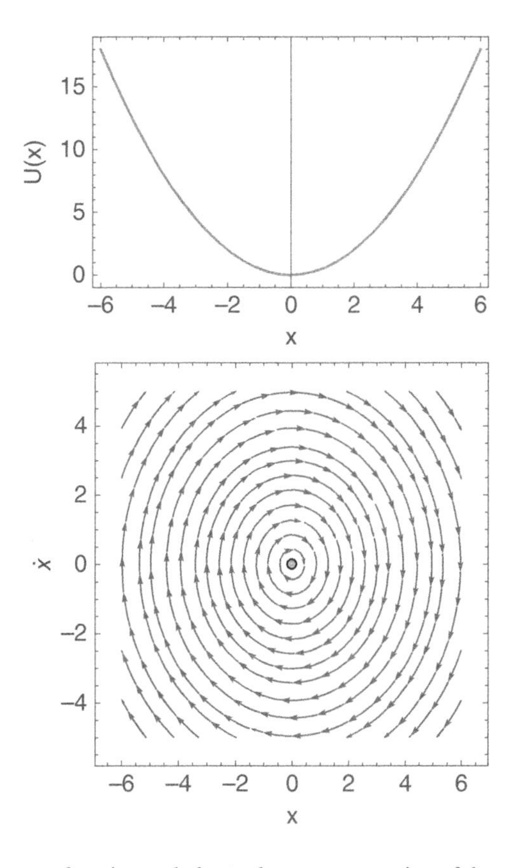


**Figure 9.8** Potential energy function and phase-plane representation of the moving wire for $\Delta = 0$.

energy function and the phase-plane representations are illustrated in Figure 9.9 for the case $\Delta = 0.2$. The phase plane reveals that the dynamic trajectories go to infinity for some initial conditions and rotate about the center for a smaller set of initial conditions.

- $\Delta = \frac{1}{4}$: When $\Delta = \frac{1}{4}$, there is only one equilibrium point: $x_{10} = \frac{1}{2}$. The eigenvalues associated with this fixed point are $\lambda_{1,2} = 0$, so the fixed point is hyperbolic and one cannot judge its nature and stability using the linearized equations. Note that at $\Delta = \frac{1}{4}$, $\frac{\mathrm{d}\mathcal{U}}{\mathrm{d}x} = \frac{\mathrm{d}^2\mathcal{U}}{\mathrm{d}x^2} = 0$, meaning the fixed point represents an inflection point in the vector field. In this case, the fixed point is called a cusp (see Figure 9.10).
- $\Delta > \frac{1}{4}$: When $\Delta > \frac{1}{4}$, the fixed points do not exist.

### *9.4.2 Non-conservative Systems*

For non-conservative systems, the sum of the kinetic and potential energy of the system is not constant. As such, it is often very difficult to construct the phase-plane representation by using the energy-levels approach discussed previously for conservative systems. In such cases,

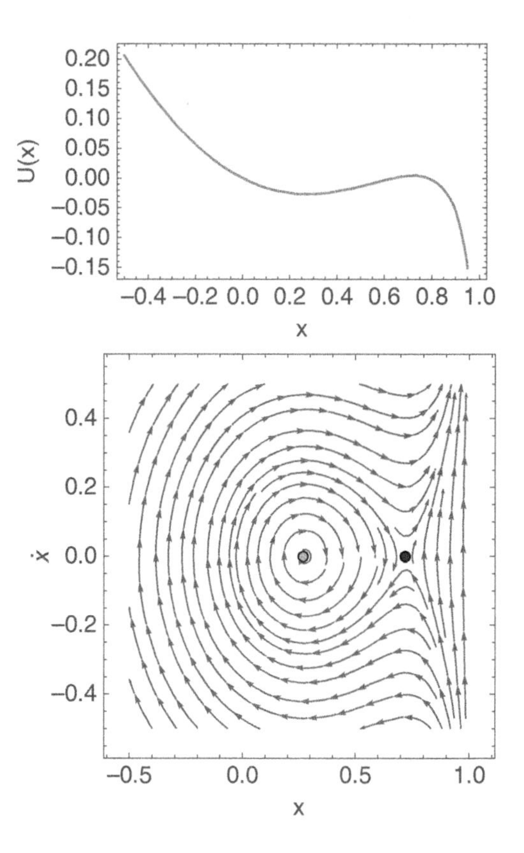


**Figure 9.9** Potential energy function and phase plane representation of the moving wire for $\Delta = 0.2$.

one reverts to using numerical techniques in order to generate the phase space. One powerful open-access numerical tool is called *pplane9* [3]. This graphical user interface system runs under Matlab and provides a numerical tool to generate the phase plane of dynamical systems. Another approach is based on using the *StreamPlot* function in the software *Mathematica*.

### Example 9.8 Phase-plane of a Damped Simple Pendulum

Construct the phase-plane representation of the damped simple pendulum whose equation of motion can be written as

$$\ddot{\theta} + \frac{c}{ml^2}\dot{\theta} + \frac{g}{l}\sin\theta = 0.$$

Use $g/l = 1$ rad/sec, $c = 1$ kg/sec, and $m = 1$ kg.

We use *Mathematica* to plot the phase-plane representation. To this end, we first express the dynamics in the general form:

$$\begin{cases} \dot{x}_1 = x_2 \\ \dot{x}_2 = -\frac{c}{ml^2}x_2 - \frac{g}{l}\sin x_1 \end{cases}$$

Subsequently, we use the following simple code in *Mathematica* to plot the phase plane.

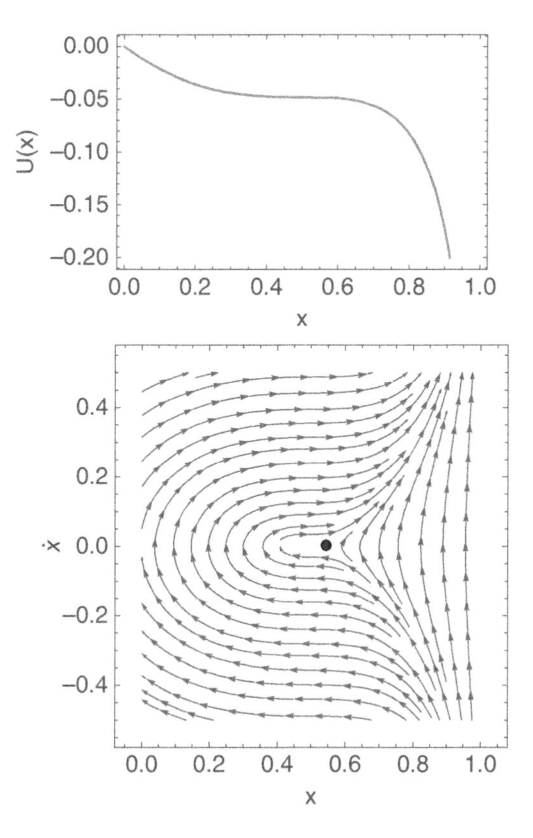


**Figure 9.10** Potential energy function and phase-plane representation of the moving wire for $\Delta = 0.25$.

```
In1=: StreamPlot[{x2, -Sin[x1] - x2}, {x1, -2 Pi,
2 Pi}, {x2, -2, 2 }, FrameLabel → {"x", "ẋ"},
LabelStyle → (FontSize → 18), StreamColorFunction
→ ColorData["HTML", "MidnightBlue"], StreamPoints →
Fine]
```

This yields the phase plane shown in Figure 9.11.

Note that the location of the fixed points does not change when adding damping to the system; only the nature of the trajectories around them changes. For the pendulum, the marginally-stable center changes into a stable node and the saddles remain unchanged by the addition of damping. The stable mainfolds entering the two saddles (red lines) create a region in the phase plane where all initial conditions lead to the stable node. This region is called the *basin of attraction* of the stable node.

### Example 9.9 Rotating T-shaped Rod with Damping

Reconsider Example 3.10, but this time we include a torsional damper of damping coefficient $c$ at the pivot. Construct the phase-plane representation of the dynamics for the following two cases: $\Omega^2 < \frac{48}{79}\frac{g}{l}$ and $\Omega^2 < \frac{48}{79}\frac{g}{l}$.

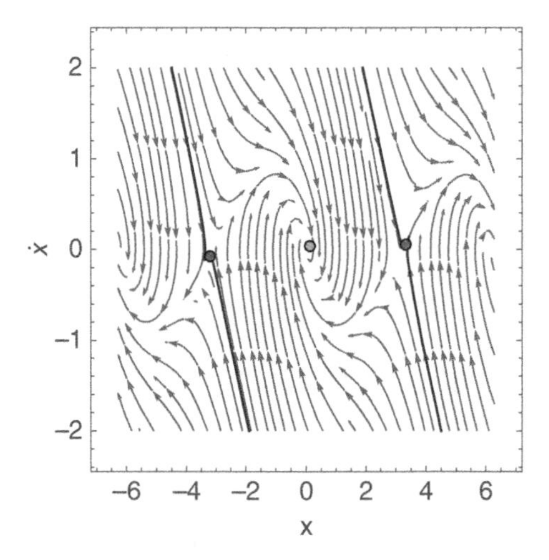


**Figure 9.11** Phase-plane representation of the damped simple pendulum. (*See color plate section for the color representation of this figure.*)

When adding a torsional damper at the pivot, the equation of motion is modified as

$$\ddot{\theta} + \frac{c}{ml^2}\dot{\theta} + \left(\frac{16}{27}\frac{g}{l} - \frac{79}{81}\Omega^2 \cos\theta\right)\sin\theta = 0.$$

Note that the addition of the damper does not change the equilibrium points, but only changes their stability characteristics. To draw the phase-plane representation for the case $\Omega^2 < \frac{48}{79}\frac{g}{l}$, we choose $g/l = 1$, $c = 1$ $m = 1$, and $\Omega^2 = 0.5$. Using these numerical values, we construct the phase-plane representation using *Mathematica's* StreamPlot command as follows:

```
In1=: StreamPlot[{x2, (16/27 - 0.5 79/81 Cos[x1]) Sin[x1]-
x2}, {x1, -1.1 Pi, 1.1 Pi}, {x2, -2, 2 }, FrameLabel
→ {"x", "ẋ"}, LabelStyle → (FontSize → 18), Stream-
ColorFunction → ColorData {"HTML", "MidnightBlue"},
StreamPoints → Fine]
```

This yields the phase-plane representation shown in Figure 9.12, which shows that the trivial equilibrium solution occurring at $\theta_0$ switched from being a center when there is no damping (Example 1.3) to become a stable node when damping is added. The basin of attraction of this stable node lies between the stable manifolds of the saddles at $\theta_0 = \pi$ and $\theta_0 = -\pi$.

To draw the phase-plane representation for the case $\Omega^2 > \frac{48}{79}\frac{g}{l}$, we choose $g/l = 1$, $c = 1$, $m = 1$, and $\Omega^2 = 1$. Using these numerical values, we construct the phase-plane representation using *Mathematica*. The resulting phase plane is shown in Figure 9.12, which illustrates that the saddles did not change their stability characteristics when damping was applied. On the other hand, the equilibrium solutions $\theta_0 = \pm \arccos\left(\frac{48g}{79l\Omega^2}\right)$ switched from being centers when there was no damping to become stable nodes when damping was added. Again, the basin of attraction of the stable nodes is determined by the stable manifolds of the saddles.

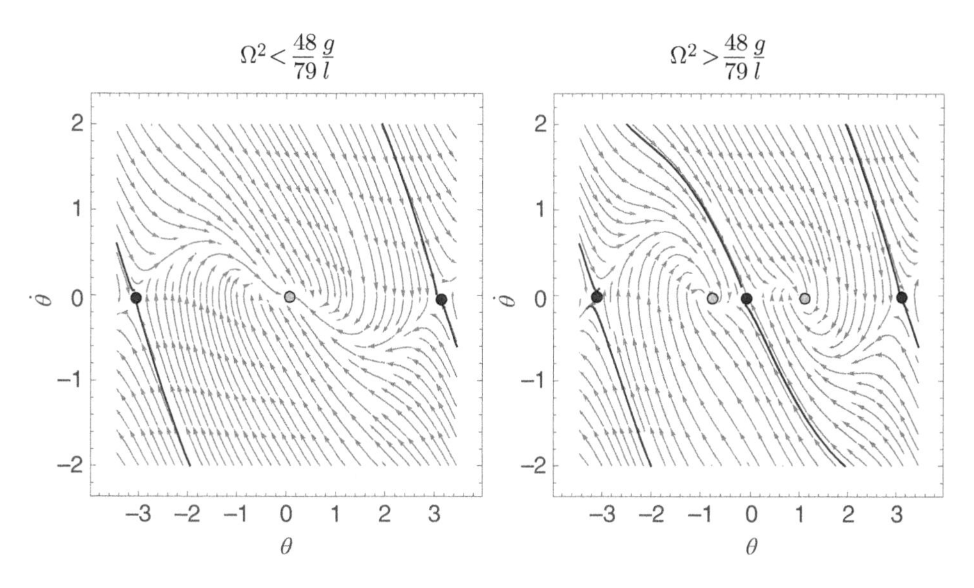


**Figure 9.12** Phase-plane representation of a damped rotating T-shaped rod. (*See color plate section for the color representation of this figure.*)

### Flipped Classroom Exercise 9.2

For the micro-mechanical switch discussed in Example 8.14, use the following numerical values: $\frac{1}{2}\frac{\varepsilon A}{k} = 4.45 \times 10^{-16}$ m$^3$/V$^2$, $d = 1$ mm, and find the equilibrium states of the switch for:

(i) $\mathcal{V} = 100$ V
(ii) $\mathcal{V} = 200$ V
(iii) $\mathcal{V} = 400$ V
(iv) $\mathcal{V} = 580$ V.

For each case, assess the stability of the equilibrium points and draw the phase-plane representation.

To solve this problem, take the following steps:

1. Use the equation of motion derived in Example 8.14.
2. Put the equation in the general form $\dot{\mathbf{x}} = \mathbf{G}(\mathbf{x})$ by letting $x_1 = y$, and $x_2 = \dot{y}$.
3. Find the equilibrium points by letting $\dot{x}_2 = 0, \dot{x}_1 = 0$. Show that, except for the fourth case involving $\mathcal{V} = 580$ V, there are three equilibrium points, one of which is always unrealizable because it is larger than the gap width $d$. For $\mathcal{V} = 580$ V, there is only one unrealizable equilibrium point.
4. Assess the stability of the resulting points and show that the equilibrium point that has the smaller magnitude is always a center, while the larger equilibrium is always a saddle.
5. Construct the phase plane for each case using the potential energy of the system.

## 9.5 Bifurcation of Equilibrium Solutions

To bifurcate is to divide or fork into two branches. The word bifurcation was first used by the French mathematician Poincaré, who was studying how equilibrium solutions change (divide and fork) as a certain parameter of interest is varied in a given dynamical system. For instance, one might be interested in understanding how changing the angular velocity $\Omega$ influences the equilibrium solutions in Example 8.9, or in understanding how the damping coefficient $c$ influences the stability of the equilibrium solutions of the simple pendulum. To this end, one would vary these parameters and inspect their influence on the system's equilibria. In such a scenario, we refer to $\Omega$ and $c$ as *bifurcation parameters*. We also refer to the point at which an equilibrium solutions changes behavior as the *bifurcation point*.

A bifurcation can also occur when more than one parameter changes simultaneously. Thus, the co-dimension of the bifurcation is dependent on the number of control parameters that need to be varied in order for the bifurcation to occur. For example, a bifurcation that occurs when $m$ bifurcation parameters are varied is called a codimension-$m$ bifurcation.

In general, bifurcations of equilibrium solutions can be classified into two major categories: *static* and *dynamic*. In a static bifurcation, one or more equilibrium solutions bifurcate into other equilibrium solutions at the bifurcation point. In other words, a static configuration gives way to another static configuration. Depending on the nature of the bifurcation, the resulting equilibrium solutions can be stable or unstable, and can have equal or different numbers of equilibrium solutions compared to the original.

In a dynamic bifurcation, however, an equilibrium solution gives way to a dynamic response; that is, a static configuration gives way to a dynamic motion at steady state. The resulting dynamic response can be periodic or aperiodic depending on the type of the bifurcation.

### 9.5.1 *Static Bifurcations*

Consider the bifurcation of the equilibrium solutions of

$$\dot{\mathbf{x}} = \mathbf{G}(\mathbf{x}, \mu), \tag{9.23}$$

under variations of the bifurcation parameter $\mu$ in the $\mathbf{x}$–$\mu$ state-control space. A static bifurcation is said to occur at $(\mathbf{x}_0, \mu_c)$ if the following conditions are satisfied:

- $\mathbf{G}(\mathbf{x}_0, \mu_c) = 0$
- the Jacobian matrix $\mathbf{D_x F}$ has a zero eigenvalue while all of its other eigenvalues have non-zero real parts at $(\mathbf{x}_0, \mu_c)$.

The first condition is used to ensure that $\mathbf{x}_0$ is an equilibrium point of the vector field $\mathbf{G}$ at $\mu_c$, while the second condition guarantees that this fixed point is non-hyperbolic. In the following, we consider the classification of static bifurcations.

***Saddle-node bifurcation*** Consider the first-order non-linear ordinary differential equation

$$\dot{x} = \mu - x^2, \tag{9.24}$$

where $\mu$ is a scalar bifurcation parameter. The equilibrium points of the system can be obtained by solving

$$\mu - x_0^2 = 0 \qquad \textit{or} \qquad x_0 = \pm\sqrt{\mu}, \qquad \mu > 0, \tag{9.25}$$

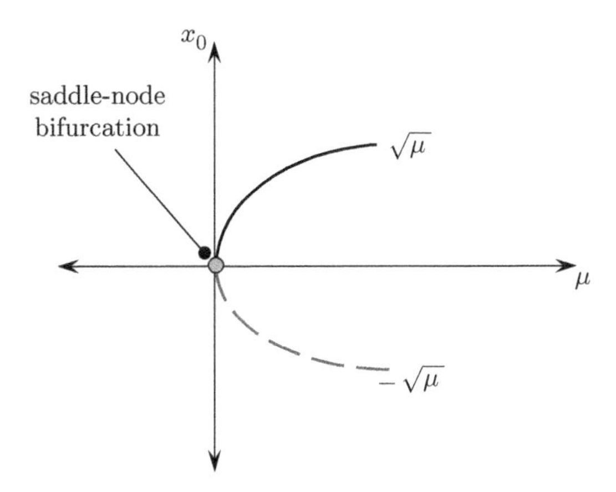


**Figure 9.13** Schematic representation of a saddle-node bifurcation. Dashed lines represent unstable solutions.

Note that when $\mu$ is less than zero, $x_0$ is undefined. The stability of the fixed points can be assessed by finding the eigenvalues of the Jacobian of Equation (9.24) evaluated at the roots, Equation (9.25). This yields

$$\lambda_{1,2} = -2x_0.$$

It follows that when $\mu = 0$, $x_0 = 0$ and $\lambda_{1,2} = 0$, meaning that $(x_0 = 0, \mu_c = 0)$ is a bifurcation point. To check the type of the bifurcation, we assess the existence and stability of the fixed points as a function of the bifurcation parameter, $\mu$.

- At $x_0 = \sqrt{\mu}$, $\lambda = -2\sqrt{\mu}$, meaning this root is stable.
- At $x_0 = -\sqrt{\mu}$, $\lambda = 2\sqrt{\mu}$, meaning this root is unstable.

Next, we plot these results in the state-control parameter space ($x$–$\mu$ space) to generate the bifurcation diagram shown in Figure 9.13. We note that at the point $(x_0 = 0, \mu_c = 0)$, there is a qualitative change in the equilibrium points of the system. To be specific, the branch of stable solutions (nodes) collides with the branch of unstable solutions (saddles) and they destroy each other, resulting in what we call a *saddle-node* bifurcation.

***Pitchfork Bifurcation*** Consider the first-order differential equation

$$\dot{x} = \mu x + \alpha x^3, \qquad \alpha > 0, \tag{9.26}$$

where $\mu$ is a scalar bifurcation parameter. The equilibrium points of the system can be obtained as $x_0 = 0$ (trivial equilibrium point) and $x_0 = \pm\sqrt{-\frac{\mu}{\alpha}}$ (non-trivial equilibrium points). Note that the trivial equilibrium point is defined everywhere, while the non-trivial equilibrium points are only defined when $\mu < 0$. Next, we find the eigenvalues

$$\lambda = \mu + 3\alpha x_0^2. \tag{9.27}$$

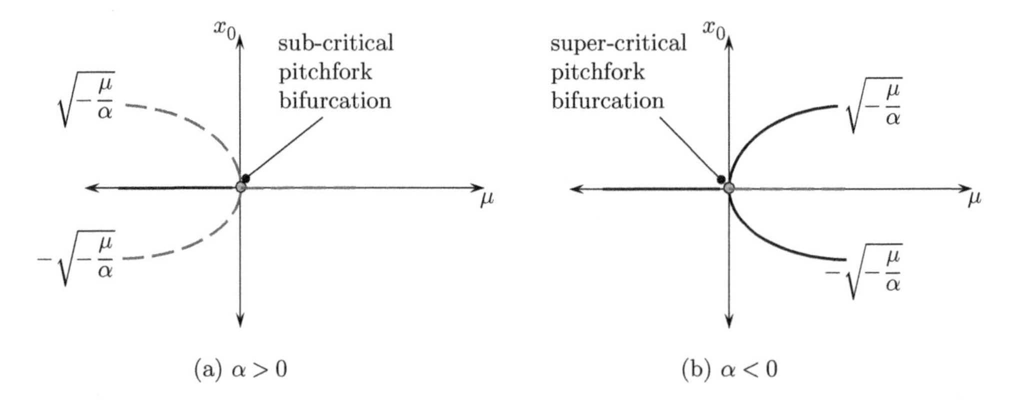


**Figure 9.14** Schematic representation of (a) a sub-critical pitchfork bifurcation and (b) a super-critical pitchfork bifurcation. Dashed lines represent unstable solutions.

Using Equation (9.27), we find

- At $x_0 = 0$, $\lambda = \mu$.
- At $x_0 = \pm\sqrt{-\frac{\mu}{\alpha}}$, $\lambda = -2\mu$.

Consequently, the trivial equilibrium point $x_0 = 0$ is stable when $\mu < 0$ and unstable when $\mu > 0$. On the other hand, the non-trivial solutions exist only when $\mu < 0$, and they are unstable. Using this understanding, we can construct the bifurcation diagram shown in Figure 9.14a. As $\mu$ is decreased towards $\mu_c = 0$, the unstable trivial solution, $x_0 = 0$, regains stability at $\mu_c = 0$. Furthermore, the unstable solutions $x_0 = \pm\sqrt{-\frac{\mu}{\alpha}}$ branch out at $(x_0 = 0, \mu_c = 0)$. These solutions remain unstable as $\mu$ is decreased further, resulting in the bifurcation behavior shown in Figure 9.14a. Such bifurcation behavior is known as a *sub-critical pitchfork* bifurcation.

When we restrict $\alpha$ to be negative, the non-trivial solutions are only defined when $\mu > 0$, and they are stable. The resulting bifurcation diagram for $\alpha < 0$ is shown in Figure 9.14b, where it can be clearly seen that as $\mu$ is increased, the stable trivial solution, $x_0 = 0$, becomes unstable at $\mu_c = 0$. Furthermore, the non-trivial stable solutions branch out at $(x_0 = 0, \mu_c = 0)$. These solutions remain stable as $\mu$ is increased further. Such bifurcation behavior is known as a *super-critical pitchfork* bifurcation.

***Transcritical Bifurcation*** Consider the system

$$\dot{x} = \mu x - x^2, \tag{9.28}$$

where $\mu$ is a scalar bifurcation parameter. For this system, the fixed points are $x_0 = 0$ (trivial), and $x_0 = \mu$ (non-trivial). The eigenvalue can be written as $\lambda = \mu - 2x_0$ :

- At $x_0 = 0$, $\lambda = \mu$.
- At $x_0 = \mu$, $\lambda = -\mu$.

In this case, both fixed points are defined everywhere in the space of $\mu$. The trivial solution is stable when $\mu < 0$ and unstable when $\mu > 0$. On the other hand, the non-trivial solution

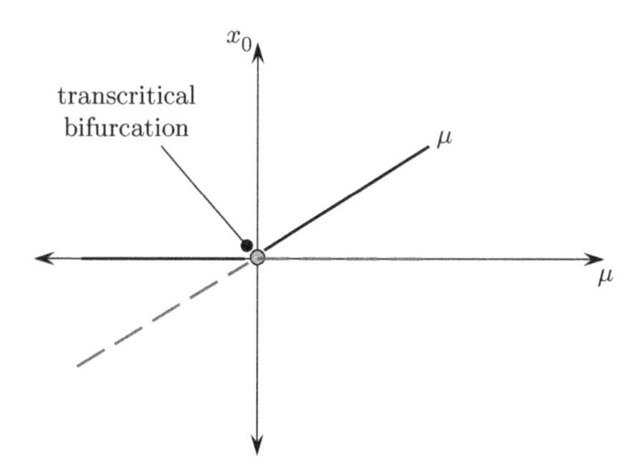


**Figure 9.15** Schematic representation of a transcritical bifurcation. Dashed lines are used to represent unstable solutions.

is stable when $\mu < 0$ and unstable when $\mu > 0$. As such, the two solutions collide at the bifurcation point and exchange stability. This bifurcation behavior is known as a *transcritical* bifurcation, and is shown in Figure 9.15.

**Example 9.10 Transcritical Bifurcation**

Consider the system

$$\dot{x} = x - \mu x(1 - x),$$

where $\mu$ is a scalar bifurcation parameter. Discuss the bifurcation which occurs at $\mu = 1$.

We find the fixed points of the system by letting $\dot{x} = 0$ and obtaining

$$x_{01} = 0, \qquad x_{02} = \frac{\mu - 1}{\mu}.$$

The eigenvalues can be obtained as $\lambda = 1 - \mu + 2\mu x_0$. Hence, for $x_{01}$, $\lambda = 1 - \mu$. It follows that $x_{01}$ is stable when $\mu > 1$ and is unstable otherwise. On the other hand, for $x_{02}$, $\lambda = 1 - \mu + 2\sqrt{\mu(\mu - 1)}$. It follows that $x_{02}$ is unstable when $\mu > 1$ and is stable otherwise. The bifurcation diagram is shown in Figure 9.16, illustrating that the two equilibrium points collide and exchange stability at $(x_0, \mu_c) = (0, 1)$. Therefore, there is a transcritical bifurcation at $\mu = 1$.

**Example 9.11 Rotating T-shaped Rod**

For the damped rotating T-shaped rod considered in Example 9.3, examine the bifurcation at $\Omega_{cr}^2 = \frac{48}{79}\frac{g}{l}$.

As mentioned before, the equilibrium points do not depend on the damping coefficient and were obtained in Example 9.3 as:

$$\begin{aligned}
(x_{10}, x_{20}) &= (\theta_0, \dot{\theta}_0) = ((i-1)\pi, 0), \qquad i = 1, 2, \ldots, \infty, \\
(x_{10}, x_{20}) &= (\theta_0, \dot{\theta}_0) = (\pm \arccos \frac{48}{79}\frac{g}{l\Omega^2}, 0), \qquad \Omega^2 \geq \frac{48}{79}\frac{g}{l}.
\end{aligned}$$

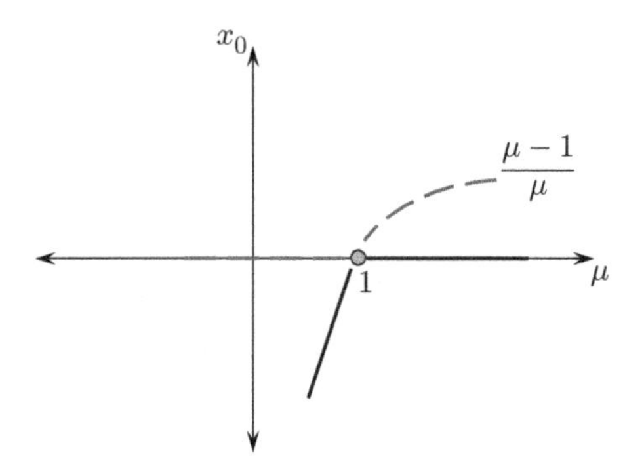


**Figure 9.16** A transcritical bifurcation at $\mu = 1$.

Since, for the first fixed point, anything beyond $i = 2$ is just a repetition of the same physical equilibrium points, we only consider the following equilibrium points:

$$\begin{aligned}(x_{10}, x_{20}) &= (\theta_0, \dot{\theta}_0) = (0, 0),\\ (x_{10}, x_{20}) &= (\theta_0, \dot{\theta}_0) = (\pi, 0),\\ (x_{10}, x_{20}) &= (\theta_0, \dot{\theta}_0) = (\pm \arccos \frac{48}{79} \frac{g}{l\Omega^2}, 0), \qquad \Omega^2 \geq \frac{48}{79}\frac{g}{l}.\end{aligned}$$

It can be clearly seen that the first two equilibrium points are independent of $\Omega$, while the third and fourth points depend on $\Omega$ and only exist when $\Omega^2 \geq \frac{48}{79}\frac{g}{l}$.

To determine the stability of the fixed points, we construct the Jacobian matrix

$$\mathbf{A} = \begin{bmatrix} 0 & 1 \\ \frac{1}{81}\left(79\Omega^2 \cos 2x_{10} - \frac{48g}{l} \cos x_{10}\right) & -\frac{c}{ml} \end{bmatrix}.$$

The eigenvalues can be obtained by solving the equation

$$|\mathbf{A} - \lambda_i \mathcal{I}| = 0.$$

The eigenvalues associated with the trivial fixed point $(0, 0)$ are

$$\lambda_{1,2} = \frac{1}{2}\left(-\frac{c}{ml} \pm \sqrt{\frac{316\Omega^2}{81} - \frac{64g}{27l} + \frac{c^2}{m^2l^2}}\right).$$

Since the damping coefficient $c > 0$, the fixed point $(0, 0)$ is

- stable when $\frac{316\Omega^2}{81} - \frac{64g}{27l} < 0$, or $\Omega^2 < \frac{48}{79}\frac{g}{l}$,
- unstable when $\frac{316\Omega^2}{81} - \frac{64g}{27l} > 0$, or $\Omega^2 > \frac{48}{79}\frac{g}{l}$.

The eigenvalues associated with the equilibrium points $(\pm \arccos \frac{48}{79}\frac{g}{l\Omega^2}, 0)$ are

$$\lambda_{1,2} = \frac{1}{2}\left(-\frac{c}{ml} \pm \sqrt{\frac{316\Omega^2}{81} - \frac{1024g}{711l^2\Omega^2} + \frac{c^2}{m^2l^2}}\right).$$

Note that when $\Omega^2 = \frac{48}{79}\frac{g}{l}$, the eigenvalues are $\lambda_{1,2} = 0$, so the equilibrium points, $(\pm \arccos \frac{48}{79}\frac{g}{l\Omega^2}, 0)$

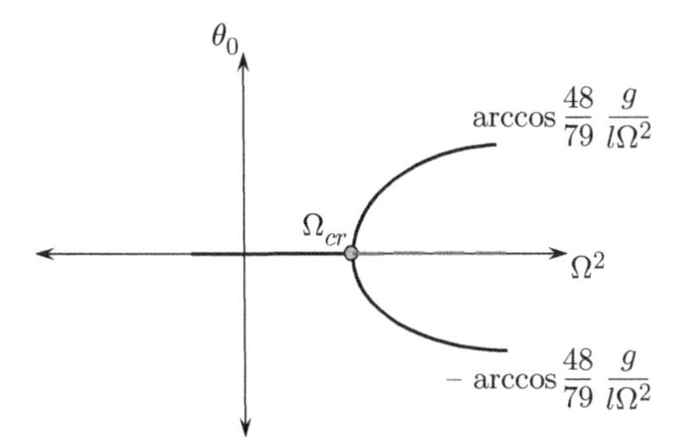


**Figure 9.17** Super-critical pitchfork bifurcation occurring at $\Omega_{cr}^2 = \frac{48}{79}\frac{g}{l}$.

- are stable when $\Omega^2 > \frac{48}{79}\frac{g}{l}$
- do not exist when $\Omega^2 < \frac{48}{79}\frac{g}{l}$.

Plotting the variation of the equilibrium points with the bifurcation parameter $\Omega^2$, as shown in Figure 9.17, it becomes evident that $\Omega_{cr}^2 = \frac{48}{79}\frac{g}{l}$ represents a super-critical pitchfork bifurcation.

**Example 9.12 Conductor Moving in the Magnetic Field of a Stationary Conductor**

Consider the equation of motion for the conductor moving in the magnetic field of another stationary wire. The equation of motion was derived in Example 9.7 as:

$$\ddot{x} + x - \frac{\Delta}{1-x} = 0.$$

Let us add a damping term $\dot{x}$ such that the equations become

$$\ddot{x} + \dot{x} + x - \frac{\Delta}{1-x} = 0.$$

Examine the bifurcation at $\Delta = \frac{1}{4}$.

We first note that when $\Delta > \frac{1}{4}$, there are no fixed points. For $0 < \Delta < \frac{1}{4}$, there are two fixed points, namely $x_{10} = \frac{1}{2} \pm \sqrt{\frac{1}{4} - \Delta}$. The stability of these two fixed points can be assessed by finding the eigenvalues of the Jacobian matrix of the system. These are given by:

$$\lambda_{1,2} = -\frac{\sqrt{4\Delta(x_{10}-1)^2 - 3(x_{10}-1)^4}}{2(x_{10}-1)^2}.$$

For the equilibrium point, $x_{10} = \frac{1}{2} + \sqrt{\frac{1}{4} - \Delta}$, $\lambda_{1,2}$ are negative and positive real numbers, so this equilibrium point is a saddle. On the other hand, for the equilibrium point $x_{10} = \frac{1}{2} - \sqrt{\frac{1}{4} - \Delta}$, $\lambda_{1,2}$ are complex conjugates with negative real parts, and so the fixed points are stable foci. The bifurcation diagram is shown in Figure 9.18 illustrating that the bifurcation is of the saddle-node type.

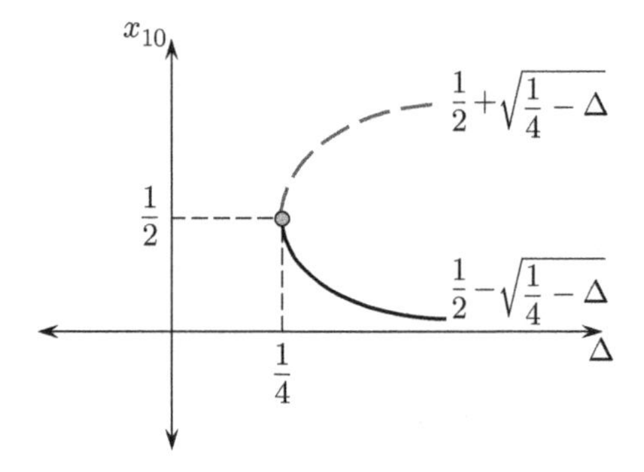


**Figure 9.18** Saddle-node bifurcation occurring at $\Delta = \frac{1}{4}$.

 **Flipped Classroom Exercise 9.3**

Examine the bifurcation that the system

$$\dot{x} = \mu + x - \ln(1 + x)$$

undergoes as the parameter $\mu$ is varied.

To solve this problem, take the following steps:

1. As you try to find the fixed points, you will quickly realize that you will not be able to find them analytically. As such, you should revert to a graphical approach, where you plot the function $\ln(1 + x)$ against $\mu + x$ for different values of $\mu$.
2. Identify the point $\mu_c$ where the two functions are tangential to each other. How many fixed points are there when $\mu > \mu_c$ and when $\mu < \mu_c$?
3. Find the eigenvalue of the system as function of the equilibrium points. Deduce which points are stable and which points are not.
4. Show that there is a saddle-node bifurcation when $\mu_c = 0$.

 **Flipped Classroom Exercise 9.4**

Examine the bifurcation that the system

$$\dot{x} = x + \frac{\mu x}{1 + x^2}$$

undergoes as the parameter $\mu$ is varied.

To solve this problem, take the following steps:

1. Find the equilibrium points of the system. Show that there are three equilibrium points. What are they?

2. Find the eigenvalues associated with each fixed point.
3. Using the resulting eigenvalues, show that there is a sub-critical pitchfork bifurcation at $\mu = -1$.

### 9.5.2 Dynamic (Hopf) Bifurcation

At a static bifurcation point, equilibrium solutions can lose or gain stability and can appear or disappear. At a dynamic (*Hopf*) bifurcation, however, an equilibrium point gives way to a dynamic solution. To better understand what a Hopf bifurcation is, consider a physical example where a mechanical damped oscillator is subjected to a fluid stream as shown in Figure 9.19. The oscillator is given an initial condition in the presence of the moving fluid. When the flow speed past the oscillator $U$ is below a critical value, known as the galloping speed, $U_{cr}$, the oscillator returns to its equilibrium state no matter how it is perturbed (Figure 9.19a). We can therefore say that the equilibrium state of the oscillator is stable. On the other hand, when the flow speed increases beyond the critical galloping speed, $U > U_{cr}$, any initial condition causes the oscillator to undergo large-amplitude sustained periodic motions, as shown in Figure 9.19b. Therefore, we can say that, at the critical galloping speed $U_{cr}$, the stable equilibrium state loses stability and gives way to a periodic response. This phenomenon, also known as the galloping instability, represents a Hopf bifurcation point. Hopf bifurcations are known to occur in many dynamical systems, including aeroelastic structures such as aircraft wings, where they produce flutter. They also produce galloping in cables and undesirable vibrational "chatter" in rotating machine tools.

For the dynamical system $\dot{\mathbf{x}} = \mathbf{G}(\mathbf{x}, \mu)$, a Hopf bifurcation is said to occur at $(\mathbf{x}_0, \mu_c)$ when the following conditions are satisfied:

- $\mathbf{F}(\mathbf{x}_0, \mu_c) = 0$.
- The Jacobian has a pair of purely imaginary eigenvalues $\pm\imath\omega$, while all of its other eigenvalues have non-zero real parts at $(\mathbf{x}_0, \mu_c)$.
- The eigenvalues must cross the imaginary axis and not only arrive there and reverse direction. In other words, they must have a non-zero speed when crossing the imaginary axis. This can be guaranteed by letting the analytic continuation of the pair of imaginary eigenvalues be $\zeta \pm \imath\omega$. For $(\mathbf{x}_0, \mu_c)$ to be a Hopf bifurcation, $(\frac{d\zeta}{d\mu}) \neq 0$ at $\mu = \mu_c$. This condition is commonly known as the *transversality condition.*

Figure 9.20 illustrates the three conditions necessary for a Hopf bifurcation to occur.

Mathematically, a Hopf bifurcation can be represented using the dynamics of the system given by the following equations:

$$\dot{r} = \mu r - r^3, \qquad \dot{\theta} = \omega + br^2,$$

where $r$ and $\theta$ are generalized polar coordinates. Here, the dynamics of $r$ are independent of those associated with $\theta$, so one can examine the equilibrium points associated with the $r$-dynamics independently. These equilibrium solutions are given by $r_{01} = 0$ and

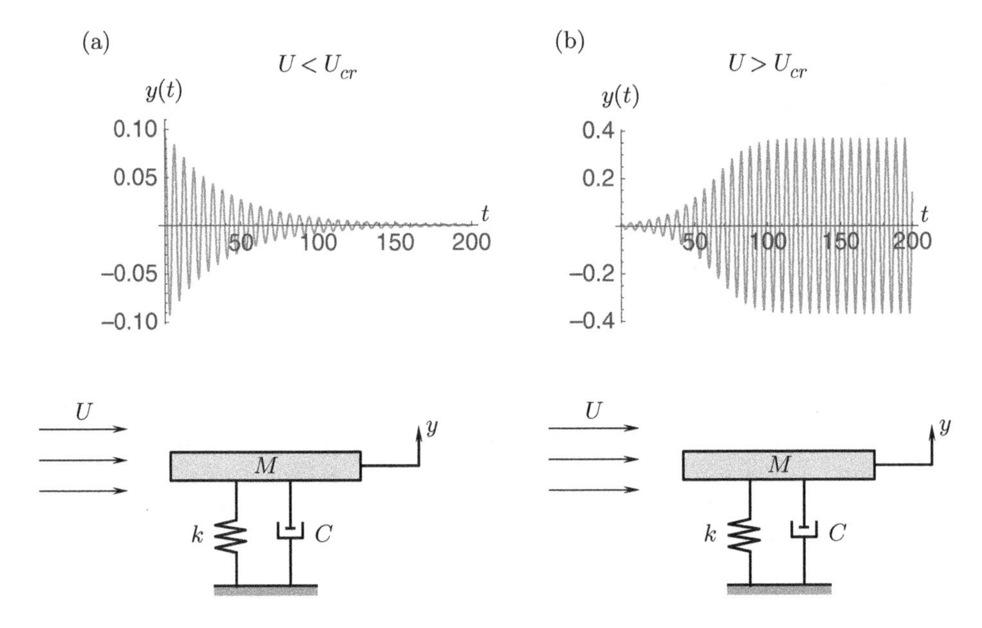


**Figure 9.19** Galloping instability is an example of a Hopf bifurcation.

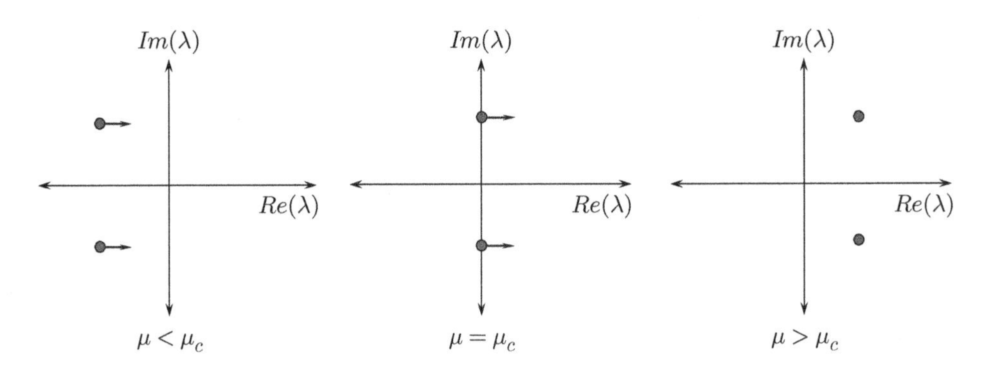


**Figure 9.20** Location of the eigenvalues associated with the equilibrium point as a function of the bifurcation parameter $\mu$ for a Hopf bifurcation.

$r_{02,03} = \pm\sqrt{\mu}, \mu > 0$. The associated eigenvalues are

$$\lambda_{1,2} = \mu - 3r_0^2.$$

It follows that the equilibrium point $r_{01}$ is stable when $\mu < 0$ and unstable when $\mu > 0$. On the other hand, the equilibria $r_{02,03}$ are stable when $\mu > 0$ and do not exist when $\mu > 0$. Therefore, as shown in Figure 9.21, the equilibrium points associated with the $r$-dynamics exhibit a super-critical pitchfork bifurcation at $\mu_c = 0$.

When including the $\theta$-dynamics in the analysis, one can directly show that $\theta(t) = (\omega + br_0^2)t + \theta_0$ at the equilibrium points, which implies that $\theta$ varies linearly with time for a given

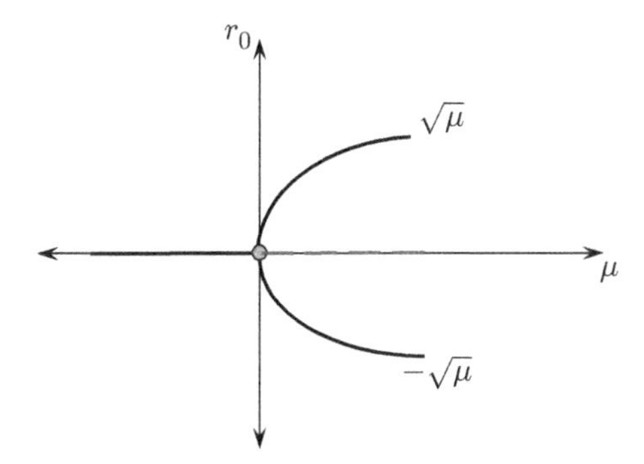


**Figure 9.21** Super-critical pitchfork bifurcation in the $(r_0, \mu)$ space.

equilibrium point $r_0$. As such, upon including the $\theta$ dynamics, $(r_0, \theta)$ is no longer a static configuration, but is actually a closed circular orbit of radius $r_0$, as shown in Figure 9.22. This means that at $\mu_c = 0$, the static equilibrium point $r_{01} = 0$ gives way to a periodic orbit whose radius is given by the magnitude of the non-trivial solutions, thereby announcing the occurrence of a super-critical Hopf bifurcation at $\mu_c = 0$.

It is also possible to arrive at a similar conclusion when projecting the dynamics into the Cartesian coordinate system by introducing the transformations $x = r\cos\theta$ and $y = r\sin\theta$. This yields

$$\dot{x} = \mu x - \omega y, \qquad \dot{y} = \omega x + \mu y,$$

for which the equilibrium point is $(x_0, y_0) = (0, 0)$. Constructing the Jacobian matrix and finding the eigenvalues, we obtain

$$\lambda_{1,2} = \mu \pm \imath\omega,$$

It follows that at $\mu_c = 0$, there rea two purely imaginary eigenvalues $\pm\imath\omega$ (the first two conditions for a Hopf bifurcation). To check for the third and final condition, we differentiate the real part of $\lambda_{1,2}$ with respect to the bifurcation parameter $\mu$ and evaluate the derivative at $\mu = \mu_c$, which yields a value of one. Therefore, the eigenvalues have a non-zero speed crossing of the imaginary axis, which means that there is a Hopf bifurcation at $(x_0, y_0) = (0, 0)$ and $\mu_c = 0$.

Similar to the sub-critical pitchfork bifurcation, there is also a sub-critical Hopf bifurcation in which an unstable periodic solution collides with a stable equilibrium solution, as shown in Figure 9.23.

**Example 9.13 Subcritical Hopf Bifurcation**

Show that the system

$$\dot{r} = \mu r + r^3 - r^5, \qquad \dot{\theta} = \omega + br^2,$$

has a sub-critical Hopf bifurcation at $\mu_c = 0$.

To solve this problem, we first note that the $r$-dynamics are decoupled from the $\theta$-dynamics. As a result, we can analyze the bifurcation in terms of the variable $r$. Once we prove that the $r$-dynamics undergo a sub-critical pirchfork bifurcation at $\mu_c = 0$, we can infer that $(r, \theta)$

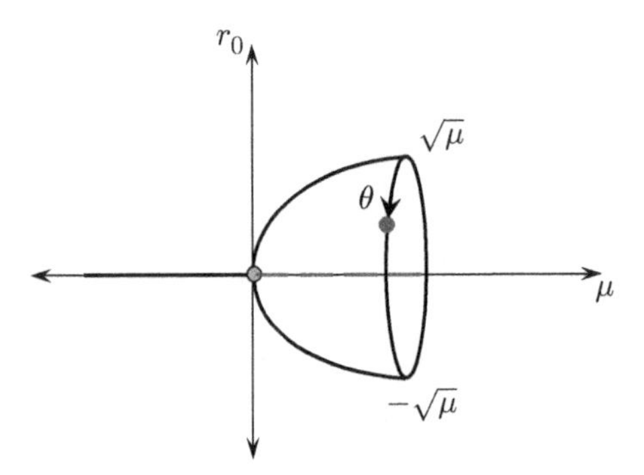


**Figure 9.22** Super-critical Hopf bifurcation in the $(r_0, \mu)$ space.

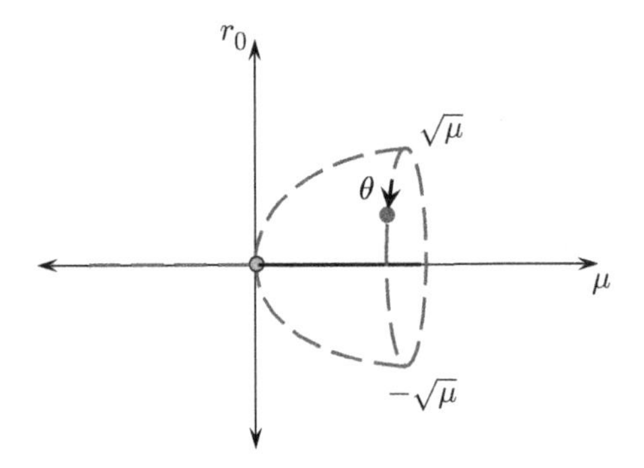


**Figure 9.23** Sub-critical Hopf bifurcation in the $(r_0, \mu)$ space.

undergoes a Hopf bifurcation at $\mu_c = 0$. To this end, we find the equilibrium points of the system by letting $\dot{r} = 0$. This yields five equilibrium points:

$$\begin{aligned} r_0^1 &= 0, \\ r_0^{2,3} &= \pm\sqrt{\frac{1-\sqrt{1+4\mu}}{2}}, \qquad -\frac{1}{4} \leq \mu \leq 0, \\ r_0^{4,5} &= \pm\sqrt{\frac{1+\sqrt{1+4\mu}}{2}}, \qquad \mu \geq -\frac{1}{4}. \end{aligned}$$

The Jacobian associated with the $r$-dynamics can be written as

$$J = \mu + 3r_0^2 - 5r_0^4.$$

For $r_0^1$, $J = \mu$, so $r_0^1 = 0$ is stable when $\mu < 0$ and unstable otherwise. For $r_0^{2,3}$, $J = -1 - 4\mu + \sqrt{1+4\mu}$, so $r_0^{2,3}$ are unstable when $-\frac{1}{4} < \mu < 0$ and do not exist otherwise. This implies that there is a sub-critical pitchfork bifurcation at $\mu_c = 0$, as shown in Figure 9.24. A sub-critical pitchfork bifurcation occurring in the space of $r$ and $\mu$ is equivalent to a sub-critical Hopf bifurcation occurring in the space of $r$, $\theta$, and $\mu$.

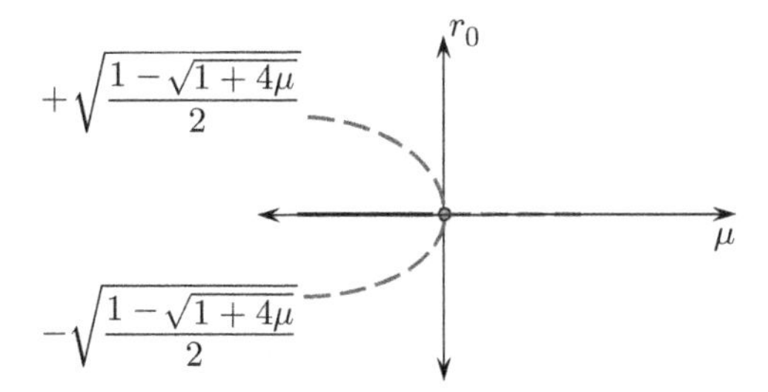


**Figure 9.24** A sub-critical pitchfork bifurcation in the ($r$-$\mu$) parameter space.

### Example 9.14 Watt Governor System

Consider the Watt governor system shown in Figure 9.25. The governor is used to control the flow rate of steam from a boiler to a steam engine so as to maintain a constant rotational speed $\Omega$ at the load. The system works as follows: the output shaft of the engine is connected through a bevel gear system of ratio $\alpha$ to the governor. This causes the governor's axle to rotate at an angular velocity $\dot{\theta} = \alpha\Omega$. When the rotational speed of the axle increases beyond a desired

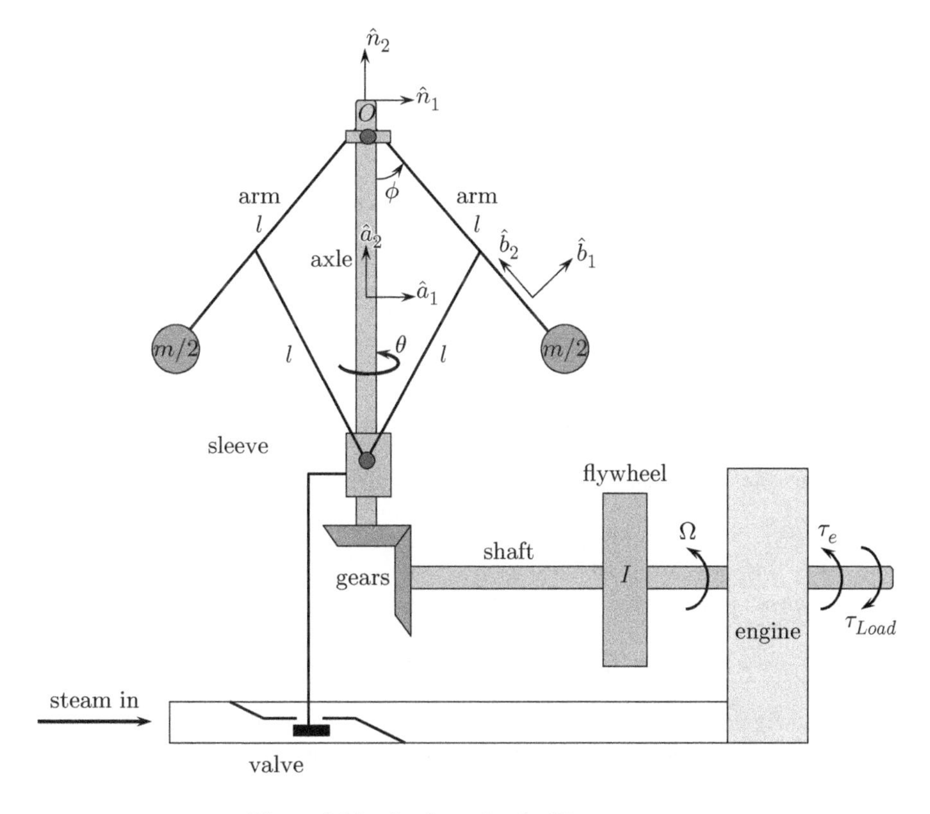


**Figure 9.25** A schematic of a Watt governor.

value, the arms of the governor, each of length $l$, rise. This pulls the valve upwards via a sleeve that supports the arms and slides along the governor's axle. This process controls the torque $\tau_e$ at the engine side, where the relation between the engine torque and the angle $\phi$ can be assumed to follow the form $\tau_e = \tau_0 \cos\phi$. Here, $\tau_e$ is zero when $\phi = \pi/2$.

Assume that the total mass of the system is dominated by the mass of the fly wheel, which has a moment of inertia $I$ and the mass of the particles $\frac{m}{2}$ at the end of the governor's arms. Also assume that when the arms rise, there is a counter-torque on the arms due to damping, which is given by $\tau_{\text{damp}} = cl^2\dot{\phi}$, where $c$ is a viscous damping coefficient.

(a) Show that the equation of motion of the system can be written as

$$\begin{aligned} \ddot{\phi} &- \alpha^2\Omega^2 \cos\phi \sin\phi + \frac{g}{l}\sin\theta + \frac{c}{m}\dot{\phi} = 0, \\ I\dot{\Omega} &= \tau_0 \cos\phi - \tau_{\text{load}}, \end{aligned}$$

where $\tau_{\text{load}}$ is the torque at the load side.

(b) Normalize the equations using $x = \phi$, $y = \sqrt{\frac{l}{g}}\dot{\phi}$, $z = \alpha\sqrt{\frac{l}{g}}\Omega$, $t = \sqrt{\frac{l}{g}}$.

(c) Show that the system exhibits a Hopf bifurcation for some combination of the design parameters.

**(a)** ***Equation of motion*** To describe the kinematics of the system, we first define the $A$-frame such that ${}^N\omega^A = \dot{\theta}\hat{a}_2$ and the $B$-frame such that ${}^A\omega^B = \dot{\phi}\hat{b}_3$ (see Figure 9.26). Subsequently, we use Euler rotational equations to find the equation of motion; that is, we use the equation

$$M_{3O} = I_{33O}\alpha_3 + (I_{11O} - I_{22O})\omega_1\omega_2,$$

where all the quantities are described in the $B$-frame such that

$$\begin{aligned} {}^N\boldsymbol{\omega}^B &= {}^N\boldsymbol{\omega}^B = \dot{\theta}\sin\phi\hat{b}_1 + \dot{\theta}\cos\phi\hat{b}_2 + \dot{\phi}\hat{b}_3, \text{ or} \\ \omega_1 &= \dot{\theta}\sin\phi, \qquad \omega_2 = \dot{\theta}\cos\phi, \qquad \omega_3 = \dot{\phi}, \qquad \alpha_3 = \ddot{\phi}. \end{aligned}$$

The moments of inertia and moment are

$$\begin{aligned} I_{11O} &= ml^2, \qquad I_{22O} = 0, \qquad I_{22O} = 0, \\ M_{3O} &= -mgl\sin\phi - cl^2\dot{\phi}. \end{aligned}$$

Note that the torque at the gear does not produce a moment in the $\hat{b}_3$ direction. It follows that

$$\ddot{\phi} - \dot{\theta}^2 \sin\phi\cos\phi + \frac{g}{l}\sin\phi + \frac{c}{m}\dot{\phi} = 0.$$

Using the gear ratio $\alpha$, one can write $\dot{\theta} = \alpha\Omega$, which yields

$$\ddot{\phi} - \alpha^2\Omega^2 \sin\phi\cos\phi + \frac{g}{l}\sin\phi + \frac{c}{m}\dot{\phi} = 0.$$

To find the equation of motion of the flywheel, we carry out a torque balance. We neglect the torque exerted by the governor on the shaft, assuming it is negligible. This yields

$$I\dot{\Omega} = \tau_0 \cos\phi - \tau_{\text{load}}.$$

**(b)** ***Normalization*** Using the normalization scheme represented in the problem statement, the equation of motion can be written as:

$$\begin{aligned} x' &= y, \\ y' &= z^2 \sin x \cos x - \sin x - \zeta y, \\ z' &= \beta(\cos x - \rho), \end{aligned}$$

where the prime is a derivative with respect to the normalized time $\tau$, $\zeta = \frac{c}{m}\sqrt{\frac{l}{g}} > 0$, $\beta = \frac{\alpha l \tau_0}{gI} > 0$, and $\rho = \frac{\tau_{\text{load}}}{\tau_0}$, $0 < \rho < 1$.

**(c)** ***Hopf bifurcation*** We first find the equilibrium solutions of the system by letting $x' = y' = z' = 0$, which yields the following equilibrium solutions:

$$(x_0, y_0, z_0) = (\cos^{-1}\rho, 0, \pm\frac{1}{\sqrt{\rho}}).$$

The Jacobian matrix evaluated at the first fixed points $(\cos^{-1}\rho, 0, \frac{1}{\sqrt{\rho}})$ yields

$$J = \begin{bmatrix} 0 & 1 & 0 \\ \frac{\rho^2-1}{\rho} & -\zeta & 2\sqrt{\rho}\sqrt{1-\rho^2} \\ -\beta\sqrt{1-\rho^2} & 0 & 0 \end{bmatrix}.$$

The eigenvalues are then obtained using the following characteristic equation:

$$\lambda^3 + \zeta\lambda^2 + \frac{1-\rho^2}{\rho}\lambda + 2\beta\sqrt{\rho}(1-\rho^2) = 0.$$

Since this is a cubic equation, it is very difficult to find insightful expressions for the eigenvalues. One way to assess the sign of the eigenvalues without actually finding them is by constructing the *Routh–Hurwitz array*. The reader can refer to the book by Ogata for more detail [4].

$$\begin{array}{lcc} \lambda^3 & 1 & \frac{1-\rho^2}{\rho} \\ \lambda^2 & \zeta & 2\beta\sqrt{\rho}(1-\rho^2) \\ \lambda^1 & \dfrac{\zeta\frac{(1-\rho^2)}{\rho} - 2\beta\sqrt{\rho}(1-\rho^2)}{\mu} & 0 \\ \lambda^0 & 2\beta\sqrt{\rho}(1-\rho^2) & \end{array}$$

According to the Routh–Hurwitz criterion, all the eigenvalues will have negative real parts when the following conditions are satisfied:

$$\zeta > 0, \quad \frac{1-\rho^2}{\rho}(\zeta - 2\beta\rho^{\frac{3}{2}}) > 0, \quad 2\beta\sqrt{\rho}(1-\rho^2) > 0.$$

The first and third conditions are satisfied by construction. This implies that when the second condition is violated, the system changes stability at the critical value

$$\zeta_c = 2\beta\rho^{\frac{3}{2}}.$$

It follows that a bifurcation must exist at $\zeta_c = 2\beta\rho^{\frac{3}{2}}$. Nevertheless, it is not straightforward to show that $\zeta_c$ corresponds to a Hopf bifurcation without actually finding the eigenvalues, and showing that there is a pair of purely imaginary eigenvalues at $\zeta_c$ while the rest have non-zero real parts.

One way to circumvent this problem is to assume that, at the critical value $\zeta_c$, there exists a Hopf bifurcation and then try to prove otherwise. To this end, we assume there are two purely imaginary eigenvalues at $\zeta_c$ while the third eigenvalue is real; that is,

$$\lambda_1 = \imath\omega, \lambda_2 = -\imath\omega, \lambda_3 = \gamma.$$

Using the assumed roots, the characteristic equation can be constructed as

$$(\lambda - \imath\omega)(\lambda + \imath\omega)(\lambda - \gamma) = 0$$

or

$$\lambda^3 - \gamma\lambda^2 + \omega^2\lambda - \alpha\omega^2 = 0.$$

Comparing the previous equation to the original characteristic equation one finds that

$$\gamma = -\zeta, \qquad \omega^2 = \frac{1-\rho^2}{\rho}, \qquad \gamma\omega^2 = -2\beta\sqrt{\rho}(1-\rho^2)$$

or

$$\omega^2 = \frac{1-\rho^2}{\rho}, \qquad \alpha\omega^2 = \frac{2\beta\sqrt{\rho}(1-\rho^2)}{\zeta}.$$

It follows that

$$\frac{1-\rho^2}{\rho} = \frac{2\beta\sqrt{\rho}(1-\rho^2)}{\zeta},$$

or

$$(\zeta - 2\beta\rho^{\frac{3}{2}})\frac{1-\rho^2}{\rho} = 0.$$

Since $0 < \rho < 1$, the previous equation can only be satisfied when $\zeta = 2\beta\rho^{\frac{3}{2}}$. As such, we have proven that at $\zeta = \zeta_c$, there is a pair of purely imaginary eigenvalues $\pm\imath\omega$ while the third eigenvalue is real.

This conclusion is not enough to prove the existence of a Hopf bifurcation. We still need to check whether the transversality condition is satisfied at $\zeta = \zeta_c$. As such, we let the analytic continuation of the eigenvalues $\pm\imath\omega$ be $\lambda_1, \lambda_2$. We differentiate all the terms in the characteristic equation with respect to $\zeta$, then substitute $\zeta = \zeta_c$ in the outcome to obtain

$$\frac{\mathrm{d}\lambda_{1,2}}{\mathrm{d}\zeta} = \frac{1 \pm \imath\zeta_c\omega}{2(1 + \frac{\zeta_c^2}{\omega^2})}.$$

Because the real parts of $\frac{\mathrm{d}\lambda_{1,2}}{\mathrm{d}\zeta}$ are non-zero, the transversality condition is satisfied. As a consequence, the Watt governor undergoes a Hopf bifurcation at $(\cos^{-1}\rho, 0, \frac{1}{\sqrt{\rho}}, \zeta = \zeta_c)$.

Next, we consider the other fixed point, namely $(x_0, y_0, z_0)$=$(\cos^{-1}\rho, 0, -\frac{1}{\sqrt{\rho}})$. In this case, the characteristic equation can be written as

$$\lambda^3 + \zeta\lambda^2 + \frac{1-\rho^2}{\rho}\lambda - 2\beta\sqrt{\rho}(1-\rho^2) = 0.$$

Using the Routh-Hurwitz criterion, the conditions for eigenvalues with negative real parts are

$$\zeta > 0, \qquad (\zeta + 2\beta\rho^{\frac{3}{2}})\frac{1-\rho^2}{\rho} > 0, \qquad -\beta\sqrt{\rho}(1-\rho^2) > 0.$$

Since $\zeta$, $\beta > 0$ and $0 < \rho < 1$, the third condition will never be satisfied. As such, the fixed point $(\cos^{-1}\rho, 0, \frac{-1}{\sqrt{\rho}})$ is always unstable.

## 9.6 Basins of Attraction

The basin of attraction of a stable equilibrium solution represents the set of all initial conditions in the phase plane that eventually lead to the equilibrium point. In the presence of multiple stable equilibrium states, a graphical representation of the basin of attraction in the phase space is very insightful in understanding the regions of initial conditions that lead to one stable state versus another.

The process of finding the basin of attraction of a non-linear dynamical system involves integrating the equation of motion for a large number of initial conditions then grouping all initial conditions that lead to each equilibrium point together. Each group of initial conditions can then be represented graphically in the phase space using different colors, resulting in a colorful map with boundaries that demarcate the different basins of attraction.

As an example, consider the phase-plane representation of the rotating T-rod with damping, as discussed in Example 1.9 for the case when $\Omega^2 > \frac{48g}{79l}$. The map is shown in Figure 9.26.

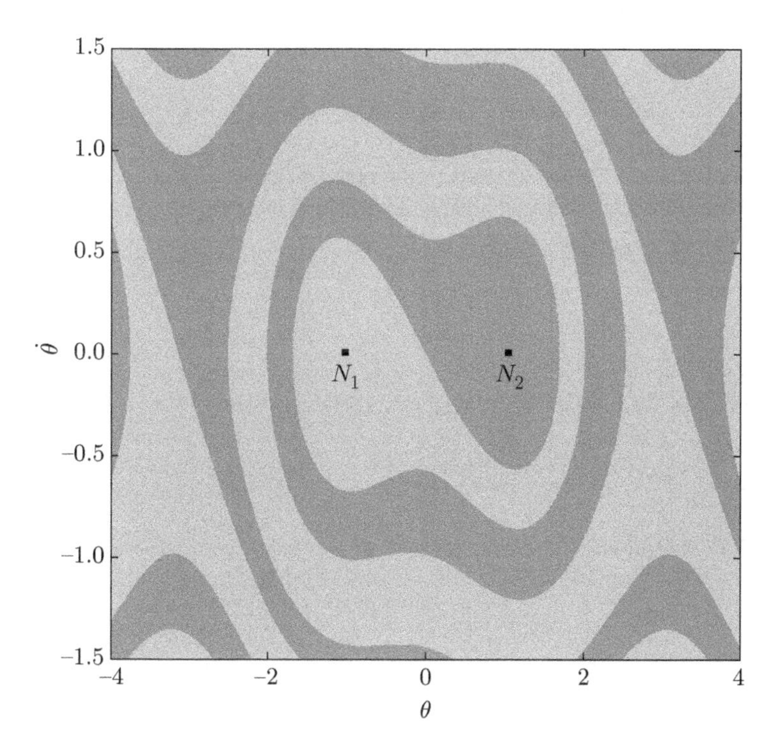


**Figure 9.26** Basins of attraction of the equilibrium solutions of the rotating T-shaped rod for $\Omega^2 > \frac{48g}{79l}$. (*See color plate section for the color representation of this figure.*)

It was generated using *Dynamics*, which is a powerful open-source software developed by Nusse and Yorke at the University of Maryland [5]. The map clearly illustrates the basin of attraction for each of the stable equilibria $(\theta_0, \dot{\theta}_0) = (\pm \arccos \frac{48}{79} \frac{g}{l\Omega^2}, 0)$, marked $N_1$ and $N_2$ on the figure.

## Exercises

9.1 Show that the system

$$\ddot{v} + 3v + 6v^2 + 4v^3 + v^4 = 0,$$

has a center at $v = 0$ and a saddle at $v = -1$.

9.2 Find the fixed points of the system

$$\dot{x} = x + e^{-y}, \qquad \dot{y} = -y,$$

and classify them.

9.3 Find the equilibrium angle of the system discussed in Example 5.4. Is it stable or unstable?

9.4 Analyze the stability of the equilibrium points of the electric dipole discussed in Example 8.11.

9.5 Analyze the stability of the equilibrium points of the electric tripole discussed in Flipped Classroom Exercise 8.2.

9.6 Sketch the phase-plane representation and classify the fixed points of the following system

$$\ddot{x} = ax + bx^3,$$

for $a = 1, b = -1$. What happens to the phase-plane representation if we add the term $-\dot{x}$ to the right-hand side of the equation? Sketch it again.

9.7 Consider the system of equations:

$$\dot{x}_1 = x_1(3 - x_1 - 2x_2), \qquad \dot{x}_2 = x_2(2 - x_1 - x_2),$$

where $x_1, x_2 \geq 0$.

(a) Find the fixed points of the system and classify them.

(b) Using the results you obtained in (a), sketch the phase-plane representation in the $x_1, x_2$ space. Remember that $x, y \geq 0$. On the sketch show the stable and unstable manifolds.

9.8 Consider the following dynamical system:

$$\dot{x} = rx - \sin x, \qquad x \geq 0,$$

Answer the following:

(a) In the limiting case when $r = 0$, find the fixed points and classify them.
(b) In the limiting case when $r \gg 1$, find the fixed points and classify them.
(c) Discuss the bifurcations that occur when $0 < r < \infty$. *Hint*: Try to graphically construct the functions $rx$ and $\sin x$ and investigate the nature of the intersections and tangents.

9.9 Discuss the bifurcation that takes place at $(x, y, \mu) = (0, 0, 0)$ for the system:

$$\dot{x} = y, \qquad \dot{y} = -x + \mu y - x^2 y.$$

9.10 Discuss the bifurcation that the system

$$\dot{x} = \mu x - \sinh x,$$

undergoes as $\mu$ is varied. *Hint*: $\sinh x = \frac{1}{2}(e^x - e^{-x})$; try to graphically construct the functions and look at the intersections and the tangents to solve the problem.

9.11 Discuss the bifurcation of the fixed points of the following system when $\mu$ is varied:

$$\dot{x} = \mu - 3x^2.$$

9.12 Show that the system

$$\dot{x} = rx - xe^x,$$

exhibits a bifurcation at $r = 1$. What type of bifurcation is it?

9.13 Show that the equation of motion for a simple pendulum forced to rotate around its axis at a fixed rate $\Omega$ can be written as

$$\ddot{\theta} + (1 - \Lambda \cos \theta) \sin \theta = 0, \qquad \Lambda = \frac{\Omega^2 l}{g},$$

where $l$ is the length of the pendulum and $g$ is the gravitational acceleration. Find the fixed points and assess their stability. Sketch the phase plane of the system for $\Lambda < 1$, $\Lambda = 1$, and $\Lambda > 1$. What type of bifurcation occurs at $\Lambda = 1$?

9.14 Show that

$$\dot{x} = x(1 - x^2) - \alpha(1 - e^{-\beta x}),$$

undergoes a transcritical bifurcation.

9.15 Show that the micro-mechanical switch of Example 8.14 with numerical parameters $\frac{1}{2}\frac{\varepsilon A}{k} = 4.45 \times 10^{-16}$ m$^3$/V$^2$, $d = 1$ mm exhibits a saddle-node bifurcation at $\mathcal{V} \approx 578$ V.

## References

1. Nayfeh, H. and Balachandran B. (1995) *Applied Nonlinear Dynamics: Analytical, Computational and Experimental Methods,* John Wiley and Sons, Inc.
2. Strogatz S. (2014) *Nonlinear Dynamics and Chaos: With Applications to Physics, Biology, Chemistry, and Engineering,* West View Press.
3. Arnold D. and Polking J.C. (2011) *Ordinary Differential Equations Using MATLAB,* Dorling Kindesley, Pearson Education.
4. Ogata K. (2010) *Modern Control Engineering,* Prentice Hall.
5. Nusse H.E. and Yorke J.A. (1997) *Dynamics: Numerical Explorations*, 2nd edn, Springer-Verlag.

# Index



*Dynamics of Particles and Rigid Bodies: A Self-Learning Approach*, First Edition. Mohammed F. Daqaq.

Companion website: www.wiley.com/go/daqaq/dynamics

## L



## M



## N



## O



## P



## R